Adaptation and Fitness in Animal Populations

Adaptation and Fitness in Animal Populations

Evolutionary and Breeding Perspectives on Genetic Resource Management

Edited by

Julius van der Werf
School of Environmental and Rural Science, University of New England, Armidale, Australia

Hans-Ulrich Graser
Animal Genetics and Breeding Unit, University of New England, Armidale, Australia

Richard Frankham
Department of Biological Sciences, Macquarie University, Sydney, Australia and Australian Museum, Sydney, Australia

and

Cedric Gondro
The Institute for Genetics and Bioinformatics, University of New England, Armidale, Australia

Springer

Editors
Julius Herman Jozef van der Werf
School of Environmental
and Rural Science
University of New England
Armidale 2351
Australia

Richard Frankham
Macquarie University
Department of Biological Sciences
Sydney 2109
Australia and Australian Museum
6 College Street
Sydney 2010
Australia

Hans-Ulrich Graser
Animal Genetics and Breeding Unit
University of New England
Armidale 2351
Australia

Cedric Gondro
The Institute for Genetics and Bioinformatics
University of New England
Armidale 2351
Australia

ISBN: 978-90-481-8050-9 e-ISBN: 978-1-4020-9005-9

Printed on acid-free paper

9 8 7 6 5 4 3 2 1

springer.com

Emeritus Professor James Stuart Flinton Barker in honour of his contributions to Population Genetics and Animal Breeding

Preface

At the 16th AAABG conference in 2005, a proposal was launched to organise a symposium to examine advances in understanding of "adaptive fitness, both in managed populations being conserved and domestic animal species being utilised for food and agriculture production". After discussion about the term "adaptive fitness" some of us decided we should organise a symposium "Adaptation and Fitness in Animal Populations – Evolutionary and breeding perspectives on genetic resource management" to be held at the 2007 AAABG meeting in Armidale.

The term 'adaptive fitness' in itself is somewhat a tautology, but fitness and adaptation are both relevant concepts when one is concerned with the long term sustainability of animal breeding programs and animal production systems in a broader sense. Fitness of farm animal populations is clearly becoming a concern, especially in programs that have achieved substantial genetic change for 'production traits' or where environmental stressors are intense. Adaptation is required to maintain fitness in new or changing environments. Adaptive mechanisms are very important in animal production systems where genotypes are used globally without being explicitly tested in all the environments where they are kept. Adaptability is also important within animal production systems experiencing large between-year variability in the environment. Understanding such mechanisms and their relationships with the production traits may help to more successfully realise sustainable productivity gains.

Evolutionary geneticists are continuously exploring the genetic mechanisms that surround fitness and adaptation in natural populations. Consequently it seemed appropriate to have a debate among animal geneticists and evolutionary biologists. The chapters from this book are a reflection of the symposium papers that were presented in September 2007 in Armidale. The book is divided into four sections, corresponding to the different sessions that we identified for the symposium. Each of these sessions addressed specific questions that we had posed before the symposium. We asked the chair of each session to summarise the discussion and this discussion summary completes the set of papers for each section of the book.

Prof. Stuart Barker's primary life's work has been directed at understanding fitness and the evolution of natural populations. He begins the book by reviewing developments over time in the definition of fitness and its components. This assisted us in arriving at a uniform set of terms for use in addressing these components.

Animal breeders have clearly been very successful in achieving selection goals. In high-input production systems in particular, dairy cows are now much more productive, sheep produce more and better wool, and cattle, broilers, and pigs grow faster and have improved meat quality. Fitness, in the sense of general health, wellbeing and reproductive ability, is commonly perceived to have declined in farm animal populations particularly of developed country production systems. Why? One straightforward model is to consider fitness as just another quantitative trait. There appears to be a negative genetic correlation with production traits and if fitness as such is not selected for, one would expect a decline. The problem could then be resolved by simply measuring fitness and including it in the selection objective. However, fitness appears to encompass complexes of characters – do these fitness characters behave like other quantitative traits? And why is the correlation with production traits negative? Can we trust that the simple quantitative genetic model will work? At this symposium, Prof. Dick Frankham explores the genetic architecture of fitness characters and shed some light on those questions. Prof. Mike Goddard discusses the same issues in the context of animal breeding programs and makes suggestions for accommodating fitness in breeding objectives.

When considering long term consequences of selection (in the farm animal context more than 10–20 generations), the main question is how genetic variation can be maintained as well as exploited, particularly for fitness traits. We are somewhat surprised by the amount of genetic variation that is still observed for traits after many generations of artificial selection. In natural populations, considerable variation exists for fitness even after millions of years of natural selection. Prof. Bill Hill and Dr. Xu-Sheng Zhang review the mechanisms that determine genetic variation and the implications for maintenance of variation in long term (artificial) selection programs. Prof. Mark Blows and Prof. Bruce Walsh discuss the constraints to selection response and adaptation, and explore whether within a multi-dimensional selection space all points are equally attainable. Dr. Piter Bijma addresses the question of how to best utilise genetic variation for fitness traits in selection programs. A relatively new approach in animal breeding is the use of a genetic model that includes effects of interactions between individuals within populations. Such models can be used to select against aggressiveness in livestock populations, an aspect that is clearly linked to fitness and animal welfare.

Adaptation is a key principle for individuals to survive in the environment they are managed in. With the possibility of environmental changes, e.g. due to global warming and perhaps increased variability of climate events, the ability to adapt to such changes has become a key issue. Dr. Kathleen Donohue discusses the ability of organisms to affect, or rather determine their environment, and how this can form a basis for genotype by environment interaction and how it can determine response to selection for phenotypic performance in different environments. Prof. John James presents principles and examples that are used in animal breeding literature to model genotype by environment interaction and interprets the meaning of different modelling approaches. An important question is how to assess fitness in different environments In *Drosophila*, fitness has been measured extensively under laboratory conditions, and Prof. Ary Hoffman explores to what extent such assessments can be extrapolated to natural conditions.

Maintaining enough variation and diversity in genetic resources is an essential component of animal breeding programs as it is a key for maintaining fitness. Prof. Brian Kinghorn et al. give an overview of the concepts and proposed strategies for maintaining genetic diversity in breeding programs. The objective of directional selection for increased merit needs to be balanced with the objective to maintain diversity. International bodies such as FAO have made considerable investments in developing strategies to assist the countries of the world to better manage their farm animal genetic resources. Dr. Louis Ollivier and Dr. Jean-Louis Foulley review measures of between and within population diversity, demonstrating also the strengths and weaknesses of this information when trying to establish conservation tactics and population adaptability. Prof. John Gibson gives an outline on the possible roles for the different institutional entities that have a responsibility to implement sustainable utilisation and conservation policies for farm animal genetic resources.

Finally, to celebrate the contributions over 55 years by Professor Stuart Barker to the research, development and teaching of population and quantitative genetics, we dedicated the symposium to him. Stuart's passion is "genetics". Fitness and adaptation have been important keywords in his research career. He has been instrumental in translating these fundamental concepts to applications in animal breeding and to the characterisation and conservation of genetic resources. This symposium brought together scientists in evolutionary genetics and animal breeding in the hope that lessons can be learned about the relevance of fitness in domesticated breeding populations and how fundamental questions about fitness and adaptation can be addressed through experimental results and observations in natural populations. Stuart has always been a strong advocate of such interactions and this book creates a unique opportunity for mutual benefits for both groups.

We thank all authors of this book for their contributions and for reviewing the contributions of their colleagues. We thank the chairpersons of the symposium for summarising the session discussions for this book. We thank Dr. Keith Hammond for raising the original idea of the symposium and the book, and his tireless efforts in keeping us on track in many ways. In relation to the program, we thank Professor Stuart Barker for his succinct and effective suggestions, Professor Frank Nicholas for his cheerful and thoughtful contributions and Professor Bill Hill for his advice on topics and speakers. We are very grateful to Kathryn Dobos who did a tremendous and meticulous job in fixing up errors in the 'final version' of the manuscript. Finally we acknowledge the financial support of the Federal Department of Agriculture, Fisheries and Forestry (DAFF) and Meat and Livestock Australia (MLA).

Armidale, NSW Australia

J. van der Werf
H.-U. Graser
R. Frankham
C. Gondro

Contents

Part I
Modelling Fitness

Defining Fitness in Natural and Domesticated Populations

James S.F. (Stuart) Barker

Abstract The term 'fitness' has been applied differently and with different definitions throughout the history of population genetics. Five concepts and definitions are presented – distinguishing (phenotypic) fitness, (genotypic) fitness, adaptedness, adaptability and durability. While the heritabilities of fitness components are low, substantial genetic change is achievable, and breeding programs should include in the breeding objectives genotypic fitness (for known QTLs) and fitness traits such as fertility and longevity, as well as production traits.

Keywords Fitness · net fitness · adaptedness · adaptability · durability

1 Introduction

Some 47 years ago, I had the temerity in the introduction to my PhD thesis (Barker 1960) to discuss the meaning of fitness. Now I am asked to come back to this question, and to consider if the concept is or should be applied differently in natural and domesticated populations.

To go back to the beginning, the concept of fitness derives from Darwin's adoption of Herbert Spencer's (1864) term 'survival of the fittest' as an alternative to 'natural selection'. Dawkins (1982, p. 179) states that Darwin adopted 'survival of the fittest' in a letter to Wallace in 1866. However, the term first appears in the various editions of 'The Origin of Species' in 1869 (5th edition), in the first paragraph of Chapter 4:

> ...Let it also be borne in mind how infinitely complex and close-fitting are the mutual relations of all organic beings to each other and to their physical conditions of life; and consequently what infinitely varied diversities of structure may be of use to each being under changing conditions of life. Can it, then, be thought improbable, seeing that variations useful to man have undoubtedly occurred, that other variations useful in some way to each being

J.S.F. Barker (✉)
School of Environmental and Rural Science, University of New England, Armidale NSW 2351, Australia
e-mail: sbarker@une.edu.au

J. van der Werf et al. (eds.), *Adaptation and Fitness in Animal Populations*,

> in the great and complex battle of life, should sometimes occur in the course of thousands of generations? If such do occur, can we doubt (remembering that many more individuals are born than can possibly survive) that individuals having any advantage, however slight, over others, would have the best chance of surviving and of procreating their kind? On the other hand, we may feel sure that any variation in the least degree injurious would be rigidly destroyed. This preservation of favourable variations, and the destruction of injurious variations, I call Natural Selection, or the Survival of the Fittest.

But it is appropriate in this Symposium to note that the prior and first use by Darwin was in the first edition of 'The Variation of Animals and Plants under Domestication' (Darwin 1868).

Fitness may refer to a genotype, an individual, a population or a species, and defining fitness has plagued evolutionary biologists for many years, partly because the term has been applied differently and with different definitions depending on which level in the hierarchy is being considered.

In the 1920s and 1930s, the theoretical foundations of population genetics were laid by Fisher, Haldane and Wright, primarily considering selection within populations and thus the relative fitness of genotypes. In his book *The Causes of Evolution*, Haldane (1932), entitled one chapter 'What is fitness?', but provided no answer (perhaps consciously and wisely for that time!). Twenty years later, Dobzhansky (1955) pointed out that 'the work in this field (*population genetics*) is severely handicapped by lack of reliable methods of comparing the fitness of populations and, of course, by the lack of clarity in the concept of fitness itself.' (italics added).

As noted here by Dobzhansky, there are two interrelated problems – defining fitness and measuring fitness.

Broadly speaking, fitness relates to the ability of organisms to pass on the genes they carry. But to illustrate the problems of definition of fitness, consider a sample:

- *The relative ability of an organism to survive and transmit its genes to the next generation (King and Stansfield 1990),*
- *The relative capacity of carriers of a given genotype to transmit their genes to the gene pool of the following generations constitutes the adaptive value, or the Darwinian fitness, of that genotype. The adaptive value is, then, a statistical concept which epitomizes the reproductive efficiency of a genotype in a certain environment (Dobzhansky 1951, p. 78),*
- *...relative fitness (is) the relative ability of different genotypes to pass on their alleles to future generations (Hedrick 1983, p. 120),*
- *Relative fitness: the fitness of a genotype, compared to another genotype, usually at the same locus (Frankham et al. 2002, p. 543),*
- *Reproductive fitness: the number of fertile offspring surviving to reproductive age contributed by an individual. Characters that contribute to fitness include male fertility, female fecundity, parental care, offspring survival and offspring fertility. Often referred to as fitness (Frankham et al. 2002, p. 543),*
- *(A numerical measure of) ability to survive and reproduce in a particular environment (Brown 1993),*

- *The probability that a unit of evolution (population) will survive for a given long period of time, such as 10^8 years, that is to say will leave descendants after the lapse of that time, is the fitness of the unit (Thoday 1953, 1958).*

Many more examples could be quoted, but these show the main variations – genotype, individual or population; one, a few or many generations; recognition of the environment or not. Dawkins (1982), Hedrick and Murray (1983) and Endler (1986) provide (from different perspectives) excellent discussions of the various ways in which the term fitness has been used, and their context (genes, genotypes, individuals, groups). Nevertheless, some of the examples above show that more than 20 years on, there are still varying definitions.

2 Fitness and Related Concepts

Fitness differences among individuals (phenotypes) are necessary for selection to lead to change in a population – whether this be selection in natural populations, or artificial selection in the laboratory or animal breeding, where some individuals are culled on phenotypic criteria and thus have a fitness of zero. Such selection will cause a change in the distribution of phenotypes *within* a generation, but will cause a changed distribution of phenotypes in the next generation only if the fitness differences are, at least partly, genetic. Failure to distinguish these two phases has been one factor contributing to confusion in defining fitness. In fact, the problem is deeper than that because, depending on context, we operate at the genetic level or at the phenotypic level, and we do not have the knowledge to transform from one to the other (Lewontin 1974). This deficiency in our understanding of the transformation from genotype to phenotype needs to be clearly recognised in relation to much of the current work attempting to identify quantitative trait loci (QTLs). A QTL found in one population in one environment may not have the same phenotypic expression in a different genetic background or in a different environment.

At the genetic level (population genetics), generation transitions are expressed in terms of change in allele frequency (Δq) of an allele at a locus after one generation:

$$\Delta q = \frac{q(1-q)}{2}\frac{\mathrm{d}\ln\bar{w}}{\mathrm{d}q} \tag{1}$$

where q is the current allele frequency and $\bar{w}$ is the mean fitness of the genotypes at this locus in the population.

At the phenotypic level (quantitative genetics), generation transitions (assuming directional selection) are expressed in terms of changes in mean phenotype:

$$\Delta P = Sh^2 \tag{2}$$

where ΔP is the change in mean phenotype in one generation, S is the selection differential (difference in mean phenotypes of the selected individuals and the whole population) and h^2 the heritability of the trait.

Given that this Symposium is titled 'Adaptation and Fitness in Animal Populations', we must recognise another aspect of the problem – that adaptation and fitness are often confused, and although related concepts, they are not identical (Endler 1986).

3 Meanings of Fitness

Endler's (1986) five concepts and definitions (Table 1) provide a basis for discussion and much of the following is adapted from his presentation.

Table 1 Concepts that have been called 'fitness' by various authors (after Endler 1986)

1. Fitness: The degree to which there is a consistent relationship between the trait and survival. Measured by the average contribution to the breeding population by a phenotype or by a class of phenotypes, relative to the contribution of other phenotypes.
2. Rate coefficient: The rate at which the process of selection proceeds. Measured by the average contribution to the gene pool of the following generation, by the carriers of a genotype or by a class of genotypes, relative to the contributions of other genotypes. May be replaced by 'net fitness' or 'genotypic fitness' to allow prediction of the dynamics of gene frequency change.
3. Adaptedness: The degree to which an organism is able to live and reproduce in a given environment or set of environments. Measured by the average absolute contribution to the next generation by a phenotype or class of phenotypes (R_0), or by the intrinsic rate of natural increase (r_m).
4. Adaptability: The degree to which an organism, population or species can remain or become adapted to a wider range of environments by physiological or genetic means.
5. Durability: Probability that an allele or genotype, a class of genotypes, a population or a species will leave descendants after a given long period of time.

3.1 *Fitness (Concept 1)*

Fitness is a measure of the degree of relationship between a trait and survival. Obviously, no relationship means that different phenotypes for a given trait will not differ in fitness, while a strong relationship means substantial fitness differences among different phenotypes for that trait. As a relative measure (i.e. relative to other phenotypes), fitness may apply to individuals, known genotypes that differ in phenotype or to populations, provided the comparison is made in the same environment. Fitness rankings when the same set of phenotypes are compared in some other environment may be quite different – phenotype x environment interaction for fitness. Although fitness is here defined in terms of a trait, that trait may simply be the overall phenotype (e.g. comparing individuals), or the overall average phenotype (e.g. comparing genotypes or comparing populations such as breeds of domesticated animals). For quantitative traits, the 'selection differential', and for polymorphic traits, the 'selection coefficient' are algebraically related measures. Robertson (1966), in a companion paper to two others on dairy cattle genetics (Barker and Robertson 1966, Robertson and Barker 1966) first noted that the selection differential is equal to the covariance of phenotype and fitness (survival – as defined here), and this relationship was extended and generalised by Price (1970, 1972).

Transforming through the genetic system gives the next concept – the selection rate coefficient.

3.2 Selection Rate Coefficient (Concept 2)

This coefficient determines the characteristic rate at which selection proceeds. It is measured by the average contribution to the gene pool of the following generation by the carriers of a genotype, or by a class of genotypes, relative to the contributions of other genotypes, and should be measured at the same life stage in two successive generations.

Consider the simplest case of a single locus with two alleles, sex-independent selection and random mating (the classical model – Prout 1969), and counting at the zygote stage in each generation. The rate coefficients are estimated by expressing the zygote frequencies in the progeny generation in terms of the parental generation allele frequencies and the selection coefficients, dividing each of these by the parental zygote frequencies, and finally expressing these ratios relative to the heterozygote genotype by dividing each homozygote ratio by the heterozygote ratio (see Endler 1986, pp. 41–42). The resulting rate coefficients for the two homozygotes are reciprocal, and they are frequency dependent, even though the selection coefficients are constant. That is, they will change in value as the allele frequencies change over successive generations within a population, or will be different in populations at different initial allele frequencies. Thus while the selection rate coefficient is of theoretical interest, it is not sufficient for application.

When considering genotypes, and possible selective differences among them, the aim is to understand the process of evolution, or in a more limited way, to predict the dynamics of gene frequency change in the short term. To do this, we need to measure the magnitude of any selective differences among the genotypes, i.e. 'net fitness' (Prout 1965). I hesitate to introduce further terminology, but this may be visualised as 'genotypic fitness' to distinguish it from fitness as in concept 1 above, i.e. 'phenotypic fitness'.

Analogous to measuring rate coefficients, estimation of net fitness involves comparing the frequencies of genotypes in one generation with the frequencies of these genotypes in the next generation. However, there are complications: Prout (1965) showed for the classical model that the estimates obtained will show spurious frequency dependence, and will not be estimates of the true net fitness *unless* the selection process has been completed at the time when the genotypes are counted in each of the two successive generations. In practice, this means counting the numbers of each genotype in each generation when all surviving individuals have reached post-reproductive age. Alternatively, mature adult numbers may be used, if it were known from separate experiments that there were no differences among genotypes in adult survival.

The four basic components of selection that contribute to net fitness are (Christiansen and Prout 2000):

- Zygotic selection: differential survival of zygotes to maturity.
- Sexual selection: differential recruitment of mature individuals to parents.
- Fecundity selection: variation in fecundity of parental pairs.
- Gametic selection: differential survival of gametes.

Prout (1969) developed a general model incorporating these components, and then reducing this to the classical model, showed that:

(i) if genotype frequencies are determined in each generation after all selection is completed, then one transition (two generations) is sufficient for the estimation of net fitness (the case discussed above),
(ii) if genotype frequencies are determined in each generation before selection has started, then two transitions are necessary and sufficient,
(iii) if genotype frequencies are determined in each generation at a partially selected stage, then both pre- and post-observational components of net fitness must be estimated, and three transitions are both necessary and sufficient.

I am not aware that net fitnesses of genotypes have been estimated for any case in domestic animal populations. However, with increasing emphasis on the identification of quantitative trait loci (QTLs) affecting the expression of economically important production traits, this should become an issue. When a QTL is identified in a particular population, the breeder may assume that the breeding objective is to bring the most favourable allele to fixation. The unstated assumption then is that the genotypes at this locus do not differ in net fitness. Perhaps more likely, there will be net fitness differences among the genotypes, and the net fitness of each genotype will need to be determined and included in the overall breeding objective. How this should best be done, I leave as an open question, but in addition, the breeder needs to recognise:

(i) the difficulties of net fitness estimation, even more so for the more complex models than the classical one,
(ii) the statistical (sampling) problems of estimation in finite populations (the classical model analysis assumes an infinite population),
(iii) that the net fitness estimates for particular genotypes refer to the population in which the estimates were made. Although estimation is being done in terms of particular identifiable genotypes, these genotypes are not divorced from the genetic background – epistatic interactions and linkage disequilibria that affect the net fitness in one population may be different in some other population,

(iv) as a further development of (iii), these net fitnesses strictly should be applied within the environment in which the estimation was made, as genotype x environment interactions may well exist.

As more QTL are identified in domesticated species, the problem becomes even more complex, as the question then will relate to the fitnesses of multi-locus genotypes. For an n-loci, two allele model, where cis and trans heterozygotes are distinguished, $2^{n-1}(2^n + 1)$ genotypes exist, e.g. two loci – 10 genotypes, three loci – 36 genotypes.

As noted previously, (phenotypic) fitness may be defined and measured for populations, in terms of their relative survival to enter the breeding population. Thus for example, comparing two livestock breeds and starting with equal numbers of new-borns in each, which are then maintained in the same environment (including the same management, disease control, etc), the numbers of each breed surviving to enter the breeding population are counted. However, this is an incomplete measure as it takes no account of survival during the reproductive period. Further, the aim of breed comparisons will be to determine the 'suitability' of those breeds for production in a given environment – where suitability will mean not only production trait criteria, but also ability to survive and reproduce in that environment. Thus for population comparisons, we need a measure analogous to the net fitness of genotypes. However, there is an important difference – genotypes are compared to one another *within* an interbreeding population, while populations are reproductively isolated from one another. The appropriate measure for populations then is adaptedness.

3.3 Adaptedness (Concept 3)

Although defined by Endler (1986) as 'the degree to which an organism is able to live and reproduce in a given set of environments; the state of being adapted', the measure proposed 'the average absolute contribution to the breeding population by a phenotype or a class of phenotypes' appears inadequate in not including reproduction. The apparent problem hinges on the use of the term 'breeding population', which could be inferred as that of the current generation.

In essence, the 'average absolute contribution' is the number of individuals born in the next generation, which could be expressed relative to the number at birth in the current generation as the rate of change in population size per generation. This ratio is referred to as the net reproductive rate, usually designated as R_0. As a rate per unit of time, it may be measured as the statistic (r_m) variously referred to as the intrinsic rate of natural increase (Lotka 1925), the innate capacity for increase (Andrewartha and Birch 1954) or the Malthusian parameter (Fisher 1930).What this measures is the rate of growth of the population in a given environment, provided that the food supply and space are not limiting, and that the population has a stable age distribution. Two sets of data are required to estimate r_m – the age-specific birth rates (m_x) and survival rates (l_x).

Given these data, r_m may be estimated by solving the equation:

$$\int_0^\infty e^{-r_m x}\, l_x m_x \delta x = 1 \tag{3}$$

While this approach has been used for some laboratory populations (Baldwin and Dingle 1986) and for some natural populations (Berryman and Lima 2006), adaptedness in domestic animal populations may be estimated more simply as the net reproduction rate (R_0). Thus for population (breed) comparisons of adaptedness, the example given above for (phenotypic) fitness is simply extended to a count of the number of newborn in each population in the next generation. One should note, however, that this gives, for the populations being compared, a measure of adaptedness to the specific set of environments during this one generation transition.

In recent years, the importance of adaptedness and the use of adapted breeds (or varieties) in genetic improvement programs have been recognised (e.g. Hammond (2000) for livestock and Namkoong (1998) for crop plant and forest tree breeding). In this context, however, adapted breeds are considered simply as those that live in some particular, usually stressful, environment, and the relative adaptedness of different breeds has not been measured. Choice of the breed (or breeds) for the foundation population of any breeding program is the basic first step (Barker 1967). Consequently, adaptedness should be a component of the characterisation and evaluation of breeds for conservation and for breeding programs.

3.4 Adaptability (Concept 4)

An individual, a population or a species may exhibit adaptability to some range of environments, or may have the capacity to become adapted. Adaptability then is a measure of actual or potential capacity, while adaptation is the process of becoming adapted or more adapted. Present (existing) adaptability could be tested, but potential adaptability cannot be predicted. Natural populations or species that lack appropriate adaptability may become extinct if the environment changes (e.g. global warming), particularly if they cannot respond (or respond fast enough) to track the changes. For domesticated populations, actual adaptability would be important if, for example, a particular breed needed to be used in some range of environments. More usually, however, the necessary criterion will be adaptedness – choice of particular breed(s) for specific environments (horses for courses or cows for climates).

3.5 Durability (Concept 5)

As noted earlier, Thoday (1953, 1958) defined fitness in terms of durability. It could be estimated as the expected time to extinction, discussed in detail by Endler (1986). However, it seems that little attention has been paid to this concept in the evolutionary literature, and it is not relevant to breeding programs in domesticated populations. However, it is a component that needs to be considered in setting priorities for conservation of endangered breeds (Ollivier and Foulley 2005).

4 Genetics of Fitness

In domestic animal populations, can fitness be improved, or at least maintained if there are negative genetic correlations between fitness and any of the production traits? That is, does fitness show genetic variation? Fisher's (1930) fundamental theorem of natural selection states 'The rate of increase in fitness of any organism at any time is equal to its genetic variance in fitness at that time'. A naïve interpretation of the theorem would be that there should be no genetic variation in fitness as alleles conferring high fitness would rapidly increase to fixation under natural selection. However, the theorem depends on many assumptions which generally will not apply in natural populations (see, for example, Charlesworth 1987).

While there are no estimates of the heritability of net fitness or adaptedness in domestic animals, the heritabilities of various components of fitness (survival, fertility, fecundity) have been estimated to be around 5–15% (Nicholas 1987, p. 405). For laboratory and natural populations, Roff and Mousseau (1987) and Mousseau and Roff (1987) summarised estimates of heritability for four categories of traits – morphology, behaviour, physiology, and life-history (i.e. fitness components). Life-history and behavioural traits had low heritability, while morphological and physiological traits had high heritabilities. But in a later analysis including more data, Hoffmann (2000) found no differences in the heritabilities of morphological, physiological or life-history traits, although life-history traits tended to exhibit a greater range and higher variance of the estimates. In an extensive analysis of a natural population, Merilä and Sheldon (2000) found that while lifetime reproductive success (as a surrogate for fitness) and various fitness traits had low heritabilities, they had higher levels of additive genetic variance than traits more distantly associated with fitness. But in addition, fitness and the fitness traits had much higher levels of residual variance (environmental and/or nonadditive genetic variance), leading to the low heritabilities. Similarly, in a domestic animal population (Norwegian dairy cattle), Andersen-Ranberg et al. (2005) found low heritabilities for fitness components, and again the residual variance was high, but substantial desirable genetic changes had been obtained.

However, while individual fitness components show significant, albeit low, heritability, there are two other aspects to be considered in developing breeding programs that include fitness – genetic correlations among the components and the genetic architecture of fitness. If individual fitness components were to be included in the breeding objectives, any negative genetic correlations among them would need to be known and taken into account. Equally, we need to acknowledge the ample evidence (briefly reviewed by Barker 2001) that epistatic interactions and dominance are substantial components of the genetic architecture of fitness. Can we identify and select for specific gene complexes?

5 Quantitative Genetics, Animal Breeding and Fitness

Animal breeding seeks to change the phenotypic value of traits that are of economic importance. Assuming a positive correlation between production traits that are being selected for and fitness (phenotypic or genotypic), there will be indirect selection for fitness. However, a negative correlation means that there will be indirect selection against fitness. In either case, an optimum breeding program would include genotypic fitness (for known QTL) and fitness traits such as fertility and longevity, together with production traits, in the breeding objectives.

Barker and Thomas (1987) noted that workers in evolutionary biology had rediscovered quantitative genetics, with a substantial development of quantitative genetic theory in relation to natural populations and evolution (e.g. Lande 1976, 1980). They also urged that empirical knowledge from selection in laboratory organisms and domestic plants and animals should provide part of the background to any consideration of quantitative traits in natural populations. I would now emphasise the reverse – that animal breeders would be well advised to include in their armoury (particularly in relation to fitness) recent developments in relation to microevolution and quantitative genetics in natural populations. This literature is quite vast, but recent papers (primarily reviews) that provide an introduction to these developments include papers in the special issue of *Genetica* (Hendry and Kinnison 2001 – particularly Agrawal et al. 2001, Arnold et al. 2001 and Wade 2001), Blows (2007), Blows and Hoffmann (2005) and Carter et al. (2005).

Finally, the rationale for this Symposium was to promote a better understanding of common problems, and better appreciation of the complementary studies in natural populations and in animal breeding. My hope is that this introduction to the Symposium will contribute in some small way to that objective.

Acknowledgments I was not in receipt of grant support during the preparation of this paper, but I am most grateful for the support of Mike Goddard and Keith Hammond, who provided critical comments on an earlier version, and Julius van der Werf for assistance with a reference.

References

Agrawal, A.F., Brodie, E.D. III. and Rieseberg, L.H. 2001. Possible consequences of genes of major effect: transient changes in the G-matrix. *Genetica* **112–113**: 33–43.

Andersen-Ranberg, I.M., Klemetsdal, G., Heringstad, B. and Steine, T. 2005. Heritabilities, genetic correlations, and genetic change for female fertility and protein yield in Norwegian dairy cattle. *J. Dairy Sci.* **88**: 348–355.

Andrewartha, H.G. and Birch, L.C. 1954. *The Distribution and Abundance of Animals*, The University of Chicago Press, Chicago, USA.

Arnold, S.J., Pfrender, M.E. and Jones, A.G. 2001. The adaptive landscape as a conceptual bridge between micro- and macroevolution. *Genetica* **112–113**: 9–32.

Baldwin, J.D. and Dingle, H. 1986. Geographic variation in the effects of temperature on life-history traits in the large milkweed bug *Oncopeltus fasciatus*. *Oecologia* **69**: 64–71.

Barker, J.S.F. 1960. PhD Thesis *The estimation of relative fitness of* Drosophila *populations and related studies in population genetics*. University of Sydney, Sydney, Australia.

Barker, J.S.F. 1967. Modern problems of population genetics in animal husbandry. *Der Zuchter* **37**: 309–323.

Barker, J.S.F. 2001. Conservation and management of genetic diversity: a domestic animal perspective. *Can. J. For. Res.* **31**: 588–595.

Barker, J.S.F. and Robertson, A. 1966. Genetic and phenotypic parameters for the first three lactations in Friesian cows. *Anim. Prod.* **8**: 221–240.

Barker, J.S.F. and Thomas, R.H. 1987. A quantitative genetic perspective on adaptive evolution. In *Genetic Constraints on Adaptive Evolution* (V. Loeschcke, ed.), Springer-Verlag, Berlin, Germany, pp. 3–23.

Berryman, A. and Lima, M. 2006. Deciphering the effects of climate on animal populations: Diagnostic analysis provides new interpretation of Soay sheep dynamics. *Amer. Nat.* **168**: 784–795.

Blows, M.W. 2007. A tale of two matrices: multivariate approaches in evolutionary biology. *J. Evol. Biol.* **20**: 1–8.

Blows, M.W. and Hoffmann, A.A. 2005. A reassessment of genetic limits to evolutionary change. *Ecology* **86**: 1371–1384.

Brown, L. (ed.) 1993. *The New Shorter Oxford English Dictionary*, Clarendon Press, Oxford, UK.

Carter, A.J.R., Hermisson, J. and Hansen, T.F. 2005. The role of epistatic gene interactions in the response to selection and the evolution of evolvability. *Theor. Pop. Biol.* **68**: 179–196.

Charlesworth, B. 1987. The heritability of fitness. *In Sexual Selection: Testing the Alternatives* (J. W. Bradbury and M. B. Andersson, eds.), Wiley, Chicester, UK. pp. 21–40.

Christiansen, F.B. and Prout, T. 2000. Aspects of fitness. In *Evolutionary Genetics from Molecules to Morphology* (R.S. Singh and C.B. Krimbas, eds.), Cambridge University Press, Cambridge, UK, pp. 146–156.

Darwin, C.R. 1868. *The Variation of Animals and Plants Under Domestication*, John Murray, London, UK. 1st edition, first issue. Volume 1.

Darwin, C.R. 1869. *On the Origin of Species by Means of Natural Selection, or the Preservation of Favoured races in the Struggle for Life*, John Murray, London, UK. 5th edition.

Dawkins, R. 1982. *The Extended Phenotype*, Oxford University Press, Oxford, UK.

Dobzhansky, Th. 1951. *Genetics and the Origin of Species*, 3rd edition. Columbia University Press, New York, USA.

Dobzhansky, Th. 1955. A review of some fundamental concepts and problems of population genetics. Cold Spr. Harb. Symp. on Quant. Biol. **20**: 1–15.

Endler, J.A. 1986. *Natural Selection in the Wild*, Princeton University Press, Princeton, NJ, USA.

Fisher, R.A. 1930. *The Genetical Theory of Natural Selection*, Clarendon Press, Oxford, UK.

Frankham, R., Ballou, J.D. and Briscoe, D.A. 2002. *Introduction to Conservation Genetics*, Cambridge University Press, Cambridge, UK.

Haldane, J.B.S. 1932. *The Causes of Evolution*, Longmans, Green & Co. Ltd, London, UK.

Hammond, K. 2000. Sustainable intensification and the conservation of farm animal genetic resources. *Asian-Aust. J. Anim. Sci.* **13**(Suppl. A): 240–246.

Hedrick, P.W. 1983. *Genetics of Populations*, Science Books International, Portola Valley, CA, USA.

Hedrick, P.W. and Murray, E. 1983. Selection and measures of fitness. In *The Genetics and Biology of Drosophila*, Vol. 3d (M. Ashburner, H.L. Carson and J.N. Thompson, Jr., eds.), Academic Press, New York, USA pp. 61–104.

Hendry, A.P. and Kinnison, M.T. 2001. An introduction to microevolution: rate, pattern, process. *Genetica* **112–113**: 1–8.

Hoffmann, A.A. 2000. Laboratory and field heritabilities: Some lessons from *Drosophila*. In *Adaptive Genetic Variation in the Wild* (T.A. Mousseua, B. Sinervo and J.A. Endler, eds.), Oxford University Press, Oxford, U.K. pp. 200–218.

King, R.C. and Stansfield, W.D. 1990. *A Dictionary of Genetics*, 4th edition. Oxford University Press, Oxford, UK.

Lande, R. 1976. The maintenance of genetic variability by mutation in a polygenic character with linked loci. *Genet. Res.* **26**: 221–235.

Lande, R. 1980. The genetic covariance between characters maintained by pleiotropic mutations. *Genetics* **94**: 203–215.

Lewontin, R.C. 1974. *The Genetic Basis of Evolutionary Change*, Columbia University Press, New York, USA.

Lotka, A.J. 1925. *Elements of Physical Biology*, Williams and Wilkins, Baltimore, MD, USA.

Merilä, J. and Sheldon, B.C. 2000. Lifetime reproductive success and heritability in nature. *Amer. Nat.* **155**: 301–310.

Mousseau, T.A. and Roff, D.A. 1987. Natural selection and the heritability of fitness components. *Heredity* **59**: 181–197.

Namkoong, G. 1998. The place of adaptation and production in genetic improvement programs for sustainable development. *Proc. 6th World Congr. Genet. Applied to Livest. Prod.* **6**(28): 69–79.

Nicholas, F.W. 1987. *Veterinary Genetics*, Clarendon Press, Oxford, UK.

Ollivier, L. and Foulley, J-L. 2005. Aggregate diversity: New approach combining within- and between-breed genetic diversity. *Livestock Prod. Sci.* **95**: 247–254.

Price, G.R. 1970. Selection and covariance. *Nature* **227**: 520–521.

Price, G.R. 1972. Extension of covariance selection mathematics. *Ann. Hum. Genet., Lond.* **35**: 485–490.

Prout, T. 1965. The estimation of fitnesses from genotypic frequencies. *Evolution* **19**: 546–551.

Prout, T. 1969. The estimation of fitnesses from population data. *Genetics* **63**: 949–967.

Robertson, A. 1966. A mathematical model of the culling process in dairy cattle. *Anim. Prod.* **8**: 95–108.

Robertson, A. and Barker, J.S.F. 1966. The correlation between first lactation milk production and longevity in dairy cattle. *Anim. Prod.* **8**: 241–252.

Roff, D.A. and Mousseau, T.A. 1987. Quantitative genetics and fitness: lessons from *Drosophila*. *Heredity* **58**: 103–118.

Spencer, H. 1864. *The Principles of Biology*, Vol. 1, Williams and Norgate, London and Edinburgh, UK.

Thoday, J.M. 1953. Components of fitness. *Symp. Soc. Exp. Biol.* **7**: 96–113.

Thoday, J.M. 1958. Natural selection and biological progress. In *A Century of Darwin* (S.A. Barnett, ed.), Harvard University Press, Cambridge, MA., USA, pp. 313–333.

Wade, M.J. 2001. Epistasis, complex traits, and mapping genes. *Genetica* **112–113**: 59–69.

Genetic Architecture of Reproductive Fitness and its Consequences

Richard Frankham

Dedicated to Stuart Barker in recognition of his many important contributions to evolutionary genetics and for his support as teacher, supervisor, mentor, and colleague

Abstract Reproductive fitness characters are crucial in animal and plant breeding, evolutionary genetics and conservation biology. However, the nature of their variation is not well understood. I review evidence on the comparative architecture of genetic variation for fitness and peripheral characters for both wild populations and domestic livestock. Fitness characters differ from peripheral characters in typically having lower heritabilities, directional dominance, higher levels of gene interactions, asymmetry of selection response, inbreeding depression and heterosis, declines in mean from mutation, and non-linear changes in genetic variation with inbreeding. These differences are a reflection of the different forces of natural selection operating on the two classes of traits. Genetic diversity for fitness in wild populations is partially due to rare, deleterious, partially recessive alleles in mutation-selection balance, with a further component due to alleles at equilibrium due to balancing selection, but there is no consensus about the relative contributions. In contrast, livestock populations are unlikely to be in equilibrium, due to changes in their environments, artificial selection, small effective population sizes and in some cases to crossing of populations. These have major effects on the genetic architecture of fitness and especially on the proportion of polymorphic loci exhibiting overdominance. In general, long-term directional artificial selection on peripheral traits is expected to move their genetic architecture towards that of fitness traits. Whilst the breeders' equation provides good predictions of selection response for peripheral characters, it does not predict asymmetrical responses to selection as observed for fitness characters, and levels of inbreeding depression and heterosis remain unpredictable. Given the importance of fitness characters and the uncertainties on many important issues, they deserve a higher priority in quantitative genetics research.

R. Frankham (✉)
Department of Biological Sciences, Macquarie University, NSW 2109, Australia, and
Australian Museum, 6 College Street, Sydney, NSW 2010, Australia
e-mail: rfrankha@els.mq.edu.au

J. van der Werf et al. (eds.), *Adaptation and Fitness in Animal Populations*,

Keywords Balancing selection · mutation-selection equilibrium · non-equilibrium · reproductive fitness · selective sweeps

1 Introduction

Reproductive fitness traits represent the most difficult to study and least understood traits in quantitative genetics, yet they are crucial in animal and plant breeding, evolutionary genetics and conservation biology. In layer chickens, the commercial output (eggs) is a reproductive fitness trait. In other domestic animals raised for meat, milk, or fibre, reproductive output contributes to the cost of producing market animals, while survival rates are crucial to the cost of production in all livestock.

In spite of the importance of reproductive fitness, the nature of the genetic variation for fitness characters is not well understood and they have received less scientific investigation than peripheral characters (ones not closely related to reproductive fitness). In this contribution, I review evidence on the comparative architecture of genetic variation for fitness and peripheral characters for both wild populations and domestic livestock and consider the implications of this for animal breeding. I use genetic architecture here to mean the array of loci, alleles, frequencies, effects, dominance, pleiotropy and interactions affecting a quantitative character. The perspectives I bring to this are derived from having worked in quantitative genetics and animal breeding for the first 26 years of my career, but having worked predominantly in conservation genetics since 1990. While it is often assumed that genetic architecture is similar in wild populations and captive/domestic ones, I will argue that there are often substantial differences between them.

1.1 Different Characteristics of Fitness and Peripheral Quantitative Characters

Reproductive fitness and peripheral quantitative traits differ in a range of characteristics that are a reflection of the different architectures of their genetic variation (Table 1). Fitness characters exhibit strong inbreeding depression, while peripheral characters show little (DeRose and Roff 1999). Fitness characters typically show lower heritabilities than peripheral characters (Mousseau and Roff 1987; Roff and Mousseau 1987). Almost all quantitative characters exhibit genetic variation in almost all naturally outbreeding populations (Lewontin 1974). However, a few characters important to survival have essentially zero additive genetic variation in a small number of wild populations, in spite of ample genetic variation for other quantitative characters and for microsatellites (Hoffmann et al. 2003; Kellermann et al. 2006). Fitness characters typically show more selection response for low than high fitness (Frankham 1990), whilst responses are closer to symmetrical for peripheral

Table 1 Differences in characteristics of fitness and peripheral quantitative traits

	Quantitative character	
	Fitness	Peripheral
Inbreeding depression and heterosis	Strong	Weak
Asymmetry of selection response	Strong	Weak
Heritabilities	Relatively low	Intermediate
Non-additive genetic variation	Higher	Lower
Change in V_A with inbreeding	Curvilinear	Linear decline
Effect of mutation on mean	Reduces	Minimal change
Genotype x environment interactions	Stronger	Weaker
Natural selection	Directional	Stabilising

characters (Falconer and Mackay 1996). Fitness characters typically have much higher levels of non-additive genetic variation than peripheral characters (Kearsey and Kojima 1967; Barker 1979; Crnokrak and Roff 1995).

Peripheral characters show an approximately linear decline in additive genetic variation with inbreeding (Gilligan et al. 2005; Van Buskirk and Willi 2006). Conversely, a meta-analysis based on 22 studies revealed that fitness characters show a non-linear change in additive genetic variation (V_A) with inbreeding, increasing to more than double base population levels at intermediate levels of inbreeding before declining (Van Buskirk and Willi 2006). New mutations cause the mean to decline for fitness traits, but have little directional impact on means for peripheral characters (Mackay 1989; Garcia-Dorado et al. 1999; Yang et al. 2001). Finally, fitness characters generally show greater susceptibility to genotype x environment interactions than peripheral characters (Frankham and Weber 2000).

These different characteristics reflect different regimes of natural selection on peripheral and fitness characters (Falconer and Mackay 1996). Natural selection in a stable environment is directional on fitness and usually stabilising on peripheral characters (Endler 1986; Falconer and Mackay 1996; Kingslover et al. 2001; Zhang and Hill 2005).

2 Equilibrium Theory for Wild Populations

Quantitative genetic variation for fitness is maintained in two ways, either from the equilibrium between the occurrence of deleterious mutations and their removal by natural selection, or from balancing selection, whether via heterozygote advantage, rare allele advantage or selection that varies in space or time (reviewed by Charlesworth and Hughes 2000). The former prediction leads us to expect low frequencies of deleterious alleles at many loci in the genome and that most of these will be partially recessive. Models of balancing selection typically involve alleles at more intermediate frequencies than those for mutation-selection balance. Antagonistic pleiotropy may also lead to maintenance of genetic diversity if the two homozygotes at a locus have effects in different direction for different fitness components. If deleterious allelic effects on life-history traits are sufficiently recessive, such antagonism can lead to heterozygote advantage for net fitness and a stable polymorphism (Charlesworth and Hughes 2000). All mechanisms will be affected by genetic drift in small populations, but the balancing selection component will generally be less affected. The relative importance of mutation-selection and balancing selection in maintaining genetic variation is a matter of controversy, as detailed below. Hill and Zhang (2009) considers some issues concerning the maintenance of quantitative genetic variation for fitness characters in another chapter.

For peripheral characters, quantitative genetic variation can be due to the mechanisms described above for fitness characters, and in addition by (a) the balance between neutral mutation and random genetic drift, (b) the equilibrium between mutation, stabilising selection and drift, or (c) the balance between mutation, drift and natural selection operating by means of pleiotropic effects of mutations on fitness related to their deviations from the mean for peripheral characters (Barton and Turelli 1989; Falconer and Mackay 1996). Heterozygote advantage on loci affecting a quantitative character can generate apparent stabilising selection (Robertson 1956). The ability of mutation-selection balance to explain quantitative genetic variation for peripheral characters is controversial with Zhang and Hill (2005) concluding that it can, whilst Johnson and Barton (2005) are sceptical.

3 Empirical Data

Much of the information on the nature of quantitative genetic variation for wild populations comes from studies in *Drosophila*. Before discussing the conclusions, it is important to point out that there are potential flaws in essentially all the data. I know of no quantitative genetic studies on the architecture of fitness that have been done under wild conditions. Studies on recently caught wild populations have been done under different conditions in the laboratory. Even studies on long term captive populations often test fitness under different conditions to those the flies have been adapted to e.g. cage adapted flies from high densities may be tested under uncrowded conditions in vials. These differences allow genotype x environment interactions to affect the conclusions reached. Further, I will argue that there are likely to be differences between wild adapted and laboratory adapted populations in genetic architecture of fitness.

There is overwhelming evidence that mutation-selection balance contributes to genetic variation for fitness. Wild outbreeding populations of essentially all species contain loads of rare recessive morphological mutations (presumably deleterious) at many loci (see Lewontin 1974; Hedrick 2005). Wild populations of the few species that have been studied (several *Drosophila* species, plus blue krillifish and zebrafish) contain loads of rare recessive lethal mutations at many loci (see Lewontin 1974; Hedrick 2005). All species examined also contain rare deleterious mutations at many loci that we typically refer to as genetic diseases (see Lewontin 1974).

Further, chromosomal homozygotes in *Drosophila* show reduced fitness on average, indicating a load of deleterious partially recessive alleles (see Lewontin 1974).

Detailed analyses of a range of quantitative genetic data also indicate that a proportion of genetic variation for fitness is due to rare alleles in mutation-selection equilibrium, but the proportion is a matter of contention. Mukai (1988), Falconer and Mackay (1996) and Houle et al. (1996) all concluded that most fitness variation was due to deleterious alleles in mutation-selection equilibrium. Further, Rodríguez-Ramilo et al. (2004) concluded on the basis of the impacts of inbreeding and artificial selection on egg-adult viability in *Drosophila melanogaster* combined with computer simulations that all of the genetic variation was probably due to alleles in mutation-selection balance.

A very thorough analysis by Charlesworth and Hughes (2000) concluded that a substantial proportion (around one-half) of the genetic variation was due to rare, partially recessive alleles in mutation-selection balance and the remainder due to loci subject to balancing selection, but that unconditional heterozygote advantage was rejected. Many studies have rejected major contributions from unconditional heterozygote advantage (see Mukai 1988; Falconer and Mackay 1996; Lynch and Walsh 1998). Frequency dependent selection is also considered to be of little importance (Dolan and Robertson 1975; Mukai 1988). In contrast to the above conclusion from Mukai (1988) for a northern Japanese population, he concluded that selection in a heterogeneous environment was involved in addition to mutation-selection balance in a southern Japanese population. Charlesworth and Hughes (2000) concluded that many loci were contributing to genetic variation for most fitness traits.

By contrast, Charlesworth et al. (2007) concluded that much of the genetic variation for early fecundity was due to alleles with intermediate frequencies, based on responses to inbreeding and directional selection. Similarly, Kelly and Willis (2001) and Kelly (2003) have obtained results inconsistent with the predominance of genetic variation due to mutation-selection balance for both flower size and male fitness components in the monkey flower, indicating the involvement of balancing selection. The latter three studies all used a method that assumes equilibrium, no epistasis for the traits and measurement of fitness in the adapted environment. The assumption of no epistasis is highly questionable for fitness traits. Further, field adapted plants were measured in the greenhouse in the latter two studies.

The *Drosophila* populations that have been analysed were captured in the wild and brought into captivity and kept there for various periods before analyses were done under laboratory conditions. This raises two reservations about whether such populations are in equilibrium with their environment. First, as I will document below, major genetic adaptation to captivity occurs and may continue for many generations. For those populations measured in captivity shortly after capture, some of the alleles that are beneficial in captivity will have been deleterious in the wild i.e. there will be genotype x environment interactions (Frankham 2008). Further, dominance of alleles may change between the wild and captivity. Second, effective population size/census ratios are often very low in mass cultures of *Drosophila*, so effective population sizes may be low (Briscoe et al. 1992; Frankham 1995). For example, we found N_e/N ratios from 0.004 to 0.051, with effective sizes of only 185 in caged populations with census sizes of 5,000 and 16 in bottle cultures with a census size of 1000. Certainly, the nature of the genetic variation, especially for fitness, cannot be taken to represent the situation in the wild environment. Levels of genetic diversity may be depleted, allele frequency and effects distributions may be changed, mildly deleterious alleles are likely to be effectively neutral and some will reach intermediate frequencies and there may be significant linkage disequilibrium. The effects of populations evolving in the wild, but being measured in captivity and of small effective sizes in long-term captive populations are both likely to bias conclusions against the mutation-selection balance component.

Notably, the different conclusion reached by Rodríguez-Ramilo et al. (2004) and Charlesworth et al. (2007) could be an artifact of laboratory adaptation and small effective size, as the *Drosophila* population studied by Rodríguez-Ramilo et al. (2004) was a recently caught wild one, while that studied by Charlesworth et al. (2007) had been in the laboratory since 1975 (about 800 generations).

In spite of substantial efforts that have been expended, there is considerable uncertainty about the quantitative details of the architecture of genetic variation for reproductive fitness and the mechanisms maintaining that genetic diversity.

There is also great uncertainly about the mechanisms maintaining genetic diversity for peripheral characters (Frankham and Weber 2000). Reviews by Barton and Turelli (1989) and Falconer and Mackay (1996) tentatively favoured mutation-drift-pleiotropy as the main mechanism.

4 QTL Mapping and Microarray Studies

In theory, the nature of genetic variation for quantitative characters can be determined by mapping the quantitative trait loci (QTL) involved and determining the frequency, effects, dominance and interactions of alleles at different loci. In practice, this is a very coarse tool whose power to detect loci is limited by the size of experiments and the number of recombinants generated. It can yield incorrect estimates for number of loci, effects of loci and their map positions (McMillan and Robertson 1974; Beavis 1994; Bost et al. 2001; Barton and Keightley 2002). Further, an average QTL in *Drosophila* encompasses a region of 4,300 kb containing an average of 500 loci (Mackay 2004). In line with this, further analyses often separate QTL into additional QTL (Table 2). Repeat experiments on the same material do not detect identical QTL. For example, Dilda and Mackay (2002) obtained an exact match for only 56% of bristle QTL detected in two studies on the same material in the same laboratory.

The ability to detect QTLs for a given sized experiment will generally be lower for fitness than for peripheral characters, as environmental variation is typically

Table 2 Number of QTL detected in a range of experiments for fitness characters in different animal species, the numbers revealed by further analyses and notes on dominance and epistasis where measured

Species	Character	# QTL detected		Notes	Reference
		Mapping	Further analyses		
Mutation					
Drosophila	Starvation resistance	383		partially recessive, mostly deleterious	1
Drosophila	Viability			deleterious, and recessive	2
Differences between stocks or inbred lines					
Drosophila	Longevity	4	15		3
Drosophila	Ovariole number	2	34		4
Drosophila	Starvation resistance	5	13		1
Drosophila	Courtship signal	3			5
Mice	Litter size	2	10	extensive epistasis	6
Mice	Litter size	3		dominance	7
Mice	Maternal performance	2	23	extensive epistasis	8
Nematode	Longevity	8		epistasis found	9
Selection lines					
Drosophila	Longevity	5, 6 and 7, total 10 5 in > 1 line		directional partial dominance	10
Drosophila	Longevity	chromosome III only 6, 7, 7, 8 and 9, total 11 10 present in > 1 line			11

References: 1. Harbison et al. (2004); 2. Lyman et al. (1996); 3. Mackay et al. (2005); 4. Wayne and McIntyre (2002); 5. Gleason et al. (2002); 6. Peripato et al. (2004); 7. Rocha et al. (2004c); 8. Peripato et al. (2002); 9. Ayyadevara et al. (2001); 10. Forbes et al. (2004); 11. Wilson et al. (2006).

higher for fitness characters. Given these limitations, I review the available evidence from QTL mapping and microarray studies.

4.1 Number of Loci

This issue involves two questions: How many loci affect the character? How many are polymorphic? The former question is addressed by studying mutations and the latter from identifying QTL differences between different populations (usually inbred).

Table 2 lists estimates of number of QTL for a range of fitness characters. For a given sized experiment, the number of loci that can be detected for a given character is expected to be least for inbred line crosses, followed by analyses of selection response, outbred populations and to be greatest for saturation mutagenesis, and empirical data generally supports these expectations (Frankham and Weber 2000).

From P element mutagenesis experiments in an isogenic *Drosophila* stock, many loci are capable of affecting a range of quantitative characters (Mackay and Anholt 2006). For example, 383 of 933 P element insertions affected starvation resistance (Harbison et al. 2004). In mice, 650 loci have some phenotypic effect on growth (Corva and Medrano 2001). It is unclear whether there are differences in number of loci affecting peripheral versus fitness characters.

Mapping of QTL differences between stocks or selection lines often reveals only up to 11 polymorphic QTL being involved for a range of traits in *Drosophila*, but these are clearly underestimates as they are expanded by finer resolution mapping (Mackay and Anholt 2006) (Table 2). For example, 4 QTL affecting longevity in *Drosophila* were separated into at least 15 QTL, 5 QTL affecting resistance to starvation into 14 QTL (Mackay and Anholt 2006), and two QTL affecting ovariole number into possibly 34 loci (Wayne and McIntyre 2002). In mice 10–30 QTL affecting differences between two inbred mice strains have been found for a range of morphological characters, with fewer being identified for fitness traits (Cheverud 2006). QTL in mice and plants have also been separated by further analyses into additional QTL (Cheverud 2006; Laurie et al. 2004).

As the number of identified QTL depends strongly on number of recombinants generated and total number of individuals scored, I will concentrate on the most extensive studies. Weber et al. (1999, 2001) analysed wing shape (a peripheral character) differences between high and low selection lines in a *Drosophila* study involving 519 recombinant isogenic chromosome III lines and measurements on 25,950 individuals and 701 recombinants for chromosome II with 35,050 individuals measured. Following extensive model comparisons, the chromosome III results could be explained either by a very large number of loci with no interactions, or 11 QTL with additive sub-equal effects, essentially no dominance and with large positive and negative interactions that nearly cancel each other. On chromosome II 10 QTL were identified that had sub-equal effects and showed epistatic interactions, but close to additive effects at each locus (no directional dominance). Analysis of 500 recombinant inbred lines (12 replicates each) from crossing the Illinois maize high and low oil selection lines indicated that greater than 50 QTL with primarily additive effects and little epistasis were involved (Laurie et al. 2004). Some of the QTL identified in crosses involving selection lines may be new mutations that arose in the lines, rather than representing loci polymorphic in the base population. From studies of over 31 mouse intercrosses, 133 obesity QTL were recorded, with at least 75 being distinct (Snyder et al. 2004; Cheverud 2006). By extrapolation from QTL identified in dairy cattle populations, Hayes and Goddard (2001) calculated that the populations were segregating for 50–110 QTL loci affecting each milk production quantitative trait. By a combination of QTL mapping, deletion mapping and microarray analyses, Wayne and McIntyre (2002) narrowed the differences in ovariole number (a fitness trait) between two inbred lines to perhaps 34 loci showing differential gene expression. Fine structure mapping of a 1 cM region in *Arabidopsis thaliana* for effects on biomass, revealed that it contained one QTL subject to balancing selection and a second region that interacted with it and that probably contained multiple loci with small effects (Kroymann and Mitchell-Olds 2005). These results have been interpreted as indicating a more polygenic view of effects than inferred from most QTL mapping experiments (Maloof 2006).

An alternative approach to the estimation of the number of QTL is to use microarrays to test for expression differences between differentiated populations. Mackay et al. (2005) reported that more than 3,700 probe sets (21% of the total) out of 18,000 differed significantly and consistently between replicate populations selected for slow versus fast mating speed. This consists of the loci involved in causing differences between the fast and slow mating speed lines, plus those that are coregulated by them. As a single locus can affect > 100 others, the above data could reduce to around 40 QTL determining selection response for mating speed, a number that is of the same order of magnitude as detected in large QTL mapping experiments. Yet, even this methodology will miss QTL of small effect, as the size of experiment sets a limit on the size of differences in transcription rates that can be detected.

In total, the above results indicate that many loci can affect quantitative loci (probably hundreds) and a substantial number of these are polymorphic (~ 50) in outbred populations. These will be minimum estimates, given the limitations of the analyses.

4.2 Distribution of Effects

Mather (1941, 1944) viewed polygenes (quantitative trait loci) as having small and approximately equal effects, while the infinitesimal model encompasses alleles of small but not necessarily equal effects (Fisher 1918). Conversely, models that are based on loci operating in biochemical pathways predict an exponential or L-shaped distribution of effects (Bost et al. 2001). A range of QTL studies have reported exponential distributions of effects of QTL for characters in both animals (Shrimpton and Robertson 1988; Hayes and Goddard 2001; Dilda and Mackay 2002; Rocha et al. 2004a; Mackay and Anholt 2006) and plants (Tanksley 1993) i.e. rare alleles of large effect and an increasing proportion of alleles of smaller effects. These are primarily for peripheral characters or mixtures of both peripheral and fitness characters and I am unaware of comparative data for fitness characters. Reports of exponential distributions of effects must be treated with caution, as QTL mapping experiments can produce L-shaped distributions regardless of the true shape of the underlying curve (Beavis 1994; Bost et al. 2001). Some QTL analyses identify loci of large effect and L-shaped distributions of effects (see above), while others do not (Weber et al. 1999, 2001; Laurie et al. 2004; Cheverud 2006; Hill 2005) and some of the disparate results involve the same trait. Notably, several of the largest studies have not yielded exponential distributions of effects and these are the ones that are more likely to be reliable.

4.3 Dominance

The full range of dominance has been found for QTL, including additive, dominant, recessive, overdominant and underdominant (Tanksley 1993; Kearsey and Farquhar 1998; Mackay and Lyman 2005). Generally, peripheral characters show more symmetrical dominance (average nearer additivity) of QTL (Weber et al. 1999, 2001; Dilda and Mackay 2002; Laurie et al. 2004; Rocha et al. 2004a, b), while fitness characters typically show considerable directional dominance (Rocha et al. 2004c; Table 2).

The relative contributions of dominance and overdominant loci to heterosis are a reflection of the genetic architecture of fitness. Using expression microarrays containing about 14,000 cDNAs and RNA from maize seedlings of two inbred parents and their F_1, Swanson-Wagner et al. (2006) found that 9.8% of transcripts differed among genotypes, with 78% showing additive gene action and 22% non-additive action. Of the non-additive cases whose gene action could be distinguished, 70% showed high parent dominance, 13% showed overdominance and the remainder showed low parent dominance, partial dominance or underdominance. As expected, dominance was the main form of gene action contributing to heterosis, but overdominance was also a significant contributor. This is consistent with the great majority of quantitative genetic analyses, which indicate that dominance of beneficial alleles is the major contributor to heterosis (see Falconer and Mackay 1996; Lynch and Walsh 1998). By contrast, Semel et al. (2006) concluded from QTL analyses in rice that overdominance is a major cause of heterosis, as introgression lines showed overdominance for fitness and yield, but it was virtually absent for peripheral characters. However, 46.4% of QTL contributing to heterosis for reproductive traits were favourable dominants, or deleterious recessives, and 14.7% were favourable overdominants, proportions that also favour dominance as the major mechanism.

4.4 Allele Frequencies

Information on allele frequencies for QTL is limited. Both common and rare molecular polymorphisms at individual candidate QTL of large effect have been detected in *Drosophila* (Mackay 2001). Wilson et al. (2006) found that of 11 chromosome III QTL loci involved in selection responses for longevity, all except 1 were found in more than one of 5 replicate populations, implying that most alleles were not rare. Analyses of 31 mouse intercrosses indicate that QTL alleles affecting obesity were generally rare, with 56% QTL being found in only one of the 31 studies (frequency < 3%) and the most common QTL being only detected in seven studies (23% frequency) (Snyder et al. 2004). However, these studies are subject to detection biases.

4.5 Epistasis and Sex Specific Effects

Epistasis is expected from the way loci act in biochemical pathways (Kacser 1989). Interactions among QTL loci (epistasis) have been a common feature of studies in *Drosophila* (Mackay and Anholt 2006), birds, mammals and plants, but are not detected in all cases (Carlborg and Haley 2004). Epistatic effects are also common in mice, especially for litter size where they account for a majority of the genetic variation in contrast to the situation for peripheral characters where their contribution was much less (Peripato et al. 2004).

Sex specific effects of QTLs are very common in *Drosophila* and occur in mice, but they do not appear to be as important in mice as in *Drosophila* (Cheverud 2006; Mackay and Anholt 2006).

4.6 Pleiotropy

Pleiotropic effects of QTL on different traits have been reported or inferred for many characters in *Drosophila* (Mackay and Anholt 2006), as predicted from the way loci operate in biochemical pathways (Keightley and Kacser 1987). Pleiotropy has also been reported for some mice QTL (Cheverud 2006). Weber et al. (2001) reported only minor pleiotropy of wing shape QTL on leg shape. Since the extent of pleiotropy depends upon the range of other characters investigated, it is not possible to compare the extent of pleiotropy across studies, characters and species.

In practice it is usually not possible to distinguish close linkage from pleiotropy for QTL, unless the QTL is a single locus, so caution is needed in interpreting the above studies.

4.7 Relation Between QTL and Major Effect Loci

There are now no reasons to believe that QTL and major genes differ qualitatively as predicted by Mather (1944). QTL have been found to be less extreme alleles at loci with known major effects in some, but not all cases (Frankham and Weber 2000; Mackay 2001; Mackay et al. 2005; Kearsey and Farquhar 1998; Mackay and Anholt 2006). For example, bristle QTL have been mapped to the known major effect loci *achaete-scute, bobbed, hairy, Hairless* and *scabrous* (Frankham 1980; Mackay and Lyman 2005).

Both structural and regulatory regions contribute to quantitative genetic variation, and the regulatory effects may be larger (Mackay and Anholt 2006). However, I am not aware of comparative data on control versus regulatory effects for fitness versus peripheral traits.

Overall, these analyses reveal many uncertainties in knowledge about the genetic architecture of both peripheral and fitness traits and several cases of conflicting information. Further, the published information in Table 2 cannot be relied upon to reflect reality. In general, there is more uncertainty than when we last reviewed this field (Frankham and Weber 2000).

5 Domestic Animals are Unlikely to be in Equilibrium

In contrast to wild populations where approximate equilibrium with their environment and mutation-selection equilibrium are credible, domestic animals are unlikely to be in equilibrium with their environment, or to be in mutation-selection equilibrium. Domestic animals have been spread around the globe from their ancestral homes into environments far removed from wild environments. Captive environments have changed several times for most domestic animals, artificial selection has been applied, some populations are derived from breed crosses and all have relatively small effective population sizes. These are likely to have wrought substantial changes on their genetic diversity for fitness.

5.1 Impact of Environmental Changes

Animals were first domesticated within the last 10,000 years. Whilst this may be long enough to reach new equilibria, most have experienced changes in environment within the last 50 years (Table 3). For example, domestic chickens derived from jungle fowl in South East Asia were first free ranging in villages at low density, followed by containment in enclosures at medium density, with most being used as dual purpose breeds for eggs and meat. Around the 1950s (about 50 generations ago), chickens were moved into high density housing or battery cages with specialisation of egg and meat strains. In Europe, layers are now moving back from cages to floor. Since domestication, chickens have been spread from South East Asia throughout the world, and their diet changed from self chosen seeds and green pick, to mash and to commercial pellets with changes in constituents.

As captive environments differ from wild ones, the genetic variants favoured in captivity differ somewhat from those favoured in natural environments. Darwin (1868) pointed out that natural selection for tameness and other adaptations to the captive environment were inevitable, especially in the context of domestication. Genetic adaptations to captive environments have been documented in mammals, fish, insects, plants and bacteria (see Zouros et al. 1982; Allard 1988; Frankham and Loebel 1992; Latter and Mulley 1995; Levin et al. 2001; Lewis and Thomas 2001; Nunney 2001; Woodworth et al. 2002; Heath et al. 2003). The extent of adaptive change may be very large, as illustrated in Table 4. For example, fecundity increased 12.6 fold over 100–150 generations in a captive population of large white butterfly (Lewis and Thomas 2001).

The time taken for adaptation to captivity to be completed is substantial. Wild *Drosophila* put into bottle cultures took about 80 generations to attain maximum captive fitness in large populations (Gilligan et al. 2003). Large captive *Drosophila melanogaster* populations in single pair vials showed an approximately linear decline in 'wild' fitness over 50 generations, so they had clearly not yet equilibrated with the captive environment (Woodworth et al. 2002). The time taken for initially homozygous populations to reach mutation-selection-drift equilibrium appears to be of the order of 250 generations in populations with N_e of 500 (García-Dorado et al. 2007).

Adaptations to captivity are overwhelmingly deleterious when species are returned to wild environments (reviewed by Frankham 2008). This effect has been

Table 3 Environmental changes in domestic animals

Domestic animal	Wild ancestor	Area of origin	Domestication (years ago)	Environmental changes		Generations* since most recent env change
				Density	Diet	
Domestic chicken	Red Jungle Fowl	South East Asia	8,000	Low to high	self choice seeds and green pick to mash to pellets	50
Pigs	Wild boar	Eurasia	9–10,000	Low to high	self choice to pellets	50
Turkeys	Wild turkeys	N and Central America	2,500	Low to high	self choice to pellets	50
Beef cattle	Auroch	Eurasia	8–10,000	Low to higher	from pasture to feedlot	~ 10
Dairy cattle	Auroch	Eurasia	8–10,000	Low to higher	pasture to pasture plus supplementary feeding to synthetic diet in winter in Northern hemisphere	~ 10
Sheep	Mouflon	Asia and Middle East	10,000	Low to higher	native pasture to improved ones with grasses and legumes (estrogenic)	
Rabbits	European rabbits	Europe	600	Free ranging to enclosures	self choice to human defined	
Salmon	Wild salmon	Atlantic ocean	recent	Low to high	self choice to human defined	~ 40

* Approximate generations since most recent major environmental change

Table 4 Extent of genetic adaptation to captive environments due to natural selection for reproductive fitness in a range of taxa (Frankham 2008). X refers to the magnitude of change, with 2x reflecting a doubling in fitness in captivity due to adaptation

Species	Adaptation	Generations
Rats	∼ 3x (fecundity)	25
Drosophila melanogaster		
extreme crowding	2x	8
modest crowding (bottles)	3.3x	80
Large white butterfly	12.6x (fecundity)	100–150
Barley	2x	∼ 50

reported for turkeys, amphibians, plants and many species of fish and biocontrol insects. The evidence from fish is extensive and informative (Allendorf and Luikart 2006; Araki et al. 2007a, 2007b). For example, lifetime reproductive success of hatchery fish stocks when returned to the wild declined at 37.5% per generation compared to that for wild fish (Araki et al. 2007b). Farmed fish, that have been domesticated by adapting to captivity plus subjected to artificial selection had a fitness only 16% that of wild fish in one study (Fleming et al. 2000) and 2–4% in another (McGinnity et al. 2003). Genetic adaptation to captivity is also strongly deleterious in biological control programs (Myers and Sabath 1980; Nunney 2001).

Since a considerable proportion of the genetic variation for fitness in the wild is probably due to rare, deleterious, partially recessive alleles in mutation-selection equilibrium, adaptation to captivity is expected to be partially due to such alleles, some of which are deleterious in the wild but beneficial in captivity. Evidence from *Drosophila*, screwworm flies, olive fruit flies and fish all points to genetic adaptation to captivity involving rare alleles, deleterious in the wild and partially recessive there (Frankham 2008). For example, an α-GPDH allele with low enzyme activity and an initial frequency of around 1% rose to high frequencies in 3 replicates in captive screwworm flies (Bush et al. 1976). The rise in frequency of initially rare alleles is expected to result in selective sweeps for linked neutral variation, and this has been observed for 7 of 8 microsatellite loci spread throughout the *Drosophila* genome (Montgomery, M.E., Woodworth, L.M., England, P.R., Briscoe, D.A. and Frankham, R., unpublished data). Until adaptation to new environments is completed, allele frequency distributions will be distorted compared to wild populations and transient polymorphism for newly favoured alleles are expected to be common.

Relatively recent environmental changes may be the reason that selection on fitness characters in domestic animals is rather more effective than would be expected from Fisher's fundamental theorem of natural selection which predicts that additive genetic variation for fitness will be rapidly depleted (Fisher 1930; Crow and Kimura 1970).

5.2 Impacts of Artificial Selection

Directional selection on a fitness trait will alter both allele frequencies and the proportion of segregating loci exhibiting heterozygotes advantage. If a wild population has three kinds of alleles affecting fitness, beneficial (+), deleterious (−) and ones showing heterozygote advantage (h), then directional high selection on fitness will tend to fix + alleles, eliminate − alleles and retain segregation at loci showing heterozygote advantage for the selected trait.

Equilibria for fitness alleles will be upset by artificial selection, even when this is applied to peripheral traits. Alleles can have effects on peripheral traits and fitness that are both beneficial (++), beneficial on the peripheral trait and neutral on fitness (+0), 0+, +−, −+, −−, beneficial on the peripheral trait and showing heterozygote advantage for fitness (+h), −h, h+, or h−. High selection in a large population on the peripheral trait will fix ++, +0, may fix +− depending upon the intensities of selection on peripheral trait and fitness and the fitness impact of − alleles, and eliminate −− and possibly −+ and −h. In time this will lead the population to retain variation predominantly due to +h, h+, and h−. In populations with smaller *Ne*, drift will have a greater impact and lead to chance fixations, but the loci subject to balancing selection will typically be less subject to drift than neutral loci. Thus, the proportion of genetic variation due to heterozygote advantage versus mutation-selection balance is likely to be much higher in long-term artificially selected populations than in wild populations at equilibrium. In line with this, many *Drosophila* selected populations have been found to have lethal alleles segregating at relatively high frequencies, presumably as the lethal allele is showing heterozygous advantage on the selected trait (Frankham et al. 1968b; Hollingdale 1971; Frankham 1980; Yoo 1980). These may have been pre-existing in the base population, but many were probably mutations arising during the course of selection (Frankham 1980). To my knowledge, lethals have not been recorded in mouse selection lines, but it is not clear whether this is due to low power to detect them in mice, or to generally lower selection pressure in mice (where selection is often done within families) compared to *Drosophila*.

Domestic livestock breeds and strains have been described with alleles showing heterozygote advantage in the selection scheme practised, including meatiness-stress sensitivity e.g. Halothane sensitivity in pigs (Webb et al. 1982), snorter dwarfism in beef cattle selected for chunky conformation (Marlowe 1964), and myostatin induced doubling muscling in racing whippet dogs (Mosher et al. 2007).

Given that the cause of difference in characteristics of fitness and peripheral characters is due to directional versus stabilising selection, long-term directional artificial selection on peripheral traits is expected to move their genetic architecture towards that of fitness traits.

Goddard (2009) discusses reasons to expect declines in fitness in populations selected for peripheral quantitative traits and for other fitness components to decline in populations selected for a fitness component (e.g. fertility and hatchability to decline in lines selected for egg production). Fitness declines have been observed widely in almost all long-term laboratory selection experiments and in most long-term livestock selection lines (Frankham et al. 1988; Rauw et al. 1998), with the study of Weber (1996) being an exception. In livestock, adverse changes have been observed in broiler chickens, turkeys, pigs and dairy cattle. For example, broiler chickens selected intensely for high growth rates for many generations show substantial declines in fitness (increased frequency of double yolker eggs, poor hatchability and poor fertility) and typically require changed management, such as restricted feeding to produce tolerable numbers of offspring. They also suffer higher mortality, elevated levels of sudden deaths and leg problems and have poorer immune function. There are currently concerns about declines in reproductive fitness of dairy cattle. These adverse fitness changes in selection lines can be due to pleiotropic effects of loci favourable to the direction of selection, to chance fixation (inbreeding depression), or to linkage disequilibrium and selective sweeps. The above observations indicate changes in the architecture of fitness in directionally selected lines for both fitness components and peripheral characters.

Artificial selection may also change the selective value of alleles with respect to fitness i.e. to generate epistatic effects.

5.3 Crossing and Introgression

Crossing of differentiated populations generates linkage disequilibrium. Some current commercial breeds or strains have been created from breed or strain crosses, including broiler chickens, some cattle breeds (e.g. Santa Gertrudis, Brangus, Australian milking zebu), some pig breeds (e.g. Lacombe), Australian merino sheep, some meat sheep (Texel) and thoroughbred race horses. In other cases single loci have been introgressed into breeds and strains, as in the cases of polled cattle and sheep and white feathers in broiler chickens and turkeys. In addition to these acknowledged hybridizations, illicit introgression has also occurred. For example, the leanness in pigs associated with Halothane anaesthetic and stress sensitivity increased in frequency in several breeds in a manner suggesting introgression. Further, I am aware of the use of 'topping up' by a broiler breeder, where very strong selection was applied to competitor's commercial stock and the resulting individuals added to a nucleus broiler strain.

Hybridization and introgression imply that such breeds and strains are not in linkage equilibrium and probably not in mutation-selection equilibrium. Further, patterns of gene interactions may be altered in hybrid populations if there is coadaptation in function of alleles across loci, as is likely for fitness characters.

5.4 Effects of Small Ne

Low N_e in domestic livestock leads to reduced genetic diversity and to build up of linkage disequilibrium and presumably to associative overdominance (Ohta and Kimura 1969; Latter 1998). Low N_e/N ratios are a general feature of unmanaged populations across a broad range of taxa, with 0.1 being the average (Frankham 1995). While breeding is now managed in domestic breeds, for much of the time since domestication it will have been essentially unmanaged, often with low N_e (Brotherstone and Goddard 2005). In addition, N_e/N ratios are lowered by artificial selection, especially selection based on progeny testing and use of artificial insemination (Robertson 1961; Wray and Thompson 1990). For example, Holstein dairy cattle number in the millions, but only have $N_e \sim 50$ (Brotherstone and Goddard 2005), while Jerseys also have $N_e \sim 50$.

In small populations selective sweeps will encompass a greater length of chromosome than in large populations due to greater linkage disequilibrium. Equilibrium frequencies for recessives alleles in mutation-selection balance are lower in small than large populations and there is also modest reduction in equilibrium frequencies for partially-recessive deleterious alleles (Crow and Kimura 1970).

6 Discussion and Implications for Livestock Breeding

The $R = Sh^2$ breeders equation has been used for about 70 years. It was validated for peripheral characters about 50 years ago by Clayton et al. (1957) and subsequently in Stuart Barker's laboratory (Frankham et al. 1968a) and elsewhere. Its predictions have generally been reasonable for base populations, but predictions for response in selection lines have been relatively poor (Frankham et al. 1968b; Sheridan 1988). For example, realised heritabilities in 27 abdominal bristle selection lines were only approximately one-half the heritabilities estimated in those lines (Frankham et al. 1968b). Further, lines at selection plateaux typically show non-zero heritabilities (Sheridan 1988). Increases in the proportion of loci showing overdominance in selection lines, as predicted above, could explain the poor ability to predict selection response from populations subject to prior selection. The Bulmer effect (Bulmer 1980) should lead to better agreement between predicted and observed response in populations subjected to prior selection than in lines selected from a base population subject to no prior selection, opposite to the observations (Sheridan 1988). Inbreeding can also lead to poor agreement between observed and predicted selection response for traits subject to inbreeding depression, but this does not apply to traits that do not suffer inbreeding depression, such as abdominal bristle number. In contrast to the above studies, Martinez et al. (2000) found good agreement between estimated and realised heritabilities in mouse selection lines.

Fitness traits typically show asymmetrical response to bi-directional selection (Frankham 1990) that is not predicted by the breeders equation. Asymmetry in fitness characters and its practical implications have been recognised, largely through arguments made by Bob Gowe in Canada. Gowe (1983 and *pers. comm.*) predicted that culling on fitness will be effective, even though $V_A \sim 0$. His argument was essentially one of a non-linear heritability due to rare, deleterious recessive alleles in mutation-selection balance, with heritability being higher at the low than the high end of the distribution. Further, by culling on fertility and hatchability in layer stocks of chickens Gowe prevented these declining, but he did not do a controlled experiment. Subsequently a controlled, replicated test in *Drosophila* showed culling on fitness to be effective and to yield similar response in the directionally selected trait in the fitness culling and non-culling treatments (Frankham et al. 1988). Anecdotal evidence indicates that several practical commercial chicken breeders used such culling on fitness in the 1980–1990s, but I do not know if it continues. There has been sporadic work on non-linear heritabilities and some cases reported (Abplanalp 1961; Nishida and Abe 1974; Robertson 1977; Bulmer 1980; Maki-Tanila 1982). However, there is no definitive evidence available to determine whether fitness traits show consistent asymmetry of offspring-parent regressions.

There has been little success in predicting the level of inbreeding depression in particular populations, and the extent of heterosis in crosses between particular populations (Falconer and Mackay 1996). Analytical theory correctly predicts the dependence of inbreeding depression and heterosis on the inbreeding level, and a linear relationship between inbreeding depression and inbreeding coefficient in the absence of epistasis. Plant breeders have had some success in predicting heterosis, based upon general combining ability from crosses between arrays of inbred lines.

Progress with improving prediction of selection response, inbreeding depression and heterosis for fitness characters in the last 50 years has been very limited. There is an urgent need for models that are more realistic and yield better practical predictions for fitness characters.

Use of equilibrium model results for livestock can rarely be justified in the light of the information above on adaptive changes in populations whose environments have changed and on the impacts of small effective sizes. If livestock are not at equilibrium with their environment, there are likely to be higher levels of additive genetic variation and higher heritabilities for fitness. The genetic architecture of fitness variation is likely to be different, with more intermediate frequency alleles due to transient polymorphisms for alleles that are headed for fixation. The asymmetry in allele frequency distribution will be lessened in rapidly adapting populations and they are expected to show less asymmetry in response to selection for fitness traits.

Linkage disequilibrium is expected due to both selective sweeps from alleles involved in adaptation to the new environment, and through low effective population size due to constrained numbers of individuals (compared to wild populations) and the inbreeding effects of artificial selection. Consequently, associative overdominance should be more important in domestic livestock than in large wild populations.

Extrapolations regarding the genetic architecture of fitness from wild populations to captive/domesticated populations for fitness traits are most unlikely to be valid. Further, extrapolations from captive populations of laboratory species to the wild are unlikely to be valid. Extrapolations from laboratory populations not subject to directional selection are unlikely to apply to domestic animals subject to artificial selection. Even extrapolations from directionally selected laboratory species to domestic livestock will be difficult unless selection is applied to related traits for similar durations and generations in captivity have been similar.

7 Conclusions

The architecture of genetic diversity for fitness in wild populations differs from that for peripheral characters in terms of dominance, symmetry of allele frequencies and dominance, levels of non-additive genetic variation, heritabilities, asymmetry of selection responses and occurrence of inbreeding depression and heterosis. In general, changes in fitness traits due to selection, inbreeding or crossing are not well predicted in animal breeding, compared to peripheral traits. Livestock populations are not likely to be in equilibrium for fitness traits due to environmental changes, limited effective population sizes and artificial selection. Given their importance and the uncertainties on many important issues, fitness traits deserve a higher priority than they are currently receiving in evolutionary and quantitative genetics and animal breeding research.

Acknowledgments I thank Bill Hill and an anonymous reviewer for many helpful comments on the manuscript.

References

Abplanalp, H. 1961. Linear heritability estimates. *Genet. Res.* **21**: 439–448.

Allard, R.W. 1988. Genetic changes associated with the evolution of adaptedness in cultivated plants and their wild progenitors. *J. Hered.* **79**: 225–238.

Allendorf, F.W. and Luikart, G. 2006. *Conservation and the Genetics of Populations*, Blackwell, Malden, MA, USA.

Araki, H., Arden, W.R., Olsen, E.M., Cooper, B. and Blouin, M.S. 2007a. Reproductive success of captive-bred steelhead trout in the wild: evaluation of three hatchery programs in the Hood River. *Conserv. Biol.* **21**: 181–190.

Araki, H., Cooper, B. and Blouin, M.S. 2007b. Genetic effects of captive breeding cause a rapid, cumulative fitness decline in the wild. *Science* **318**: 100–103.

Ayyadevara, S., Ayyadevara, R., Hou, S., Thaden, J.J. and Shmookler Reis, R.J. 2001. Genetic mapping of quantitative trait loci governing longevity of *Caenorhabditis elegans* in recombinant-inbred progeny of a Bergerac-BO x RC301 interstrain cross. *Genetics* **157**: 655–666.

Barker, J.S.F. 1979. Inter-locus interactions: A review of experimental evidence. *Theor. Popul. Biol.* **16**: 323–346.

Barton, N.H. and Keightley, P.D. 2002. Understanding quantitative genetic variation. *Nat. Rev. Genet.* **3**: 11–21.

Barton, N.H. and Turelli, M. 1989. Evolutionary quantitative genetics: how little do we know? *Ann. Rev. Genet.* **23**: 337–370.

Beavis, W.D. 1994. The power and deceit of QTL experiments: lessons from comparative QTL studies. *Proc. 49th Annu. Corn Sorghum Ind. Res. Conf.*, American Seed Trade Association, Washington, D.C., USA, pp. 252–268.

Bost, B., de Vienne, D., Hospital, F., Moreau, L. and Dillman, C. 2001. Genetic and nongenetic bases for the L-shaped distribution of quantitative trait loci effects. *Genetics* **157**: 1773–1787.

Briscoe, D.A., Malpica, J.M., Robertson, A., Smith, G.J., Frankham, R., Banks, R.G. and Barker, J.S.F. 1992. Rapid loss of genetic variation in large captive populations of *Drosophila* flies: Implications for the genetic management of captive populations. *Conserv. Biol.* **6**: 416–425.

Brotherstone, S. and Goddard, M. 2005. Artificial selection and the maintenance of genetic variance in the global dairy cow population. *Philos. Trans. R. Soc. Lond. B Biol. Sci.* **360**: 1479–1488.

Bulmer, M.G. 1980. *The Mathematical Theory of Quantitative Genetics*, Clarendon Press, Oxford, UK.

Bush, G.L., Neck, R.W. and Kitto, G.B. 1976. Screwworm eradication: Inadvertent selection for noncompetitive ecotypes during mass rearing. *Science* **193**: 491–493.

Carlborg, Ö. and Haley, C.S. 2004. Epistasis: too often neglected in complex trait studies? *Nat. Rev. Genet.* **5**: 618–625.

Charlesworth, B. and Hughes, K.A. 2000. The maintenance of genetic variation in life-history traits. In *Evolutionary Genetics from Molecules to Morphology* (R.S. Singh and C. Krimbas, eds.), Cambridge University Press, Cambridge, UK, pp. 369–392.

Charlesworth, B., Miyo, T. and Borthwick, H. 2007. Selection response of means and inbreeding depression for female fecundity in *Drosophila melanogaster* suggest contributions from intermediate-frequency alleles to quantitative trait variation. *Genet. Res.* **89**: 85–91.

Cheverud, J.M. 2006. Genetic architecture of quantitative variation. In *Evolutionary Genetics: Concepts and Case Studies* (C.W. Fox and J.B. Wolf, eds.), Oxford University Press, New York, USA, pp. 288–309.

Clayton, G.A., Morris, J.A. and Robertson, A. 1957. An experimental check on quantitative genetical theory. I. Short-term responses to selection. *J. Genet.* **55**: 131–151.

Corva, P.M. and Medrano, J.F. 2001. Quantitative trait loci (QTLs) mapping for growth traits in the mouse: A review. *Genet, Sel. Evol.* **33**: 105–132.

Crnokrak, P. and Roff, D.A. 1995. Dominance variance: associations with selection and fitness. *Heredity* **75**: 530–540.

Crow, J.F. and Kimura, M. 1970. *An Introduction to Population Genetics Theory*, Harper and Rowe, New York, USA.

Darwin, C. 1868. *The Variation of Animals and Plants Under Domestication*, John Murray, London, UK.

DeRose, M.A. and Roff, D.A. 1999. A comparison of inbreeding depression in life-history and morphological traits in animals. *Evolution* **53**: 1288–1292.

Dilda, C. and Mackay, T.F.C. 2002. The genetic architecture of *Drosophila* sensory bristle number. *Genetics* **162**: 1655–1674.

Dolan, R. and Robertson, A. 1975. The effect of conditioning the medium in *Drosophila* in relation to frequency-dependent selection. *Heredity* **35**: 311–316.

Endler, J.A. 1986. *Natural Selection in the Wild*, Princeton University Press, Princeton, NJ, USA.

Falconer, D.S. and Mackay, T.F.C. 1996. *Introduction to Quantitative Genetics*, 4th ed. Longman, Harlow, England.

Fleming, I.A., Hindar, K., Mjolnerod, I.B., Jonsson, B., Balstad, T. and Lamberg, A. 2000. Lifetime success and interactions of farm salmon invading a native population. *Proc. R. Soc. Lond. B* **267**: 1517–1523.

Fisher, R.A. 1918. The correlation between relatives on the supposition of Mendelian inheritance. *Trans. R. Soc. Edinb.* **52**: 399–433.

Fisher, R.A. 1930. *The Genetical Theory of Natural Selection*, Clarendon Press, Oxford, UK.

Forbes, S.N., Valenzuela, R.K., Keim, P. and Service, P.M. 2004. Quantitative trait loci affecting lifespan in replicated populations of *Drosophila melanogaster*. I. Composite interval mapping. *Genetics* **168**: 301–311.

Frankham, R. 1980. Origin of genetic variation in selection lines. In *Selection Experiments in Laboratory and Domestic Animals* (A. Robertson, ed.), Commonwealth Agricultural Bureaux, Farnham Royal, Bucks, UK, pp. 56–68.

Frankham, R. 1990. Are responses to artificial selection for reproductive fitness traits consistently asymmetrical? *Genet. Res.* **56**: 35–42.

Frankham, R. 1995. Effective population size/adult population size ratios in wildlife: A review. *Genet. Res.* **66**: 95–107.

Frankham, R. 2008. Genetic adaptation to captivity in species conservation programs. *Mol. Ecol.* **17**: 325–333.

Frankham, R. and Loebel, D.A. 1992. Modelling problems in conservation genetics using captive *Drosophila* populations: Rapid genetic adaptation to captivity. *Zoo Biol.* **11**: 333–342.

Frankham, R. and Weber, K. 2000. Nature of quantitative genetic variation. In *Evolutionary Genetics: From Molecules to Man* (R. Singh and C.B. Krimbas, eds.), Cambridge University Press, Cambridge, UK, pp. 351–368.

Frankham, R., Jones, J.P. and Barker, J.S.F. 1968a. The effect of population size and selection intensity in selection for a quantitative character in *Drosophila*. I. Short-term response to selection. *Genet. Res.* **12**: 237–248.

Frankham, R., Jones, L.P. and Barker, J.S.F. 1968b. The effects of population size and selection intensity in selection for a quantitative character in *Drosophila*. III. Analyses of the lines. *Gene. Res.* **12**: 267–283.

Frankham, R., Yoo, B.H. and Sheldon, B.L. 1988. Reproductive fitness and artificial selection in animal breeding: culling on fitness prevents a decline in reproductive fitness in lines of *Drosophila melanogaster* selected for increased inebriation time. *Theor. Appl. Genet.* **76**: 909–914.

Garcia-Dorado, A., Lopez-Fanjul, C. and Caballero, A. 1999. Properties of spontaneous mutations affecting quantitative traits. *Genet. Res.* **74**: 341–350.

García-Dorado, A., Ávila, V., Sánchez-Molano, E., Manrique, A. and López-Fanjul, C. 2007. The build up of mutation-selection-drift balance in laboratory *Drosophila* populations. *Evolution* **61**: 653–665.

Gilligan, D.M., Briscoe, D.A. and Frankham, R. 2003. Dynamics of genetic adaptation to captivity. *Conserv. Genet.* **4**: 189–197.

Gilligan, D.M., Briscoe, D.A. and Frankham, R. 2005. Comparative losses of quantitative and molecular genetic variation in finite populations of *Drosophila melanogaster*. *Genet. Res.* **85**: 47–55.

Gleason, J.M., Nuzhdin, S.V. and Ritchie, M.G. 2002. Quantitative trait loci affecting a courtship signal in *Drosophila melanogaster*. *Heredity* **89**: 1–6.

Goddard, M.E. 2009. Fitness traits in animal breeding programs. In *Adaptation and Fitness in Animal Populations*. Evolutionary and Breeding Perspectives on Genetic Resource Management (J.H.J. van der Werf, H.-U. Graser, R. Frankham and C. Gondro, eds.), Springer, The Netherlands.

Gowe, R.S. 1983. Lessons from selection studies in poultry for animal breeders. *Proc. 32nd. Annu. Breeders' Roundtable*, pp. 22–50.

Harbison, S.T., Yamamoto, A.H., Fanara, J.J., K.K. Norga. and Mackay, T.F.C. 2004. Quantitative loci affecting starvation resistance in *Drosophila melanogaster*. *Genetics* **166**: 1807–1823.

Hayes, B. and Goddard, M.E. 2001. The distribution of the effects of genes affecting quantitative traits in livestock. *Genet. Sel. Evol.* **33**: 209–229.

Heath, D.D., Heath, J.W., Bryden, C.A., Johnson, R.M. and Fox, C.W. 2003. Rapid evolution of egg size in captive salmon. *Science* **299**: 1738–1740.

Hedrick, P.W. 2005. *Genetics of Populations*, 3rd edition, Jones and Bartlett, Sudbury, MA, USA.

Hill, W.G. 2005. A century of corn selection. *Science* **307**: 683–684.

Hill, W.G. and Zhang, X.-S. 2009. Maintaining genetic variation in fitness. In *Adaptation and Fitness in Animal Populations*. Evolutionary and Breeding Perspectives on Genetic Resource Management (J.H.J. van der Werf, H.-U. Graser, R. Frankham and C. Gondro, eds.), Springer, The Netherlands.

Hoffmann, A.A., Hallas, R.J., Dean, J.A. and Schiffer, M. 2003. Low potential for climatic stress adaptation in a rainforest *Drosophila* species. *Science* **301**: 100–102.

Hollingdale, B. 1971. Analyses of some genes from abdominal bristle selection lines in *Drosophila melanogaster*. *Theor. Appl. Genet.* **41**: 292–301.

Houle, D., Morikawa, B. and Lynch, M. 1996. Comparing mutational variabilities. *Genetics* **143**: 1467–1483.

Johnson, T. and Barton, N. 2005. Theoretical models of selection and mutation on quantitative traits. *Philos. Trans. R. Soc. Lond. B Biol. Sci.* **360**: 1411–1425.
Kacser, H. 1989. Quantitative variation and the control analysis of enzyme systems. In *Evolution and Animal Breeding: Reviews on Molecular and Quantitative Approaches in Honour of Alan Robertson* (W.G. Hill and T.F.C. Mackay, eds.), C.A.B. International, Wallingford, UK, pp. 219–226.
Keightley, P.D. and Kacser, H. 1987. Dominance, pleiotropy and metabolic structure. *Genetics* **117**: 319–329.
Kearsey, M.J. and Kojima, K. 1967. The genetic architecture of body weight and egg hatchability in *Drosophila melanogaster. Genetics* **56**: 23–37.
Kearsey, M.J. and Farquhar, A.G.L. 1998. QTL analysis in plants: Where are we now? *Heredity* **80**: 137–142.
Kellermann, V.M., van Heerwaarden, B., Hoffmann, A.A. and Sgrò, C.M. 2006. Very low additive genetic variance and evolutionary potential in multiple populations of two rainforest *Drosophila* species. *Evolution* **60**: 1104–1108.
Kelly, J.K. 2003. Deleterious mutations and the genetic variance of male fitness components in *Mimulus guttatus. Genetics* **164**: 1071–1085.
Kelly, J.K. and Willis, J.H. 2001. Deleterious mutations and genetic variation for flower size in *Mimulus guttatus. Evolution* **55**: 937–942.
Kingslover, J.G., Hoekstra, H.E., Hoekstra, J.M., Berrigan, D., Vignieri, S.N., Hill, C.E., Hoang, A., Gilbert, P. and Beerli, P. 2001. The strength of phenotypic selection in natural populations. *Am. Nat.* **157**: 245–261.
Kroymann, J. and Mitchell-Olds, T. 2005. Epistasis and balanced polymorphism influencing complex trait variation. *Nature* **435**: 95–98.
Latter, B.D.H. 1998. Mutant alleles of small effect are primarily responsible for the loss of fitness with slow inbreeding in *Drosophila melanogaster. Genetics* **148**: 1143–1158.
Latter, B.D.H. and Mulley, J.C. 1995. Genetic adaptation to captivity and inbreeding depression in small laboratory populations of *Drosophila melanogaster. Genetics* **139**: 255–266.
Laurie, C.C., Chasalow, S.D., LeDeaux, J.R., McCarroll, R., Bush, D., Huage, B., Lai, C., Clark, D., Rocheford, T.R. and Dudley, J.W. 2004. The genetic architecture of response to long-term artificial selection for oil concentration in the maize kernel. *Genetics* **168**: 2141–2155.
Levin, P.S., Zabel, R.W. and Williams, J.G. 2001. The road to extinction is paved with good intentions: negative association of fish hatcheries with threatened salmon. *Proc. R. Soc. Lond. B: Biol. Sci.* **268**: 1153–1158.
Lewis, O.T. and Thomas, C.D. 2001. Adaptation to captivity in the butterfly *Pieris brassicae* (L) and the implications for *ex situ* conservation. *J. Insect Conserv.* **5**: 55–63.
Lewontin, R.C. 1974. *The Genetic Basis of Evolutionary Change*, Columbia University Press, New York, USA.
Lynch, M. and Walsh, B. 1998. *Genetics and Analysis of Quantitative Traits*, Sinauer, Sunderland, MA, USA.
Lyman, R.F., Lawrence, F., Nuzhdin, S.V. and Mackay, T.F.C. 1996. Effects of single *P*-element insertions on bristle number and viability in *Drosophila melanogaster. Genetics* **143**: 277–292.
Mackay, T.F.C. 1989. Mutation and the origin of quantitative variation. In *Evolution and Animal Breeding: Reviews on Molecular and Quantitative Approaches in Honour of Alan Robertson* (W.G. Hill and T.F.C. Mackay, eds.), C.A.B. International, Wallingford, UK, pp. 113–119.
Mackay, T.F.M. 2001. The genetic architecture of quantitative traits. *Annu. Rev. Genet.* **35**: 303–339.
Mackay, T.F.C. 2004. The genetic architecture of quantitative traits: lessons from *Drosophila. Curr. Opin. Genet. Devel.* **14**: 253–257.
Mackay, T.F.C. and Anholt, R.R. 2006. Of flies and man: *Drosophila* as a model for human complex traits. *Annu. Rev. Genomics and Hum. Genet.* **7**: 339–367.
Mackay, T.F.C. and Lyman, R.F. 2005. *Drosophila* bristles and the nature of quantitative genetic variation. *Philos. Trans. R. Soc. Lond. B Biol. Sci.* **360**: 1513–1527.

Mackay, T.F.C., Heinsohn, S.L., Lyman, R.F., Moehring, A.J., Morgan, T.J. and Rollmann, S.M. 2005. Genetics and genomics of *Drosophila* mating behavior. *Proc. Natl Acad. Sci., U.S.A* **102** Suppl. 1: 6622–6629.

Maki-Tanila, A. 1982. Ph.D. thesis. *The Validity of the Heritability Concept in Quantitative Genetics*. University of Edinburgh, Edinburgh, UK.

Maloof, J.N. 2006. Small but not forgotten. *Heredity* **96**: 1–2.

Marlowe, T.J. 1964. Evidence of selection for the snorter dwarf gene in cattle. *J. Anim. Sci.* **23**: 454–460.

Martinez, V., Bünger, L. and Hill, W.G. 2000. Analysis of response to 20 generations of selection for body composition in mice: fit to infinitesimal model assumptions. *Genet. Sel. Evol.* **32**: 3–21.

Mather, K. 1941. Variation and selection of polygenic characters. *J. Genet.* **41**: 159–193.

Mather, K. 1944. The genetical activity of heterochromatin. *Proc. R. Soc. Lond. B* **132**: 308–332.

McGinnity, P., Prodöhl, P., Ferguson, A., Hynes, R., Ó Maoiléidigh, N., Baker, N., Cotter, D., O'Hea, B., Cooke, D., Rogan, G., Taggart, J. and Cross, T. 2003. Fitness reduction and potential extinction of wild populations of Atlantic salmon, *Salmo salar*, as a result of interactions with escaped farmed salmon. *Proc. R. Soc. Lond. B. Biol. Sci.* **270**: 2443–2450.

McMillan, I. and Robertson, A. 1974. The power of methods for detecting major genes affecting quantitative characters. *Heredity* **32**: 349–356.

Mosher, D.S., Quignon, P., Bustamante, C.D., Sutter, N.B., Mellersh, C.S., Parker, H.G. and Ostrander, E.A. 2007. A mutation in the myostatin gene increases muscle mass and enhances racing performance in heterozygote dogs. *PLoS Genet.* **3**(5): e79.

Mousseau, T.A. and Roff, D.A. 1987. Natural selection and the heritability of fitness components. *Heredity* **59**: 181–197.

Mukai, T. 1988. Genotype-environment interaction in relation to the maintenance of genetic variability in populations of *Drosophila melanogaster*. In *Proceedings of the Second International Conference on Quantitative Genetics* (B.S. Weir, E.J. Eisen, M.M. Goodman and G. Namkoong, eds.), Sinauer, Sunderland, MA, USA, pp. 21–31.

Myers, D. and Sabath, M.D. 1980. Genetic and phenotypic variability, genetic variance, and the success of establishment of insect introductions for the biological control of weeds. *Proceedings of the V International Symposium on Biological Control of Weeds*, 91–102.

Nishida, A. and Abe, T. 1974. The distribution of genetic and environmental effects and the linearity of heritability. *Can. J. Genet. Cytol.* **16**: 3–10.

Nunney, L. 2001. Managing captive populations for release: a population genetic perspective. In *Quality Control of Natural Enemies used in Biological Pest Control: Theoretical Background and Development of Testing Procedures* (J.C. van Lenteren, ed.), Cambridge University Press, Cambridge, UK, pp. 73–88.

Ohta, T. and Kimura, M. 1969. Linkage disequilibrium due to random genetic drift. *Genet. Res.* **13**: 47–55.

Peripato, A.C., De Brito, R.A., Vaughn, T.T., Pletscher, L.S., Matioli, S.R. and Cheverud, J.M. 2002. Quantitative trait loci for maternal performance for offspring survival in mice. *Genetics* **162**: 1341–1353.

Peripato, A.C., De Brito, R.A., Matioli, S.R., Pletscher, L.S., Vaughn, T.T. and Cheverud, J.M. 2004. Epistasis affecting litter size in mice. *J. Evol. Biol.* **17**: 593–602.

Rauw, W.M., Kanis, E., Noordhuizen-Stassen, E.N. and Grommers, F.J. 1998. Undesirable side effects of selection for high production efficiency in farm animals: A review. *Livest. Prod. Sci.* **56**: 15–33.

Robertson, A. 1956. The effect of selection against extreme deviants based on deviation or on homozygosis. *J. Genet.* **54**: 236–248.

Robertson, A. 1961. Inbreeding in artificial selection programmes. *Genet. Res.* **2**: 189–194.

Robertson, A. 1977. The non-linearity of offspring-parent regression. In *Proceedings of the International Conference on Quantitative Genetics* (E. Pollak, O. Kempthorne and T.B. Bailey, eds.), Iowa State University Press, Ames, IA, USA, pp. 297–304.

Rocha, J.L., Eisen, E.J., Van Vleck, L.D. and Pomp, D. 2004a. A large-sample QTL study in mice: I. Growth. *Mamm. Genome* **15**: 83–99.

Rocha, J.L., Eisen, E.J., Van Vleck, L.D. and Pomp, D. 2004b. A large-sample QTL study in mice: II. Body composition. *Mamm. Genome* **14**: 100–113.

Rocha, J.L., Eisen, E.J., Siewerdt, F., Van Vleck, L.D. and Pomp, D. 2004c. A large-sample QTL study in mice: III. Reproduction. *Mamm. Genome* **15**: 878–886.

Rodríguez-Ramilo, S.T., Pérez-Figuero, A., Fernández, B., Fernández, J. and Caballero, A. 2004. Mutation-selection balance accounting for genetic variation for viability in *Drosophila melanogaster* as deduced from an inbreeding and artificial selection experiment. *J. Evol. Biol.* **17**: 528–541.

Roff, D.A. and Mousseau, T.A. 1987. Quantitative genetics and fitness: lessons from *Drosophila. Heredity* **58**: 103–118.

Semel, Y., Nissenbaum, J., Menda, N., Zinder, M., Krieger, U., Issman, N., Pleban, T., Lippman, Z., Gur, A. and Zamir, D. 2006. Overdominant quantitative trait loci for yield and fitness in tomato. *Proc. Natl Acad. Sci. U.S.A.* **103**: 12981–12986.

Sheridan, A.K. 1988. Agreement between estimated and realised genetic parameters. *Anim. Breed. Abstr.* **56**: 877–889.

Shrimpton, A. and Robertson, A. 1988. The isolation of polygenic factors controlling bristle score in *Drosophila* melanogaster. II. Distribution of third chromosome bristle effects within chromosome sections. *Genetics* **118**: 445–459.

Snyder, E.E., Walts, B., Perusse, L., Chagnon, Y.C., Weisnagel, S.J., Rankinen, T. and Bouchard, C. 2004. The human obesity gene map: the 2003 update. *Obes. Res.* **12**: 369–439.

Swanson-Wagner, R.A., Jia, Y., DeCook, R., Borsuk, L.A., Nettleton, D. and Schnable, P.S. 2006. All possible modes of gene action are observed in a global comparison of gene expression in a maize F_1 hybrid and its inbred parents. *Proc. Natl. Acad. Sci. U.S.A.* **103**: 6805–6810.

Tanksley, S.D. 1993. Mapping polygenes. *Annu. Rev. Genet.* **27**: 205–233.

Van Buskirk, J. and Willi, Y. 2006. The change in quantitative genetic variation with inbreeding. *Evolution* **60**: 2428–2434.

Wayne, M.L. and McIntyre, L.M. 2002. Combining mapping and arraying: An approach to candidate gene identification. *Proc. Natl. Acad. Sci. U.S.A.* **99**: 14903–14906.

Webb, A.J., Carden, A.E., Smith, C. and Imlah, P. 1982. Porcine stress syndrome in pig breeding. *Proceedings of the 2nd World Congress on Genetics Applied to Livestock Production*, pp. 588–608.

Weber, K.E. 1996. Large genetic change at small fitness cost in large populations of *Drosophila melanogaster* selected for wind tunnel flight: Rethinking fitness surfaces. *Genetics* **144**: 205–213.

Weber, K., Eisman, R., Morey, L., Patty, A., Sparks, J., Tausek, M. and Zeng, Z.-B. 1999. An analysis of polygenes affecting wing shape on chromosome 3 in *Drosophila melanogaster. Genetics* **153**: 773–786.

Weber, K., Eisman, R., Higgins, S., Morey, L., Patty, A., Tausek, M. and Zeng, Z.-B. 2001. An analysis of polygenes affecting wing shape on chromosome 2 in *Drosophila melanogaster. Genetics* **159**: 1045–1057.

Wilson, R.H., Morgan, T.J. and Mackay, T.F.C. 2006. High-resolution mapping of quantitative trait loci affecting increased lifespan in *Drosophila melanogaster. Genetics* **173**: 1455–1463.

Woodworth, L.M., Montgomery, M.E., Briscoe, D.A. and Frankham, R. 2002. Rapid genetic deterioration in captivity: Causes and conservation implications. *Conserv. Genet.* **3**: 277–288.

Wray, N.R. and Thompson, R. 1990. Prediction of rates of inbreeding in selected populations. *Genet. Res.* **55**: 41–54.

Yang, H.-P., Tanikawa, A.Y., Van Voorhies, W.A., Silva, J.C. and Kondrashov, A.S. 2001. Whole-genome effects of ethyl methanesulfonate-induced mutation on nine quantitative traits in outbred *Drosophila melanogaster. Genetics* **157**: 1257–1265.

Yoo, B.H. 1980. Long-term selection for quantitative characters in large replicate populations of *Drosophila melanogaster*. II. Lethals and visible mutants with large effects. *Genet. Res.* **35**: 19–31.

Zhang, X.-S. and Hill, W.G. 2005. Genetic variability under mutation selection balance. *Trends Ecol. Evol.* **20**: 468–470.

Zouros, E., Loukas, M., Economopoulos, A. and Mazomenos, B. 1982. Selection at the alcohol dehydrogenase locus of the olive fruit fly *Dacus oleae* under artificial rearing. *Heredity*, **48**: 169–185.

Fitness Traits in Animal Breeding Programs

Michael Goddard

Abstract Traits measuring reproduction and survival sometimes show a negative genetic trend in livestock populations despite their importance to profitability. This occurs due to inbreeding depression and selection for other traits. For many traits there are genes with an allele that increases the trait value but is initially at low frequency due to negative effects on fitness. Therefore, theory suggests that fitness traits will tend to decline due to selection for other traits even if this is not predicted by the genetic correlation in the base population. The most important recommendation to overcome this problem is simply to include fitness traits in the breeding objective, the recording scheme and the selection index. The economic weight given to fitness traits should take account of the high likelihood that they will be important in the future whereas other breeding objectives are more likely to change with time.

Keywords Breeding objectives · fitness · dairy cattle · reproduction · inbreeding depression

1 Introduction

I will define fitness traits as those measuring the survival and reproductive rate of animals. These traits are of economic value to livestock producers, so one would expect them to be included in breeding objectives that are designed to maximise profit. However, it is often claimed that selection programs, whose purpose is to increase profit, have resulted in a decline in fitness. The purpose of this paper is to consider why this might be the case and what should be done to achieve a better outcome.

Maintaining fitness in livestock does not necessarily mean maintaining traits or genotypes that occur in the wild progenitors of our domestic animals. For instance, aggressiveness in contests for females might be a fitness trait among wild males

---

M. Goddard (✉)
University of Melbourne and Department of Primary Industries, Victoria
e-mail: mike.goddard@dpi.vic.gov.au

J. van der Werf et al. (eds.), *Adaptation and Fitness in Animal Populations*,

but it is not a desirable trait in domestic livestock. Therefore I will consider fitness from a utilitarian viewpoint. I will assume that there is an overall objective that the animal breeder is trying to maximise and for brevity I will call this objective 'profit'. Fitness traits are only important if they contribute to this overall objective.

This objective should be a long term objective. This is important because economic conditions change with time and consequently some short term breeding objectives change with time, whereas fitness traits tend always to be valuable, regardless of changes in prices and costs. Goddard (1992) suggested that, to take account of this, the economic weight of traits that fluctuate with prices and costs be discounted to reflect the risk that they are no longer valuable in the future. For instance, the market premium for intramuscular fat content in meat might disappear at some time in the future. However, reproduction and survival are likely to be valuable under all future markets for meat. In general the economic weights for fitness traits are less likely to decline than the economic weights for other traits and so they should be discounted less than that of other traits.

2 Has Fitness Declined?

It is difficult to give a general answer to this question, but at least some fitness traits in some populations have declined. For instance, fertility in Holstein dairy cattle. In the USA days open increased genetically by 24 days between 1965 and 1995 (i.e. 0.8 days / year) but has not increased further since then due perhaps to selection for productive life (Van Raden pers. comm.). In the Netherlands, over the period 1982–1998, calving interval increased by 1.45 days / year (de Jong 2005). The greater change for the worse in the Netherlands compared to the USA could be due to changing from Dutch Friesian to American Holstein as well as the direct effect of selection. In Australia, Fulkerson et al. (2001) compared Holsteins of high and low ABV for fat plus protein and found the high ABV cows were open for 8 days longer. This comparison in ABV for fat and protein yield is equivalent to 15 years of genetic trend and also includes the effect of an increase in the proportion of American Holstein genes in the high ABV cows.

3 Why Has Fitness Declined?

Three explanations seem possible:

1. *This is a logical response given the economic weights and genetic parameters.* For instance, if fertility in dairy cows is of low economic importance compared to milk yield and negatively correlated with it, it might be logical to implement a selection index that drives fertility down.
   In my experience, this is seldom the case. Usually fitness traits have simply been ignored in the construction of the index because they were said to have low heritability or to be too hard to record.

2. *Fitness has declined due to an inappropriate breeding program but the underlying genetic process is well understood.*
   The two well understood processes are inbreeding depression and selection and these will be considered in the next two sections of the paper.
3. *Fitness has declined more than would have been expected from knowledge of genetic parameters in the base population.*
   Evidence for this and possible mechanisms will be discussed after the section on selection.

4 Inbreeding

Inbreeding depression occurs in most fitness traits although the estimates are not very precise or consistent. For instance, 1% inbreeding has been reported to cause an increase in days open of dairy cows of 0.3 days (Biffani et al. 2002), a decline in non-return rate of 0.16% (Wall et al. 2005) and a decline in calving rate of 0.4% (Cassell et al. 2003).The cost of inbreeding is even greater than this because many non-fitness traits also show inbreeding depression and inbreeding also reduces genetic variance and hence reduces future genetic gains. Milk production traits in dairy cows decline by about 0.3% per 1%F. The estimated effect on profit of 1%F varies widely from $3 per cow-year (Man 2004) to $24 per cow-lifetime (Smith et al. 1998).

Intense selection increases additive genetic gain and increases inbreeding, so economically optimum breeding programs do result in a moderate level of inbreeding. For instance, Goddard (1992) and Goddard and Smith (1990) found that optimum dairy cattle breeding programs had inbreeding rates of 0.2% per year. In fact the realised rate of inbreeding in dairy cattle is about 0.2% per year (Thompson et al. 2000a and b). Therefore, inbreeding depression seems to explain only a little of the decline in fertility. For instance, even the high inbreeding depression estimate of Cassell et al. (2003) of 0.4% reduction in calving rate per 1%F predicts only $0.2*0.4 = 0.08\%$ decline in calving rate per year compared with the observed trend of 0.4% per year in non-return rate.

These estimates of inbreeding depression typically come from data where the inbreeding has occurred in the recent past, so that selection has had little opportunity to prevent deleterious genes drifting up in frequency. However, the population as a whole is inbreeding much more slowly and so natural selection has an opportunity to oppose random increases in the frequency of deleterious alleles. This would imply that the actual inbreeding depression experienced by livestock populations is less than predicted above by estimating inbreeding depression per %F due to one generation of inbreeding. Experimental evidence supports this suggestion: slow inbreeding leads to less inbreeding depression than rapid inbreeding when the same inbreeding coefficient is reached (Ehiobu et al. 1989). Therefore we might expect the losses from inbreeding at 0.2% per year are less than predicted above.

In fact, slow inbreeding, due to finite population size in livestock populations, has occurred over many years. How rapid has this inbreeding been and how much inbreeding depression has it caused? The rate of inbreeding depends of course on the effective population size (Ne) in the past. For *Bos taurus* cattle, MacEachern (2007) has estimated Ne from molecular data to be >50, 000 prior to domestication, 1000–2000 after domestication, and falling to 50–100 in modern breeds such as Holstein.

How much inbreeding depression has this slow inbreeding caused over the evolution of modern cattle? One way to estimate the amount of inbreeding depression that this slow inbreeding has caused is from the heterosis that occurs when breeds are crossed. This heterosis can be compared to the inbreeding depression predicted by multiplying the inbreeding depression per %F estimated from inbreeding experiments by the inbreeding coefficient since the parent breeds diverged, estimated using genetic markers (Goddard and Ahmed 1982). Heterosis occurs in crosses between breeds and species that have diverged at different points in this history. In crosses among European *B. taurus* breeds, that have diverged at Ne = 50−2000, heterosis is not too different from that expected based on the inbreeding coefficient and the experimental rate of inbreeding depression per %F (Goddard and Ahmed 1982). This implies that natural and artificial selection since the European breeds diverged has been ineffective in preventing inbreeding depression (that is, in preventing random increases in the frequency of deleterious, recessive genes).

Crosses between *B. indicus* and *B. taurus* generate more heterosis than crosses within either sub-species indicating that even Ne = 50,000 may not prevent random increases in the frequency of some deleterious alleles. However, the heterosis was less than expected from genetic distance and inbreeding depression (Goddard and Ahmed 1982). The existence of heterosis in crosses implies that the breeds differ in gene frequency and that some detrimental recessive alleles have drifted to higher than expected frequency in some breeds, but this does not necessarily imply a decrease in the mean genetic merit of the population. A large population will eventually reach an equilibrium between mutation, selection and drift and its mean fitness will not decline further despite the inevitable slow inbreeding. At this equilibrium a number of (partially) recessive deleterious mutations have risen in frequency and heterosis occurs in crosses if these mutations are different in different breeds. However, if selection was completely effective in preventing these deleterious alleles increasing in frequency, there would be no heterosis in crosses. Thus it seems likely that inbreeding depression has been accumulating in cattle since their domestication and possibly before domestication. Therefore neither natural nor artificial selection will be effective in preventing inbreeding depression in modern breeds with Ne = 50−100.

The explanation of these results could be that inbreeding depression is caused by a large number of genes of small effect, so that the selection coefficient(s) on individual genes is so small that $N_e s < 1$. This would require 1000's of genes to explain inbreeding depression. An alternative explanation is that differential selection pressures in different breeds causes their gene frequencies to diverge and therefore heterosis occurs when the breeds are crossed. This alternative explanation implies that the difference in allele frequencies between breeds, that is necessary for heterosis to occur, is due, in part, to differential selection, not just genetic drift.

In summary, if there are a large number of genes affecting fitness, neither artificial nor natural selection will be able to completely prevent the decline in fitness caused by inbreeding.

5 Selection

Some fitness traits have low heritability but this does not imply negligible genetic variance. Often the low heritability is because the phenotypic variance is large rather than the genetic variance is small. For instance, 6-week pregnancy rate in dairy cattle has a mean of about 0.7, a phenotypic standard deviation of 0.45, $h^2 = 0.04$, genetic standard deviation of 0.09 and genetic CV of $0.09/0.7 = 13\%$ which is as great as that of milk yield.

Despite economically significant genetic variation, fitness traits have often been omitted from the selection index. This has been justified by the low heritability and, in some cases, by the practical difficulties in recording the trait. For instance, reproductive performance of dairy and beef cows was not always captured in the central database from which EBVs were calculated. In Australian dairy cattle today, there is very incomplete recording of pregnancy tests, cow and calf mortalities (Haile-Mariam et al. 2007b).

Fitness traits are expected to decline if they are not directly selected and are unfavourably correlated with a trait under selection. Van Raden (pers. comm.) calculates that most of the increase in days open in dairy cows in USA is expected due to the genetic trend in milk production and the genetic correlation between them of 0.35. This would predict a genetic decline in 6 week pregnancy rate of $< 0.5\%$ per year (calculations in appendix 1).

However, if fitness traits are recorded and included in the index, this decline caused by unfavourable correlations, can be reduced or reversed. In Swedish Red and White dairy cows, fertility has been included in the index and has not declined although the fertility of Swedish Holsteins, that have a large American influence, has declined (Philipsson et al., 2005). The genetic trend in dairy cow fertility in USA has been flat since 1995 perhaps due to selection for productive life. In Australian dairy cattle the correlation between the fertility ABV and the economic index (APR) is close to zero and so selection on the APR would be expected to cause no change in fertility.

Why are genetic correlations between production traits and fitness traits often unfavourable? Natural selection is expected to optimise the genotype with respect to fitness. For some traits we expect that fitness will be greatest for an intermediate value of the traits (so called stabilising selection). For instance, if the milk yield of cows is too low the survival and growth of the calf suffers but if milk yield is too high the fertility and survival of the cow is reduced. Under these circumstances, directional, artificial selection for a trait subject to stabilising, natural selection leads to a decline in fitness including a decline in some components of fitness.

The environment of modern livestock is different to that in the past or to that of their wild ancestors, so the optimum value for traits may be different to what it was in the past and so the mean value of a trait may not be optimum for the modern environment. Certainly if the aim is to maximise profit, then the historical mean of traits is unlikely to be optimum. However, because the optimum results from a compromise between different components of fitness, it is likely that at least some components will decline as selection moves the mean. For instance, the optimum milk yield in modern cattle might be higher than in their wild relatives but this does not mean that increasing milk yield will not decrease fertility.

6 Long Term Response

There is a suspicion that fitness of livestock has declined faster than predicted from selection theory. Experiments demonstrate that the theory of selection response, based on heritabilities and genetic correlations, gives good predictions of genetic change in the short term. However, in the longer term we expect allele frequencies to change, due in part to selection, and hence genetic parameters to change. Therefore, to predict long term response to selection requires knowledge of allele frequencies.

There are several reasons to believe that alleles that have an effect on quantitative traits and a detrimental effect on fitness are segregating in populations at low frequency. These reasons include, firstly, the genetic variance added to a typical quantitative trait by mutation (Vm) is at least 0.001 times the environmental variance (Ve) per generation (Lynch and Walsh 1998). If these mutations were neutral, an equilibrium would be reached between mutation and genetic drift when the genetic variance reaches 2NeVm where Vm is the mutational variance $= 0.001$ Ve. This means that if Ne $= 10000$ (as commonly quoted for humans) the heritability would be $20/21 = 0.95$ because $2^{*}10000^{*}.001$ Ve $= 20$ Ve. We do not observe these high heritabilities, presumably because selection eliminates many of the mutations that are causing the mutational variance. Before mutations are eliminated, they will be at low frequency but still contributing to the genetic variance.

Secondly, under the neutral model, the heritability increases as Ne of the population increases. This is not what is observed, presumably because selection is eliminating some mutations and selection is more effective in larger populations. This explanation implies that many of the selection coefficients are small. For instance, if mutations were divided into those that were highly deleterious and eliminated quickly and those that were neutral, heritability would still increase as Ne increased. Small selection coefficients imply that mutations may segregate at low frequency for many generations. Models of genetic variation in quantitative traits that assume many genes subject to small selection pressures mimic the observed genetic variance and long term response to selection reasonably well (Goddard and Hayes 2003; Zhang and Hill 2005)

Thirdly, polymorphisms within the protein coding DNA that change an amino acid are much less frequent than synonymous polymorphisms when compared to the number of non-synonymous and synonymous sites. In *B. taurus* cattle, MacEachern (2007) found the rate of non-synonymous polymorphisms per site (Dn) to be 0.0003–0.0006 compared to the rate of synonymous polymorphisms per site (Ds) of 0.003. If all these mutations were neutral, we would expect Dn=Ds. Mutations at non-synonymous, amino acid changing sites undoubtedly occur but most are eliminated by selection due to their negative effect on fitness. In comparisons between *B. taurus* and closely related species, he found the Dn/Ds ratio to be 0.1 i.e. lower than in polymorphisms within B. *taurus* (0.1–0.2). This implies that many deleterious mutations, that will never become fixed and contribute to species differences, are segregating within *B. taurus*. The proportion of non-synonymous mutations that are deleterious is probably even higher than this comparison suggests. MacEachern (2007) found 33 polymorphisms that have existed since *B. taurus* shared a common ancestor with yak, banteng and bison about 1 million years ago. These polymorphisms are probably neutral and within them Dn/Ds = 0.06. This is lower than the ratio found in species comparisons perhaps because some species differences have been driven by selection. If mildly deleterious alleles are commonly segregating in *B. taurus* and presumably in other species, we expect them to be biased towards a low frequency.

The selection that operates to limit genetic variance in quantitative traits could be simple purifying selection. That is, the mutant allele is deleterious and selected against but, if $N_e s$ is small enough, may persist in the population for some time. Alternatively, stabilising selection, in which animals near the mean for a quantitative trait are more fit than extreme animals, will have the same effect of eliminating mutations and reducing genetic variance.

Except for traits that are major components of fitness, it seems likely that these mildly deleterious alleles that segregate at low frequency in the population, sometimes increase and sometimes decrease a given quantitative trait. If, for instance, they consistently decreased the trait, then selection for high values of the quantitative trait would quickly drive them towards elimination and genetic variance would decline. This is not what we see. Long term selection experiments and experience with livestock shows that genetic variance does not sharply decline if the Ne is greater than approximately 100 (Dudley 1977; Weber and Diggans 1990).

Based on these three reasons, I conclude that there are many slightly deleterious alleles segregating at low frequency, some of which have a positive effect on almost any quantitative trait that we study. The implication of this is that selection for almost any trait will increase the frequency of some deleterious alleles and so decrease fitness. Of course, this artificial selection will also be increasing the frequency of some alleles that have a positive effect on fitness, but since these will typically be at high frequency already, this will not compensate for the increase in deleterious alleles. This process can be summarised by saying that alleles that increase both the traits under selection and fitness will be fixed quickly, leaving alleles with opposite effects on the trait and fitness still segregating. Thus selection changes gene frequencies such that the genetic correlation between the trait and fitness becomes more unfavourable over time.

There are examples of this process where by chance, an allele with a major effect is segregating and hence has been identified in experiments. *Drosophila* selection experiments sometimes have shown that selection was for a recessive lethal that increased the selected trait when heterozygous (Yoo 1980). In cattle, selection for muscling has repeatedly selected myostatin mutants that are detrimental to fitness (Grobet et al. 1998), for instance, because they increase dystocia (Arthur 1995). In pigs, the rise in frequency of the halothane mutation due, possibly, to selection for leanness, decreased fitness, for instance, through increased mortality under stress (Fàbrega et al. 2002).

In the absence of knowledge about individual genes, we might observe the consequences of this process as a change in genetic correlations and heritabilities. As a major gene such as a myostatin mutant becomes common the genetic variance of muscularity and dystocia and the genetic correlation between them is expected to increase. The difficulty of estimating genetic parameters with high precision makes this process hard to recognise. There is some suggestion that the genetic correlation between milk production and fertility in cows may have become more negative over time.

Thus it seems likely that selection will lead to decreases in fitness even if the genetic correlation in the base population does not predict this.

7 Recommended Policy

Given that fitness traits are likely to decline as a result of inbreeding depression and selection, what should we do about it?

The first obvious recommendation is to avoid inbreeding as much as possible. This means maintaining a high Ne and avoiding mating animals that are related where possible.

There is a conflict between maximising selection intensity and minimising inbreeding but this process has been well studied and tools exist to help the breeder manage the conflict. Wray and Goddard (1994) and Meuwissen (1997) showed how to maximum genetic gain adjusted for inbreeding depression (or to maximise genetic gain at a fixed inbreeding rate) by selecting the best animals while minimising the average relationship among the selected animals. A different approach, that is more suited to a company competing in an open market to sell breeding stock or semen, has recently been proposed (Haile-Mariam et al. 2007a, b).

The breeding objective should include all traits including fitness traits. The selection index should then account for fitness traits provided they are recorded. In the past the main problems have arisen where fitness traits were simply ignored. In many cases fitness would not have declined if conventional selection index theory had been applied. However, there are some policies which would decrease the risk that fitness declines and should be applied. These are discussed below.

In selection index theory the economic weights should reflect future prices and costs about which there is inevitable uncertainty. I have argued here and previously (Goddard 1998) that the economic weight of some traits should be reduced to account for this uncertainty, but that this should not apply to fitness traits because they are likely to be valuable under all future prices. Adopting this policy would not even reduce current gains greatly because the total gain in the objective is rather insensitive to small changes in index weights. Therefore if the weight given to fitness traits was higher than optimum under current prices, the reduction in overall genetic gain would be minimal.

Breeding objectives sometimes vary only slightly between populations such as countries. However, if these slight differences in objectives are recognised instead of adopting a common objective, the number of sires selected increases and hence the rate of inbreeding decreases (Goddard 1992).

Selection indices that rely heavily on estimates of genetic correlations are more risky than those that do not, because genetic correlations are hard to estimate accurately and may change with time. In particular, if many possible indirect selection criteria are available, the accuracy of the estimated ideal index is overestimated, sometimes grossly so. This occurs because the selection index uses all the estimation errors in the genetic correlations to 'increase' the accuracy of the optimum index (Goddard and Thompson 1992).

The least risky policy is to use direct selection whenever possible. For instance, if selection were based on a progeny test of the actual lifetime profit of offspring, this would be a very robust policy, that would be very insensitive to errors in estimates or changes in genetic parameters (Meuwissen and Goddard 1997). Such a policy might sacrifice too much genetic gain but it would be safer to place more emphasis than optimum on direct measures of profit, than to place less than optimum emphasis on direct measures of profit. For instance, we should record the traits that directly affect profit wherever possible. This means recording reproductive and survival traits and using them in the selection criterion. If these fitness traits were included in multi-trait BLUPs, they would counteract any unforseen negative effects of selection on other traits.

The number of possible fitness traits that could be recorded is probably too large to record all of them. In the past, the negative effects of selection have sometimes been hard to anticipate. For instance, selection of poultry has resulted in over fatness, leg weakness and ascites (McEntee et al. 2002). The important policy may be simply to be monitoring overall fitness traits such as mortality and veterinary treatment so that problems are detected early. All these problems in meat chicken breeding were detected and selection practices changed accordingly.

In the future, selection on individual genes and genetic markers is likely to become more common. Selection at a single locus should be based on the overall effect of the different genotypes on profit. If the alleles act additively, then selection should be for the most profitable allele regardless of the pattern of favourable and unfavourable effects it might have on different traits. Of course this can only be determined if all traits contributing to profit are recorded.

If a gene shows dominance it would still be best to fix the desirable allele. However, if a gene shows dominance for one trait (e.g. fertility) but additivity for another (e.g. milk yield), it may be that the heterozygote is the most profitable (i.e. overdominance for profitability). This could lead to selection maintaining the polymorphism. A better outcome might be to cross two lines fixed for alternative alleles. This implies selection for crossbred performance. This has seldom been practical in the past but with the advent of selection largely on genetic markers, it could easily be implemented in the future. It would have benefits even for genes showing dominance, rather then overdominance, because selection could concentrate on some loci in one population and other loci in the population to which the first is to be crossed. The crossbreeding might involve existing breeds (e.g. Holstein x Jersey) or it might utilise two lines of the same breed (e.g. two Holstein lines).

8 Conclusions

Economically derived breeding objectives include fitness traits such as fertility and survival. Where these traits have declined it is not necessarily due to a failure of current quantitative genetic theory but rather to a failure to utilise this theory. In many cases a decline in fitness traits has occurred because they have simply been ignored in the breeding objective or the recording scheme or the selection criterion. The first improvement in practice that can be made is to properly record and use fitness traits. Even when there is an unfavourable genetic correlation between traits this correlation is seldom 1.0. Therefore it is possible to find animals that are genetically above average for both traits but it is necessary to adequately record both traits.

Inbreeding depression will also cause a decline in fitness. There are methods that balance the gains from intense selection against the losses from inbreeding and these should be used. Recognition that breeding objectives differ between places or production systems would lead to an increase in Ne.

There are reasons to think that selection for quantitative production traits will cause a decline in fitness more often than an increase, even when the genetic parameters in the base population do not predict this. Direct selection for overall profit is a robust policy that helps to protect against these unexpected, deleterious effects. Even where an unfavourable trend occurs, good recording of major components of profit such as fertility and survival, will detect the trend so that it can be reversed.

The likely change in the future to selection based on DNA markers is both a threat and an opportunity. Selection for an allele that improves some traits but decreases overall profit could easily occur unless we estimate the effects of DNA markers on a complete breeding objective. The use of selection on DNA markers will make it more practical to develop lines selected to maximise the performance of a cross between them. This will reduce the losses in fitness due to inbreeding and possibly some losses due to selection as well.

References

Arthur, P.F. 1995. Double muscling in cattle: A review. *Aust. J. Agri. Res.* **46**:1493–1515.

Biffani, S., Samoré, A.B. and Canavesi, F. 2002. Inbreeding depression for production, reproduction and functional traits in Italian Holstein cattle. *Proc 7th World Congr. Genet. Applied to Livest. Prod.* Comm. No. 09–44.

Cassell, B.G., Adamec, V. and Pearson, R.E. 2003. Maternal and fetal inbreeding depression for 70-Day nonreturn and calving rate in Holsteins and Jerseys. *J. Dairy Sci.* **86**: 2977–2983.
Dudley, J.W. (1977). 76 generations of selection for oil and proteinpercentage in maize. In *Proc. Int. Conf. Quant. Genet* (E. Pollak, O. Kempthorne and T.B. Bailey, eds.), Iowa State University, Ames, IA, pp. 459–473.
Ehiobu, N.G., Goddard, M.E. and Taylor, J.F. (1989). Effect of rate of inbreeding on inbreeding depression in *Drosophila melanogaster. Theor. Appl. Genet.* **77**: 123–127.
Fàbrega, E., Diestre, A., Carrión, D., Font, J. and Manteca, X. (2002). Effect of the halothane gene on pre-slaughter mortality in two Spanish commercial pig abattoirs. *Anim. Welf.* **11**: 449–452.
Fulkerson, W.J., Wilkins, J., Dobos, R.C., Hough, G.M., Goddard, M.E. and Davison, T. (2001). Reproductive performance in Holstein-Friesian cows in relation to genetic merit and feeding when grazing pasture. *Anim. Sci.* **73**: 397–406.
Goddard, M.E. (1998). Consensus and debate in the definition of breeding objectives. *J. Dairy Sci.* **81** (suppl. 2): 6–18.
Goddard, M.E. (1992). Optimum effective population size for the global population of black and white dairy cattle. *J. Dairy Sci.* **75**: 2902–2911.
Goddard, M.E. and Ahmed, A.M. (1982). The use of the genetic distance between cattle breeds to predict the heterosis in crosses. *Proc. 2nd World Congr. Genet. Appl. Livest. Prod.* **8**: 377–382.
Goddard, M.E. and Hayes, B.J. (2003). The effect of artificial selection on the distribution of QTL effects. *Proc. Assoc. Advmt. Anim. Breed. Genet.* **15**: 35–38.
Goddard, M.E. and Smith, C. (1990). Optimum number of bull sires in dairy cattle breeding. *J. Dairy Sci.* **73**: 1113–1122.
Goddard, M.E. and Thompson, R. (1998). Use of Prior Information in forming selection indices. *J. Anim. Breed. Genet.* **115**: 327–340.
Grobet, L., Poncelet, D., Royo, L.J., Brouwers, B., Pirottin, D., Michaux, C., Ménissier, F., Zanotti, M., Dunner, S. and Georges, M. (1998). Molecular definition of an allelic series of mutations disrupting the myostatin function and causing double-muscling in cattle. *Mamm. Genome* **9**: 210–213.
Haile-Mariam, M., Bowman, P.J. and Goddard, M.E. 2007a. A practical approach for minimising inbreeding and maximising genetic gain in Australian dairy cattle. *Genet. Sel. Evol.* **39**: 369–389.
Haile-Mariam, M., Schelfhorst, E. and Goddard, M.E. 2007b. Effect of data collection methods on the availability of calving ease, fertility and herd health data for evaluating Australian dairy cattle. *Aust. J. Exp. Agric.* **47**: 664–671.
de Jong, G. (2005). Usage of predictors of fertility in the genetic evaluation, application in the Netherlands. *Interbull Bull.* **33**: 69–73.
Lynch, M. and Walsh, B. (1998). *Genetics and Analysis of Quantitative Traits*, Sinauer Associates, Sunderland, UK.
MacEachern, S. (2007). PhD thesis. *Molecular Evolution of the Domesticated Cow (Bos taurus).* Latrobe University, Melbourne, Australia.
McEntee, G.M., McDevitt, R.M. and Rance, K.A. (2002). Sustainable broiler breeding: Achieving the balance between support and demand tissues. *Proc 7th World Congr. Genet. Appl. Livest. Prod.* Comm. No. 4–38. **30**: 349–352.
Man, W.Y.N. 2004. PhD Thesis. *Inbreeding in Australian Holstein Friesian Cattle*. The University of Sydney, Sydney, Australia.
Meuwissen, T.H.E. 1997. Maximizing response to selection with a predefined rate of inbreeding. *J. Anim. Sci.* **75**: 934–940.
Meuwissen, T.H.E. and Goddard, M.E. (1997). Selection of farm animals for non-linear traits and profit. *Anim. Sci.* **65**: 1–8.
Philipsson, J., Eriksson, J-Å. and Stålhammar, H. 2005. Know-how transfer in animal breeding – the power of integrated cow data bases for farmer's selection of bulls to improve functional traits in dairy cows. In *Knowledge Transfer in Cattle Husbandry. EAAP Publication* **117**: 85–95. Wageningen Academic Publishers, The Netherlands.

Smith, L.A., Cassell, B.G. and Pearson, R.E. 1998. The effects of inbreeding on the lifetime performance of dairy cattle. *J. Dairy Sci.* **81**: 2729–2737.

Thompson, J.R., Everett, R.W. and Hammerschmidt, N.L. 2000a. Effects of inbreeding on production and survival in Holsteins. *J. Dairy Sci.* **83**: 1856–1864.

Thompson, J.R., Everett, R.W. and Wolfe, C.W. 2000b. Effects of inbreeding on production and survival in Jerseys. *J. Dairy Sci.* **83**: 2131–2138.

Wall, E., Brotherstone, S., Kearney, J.F., Woolliams, J.A. and Coffey, M.P. 2005. Impact of non-additive genetic effects in the estimation of breeding values for fertility and correlated traits. *J. Dairy Sci.* **88**: 376–385.

Weber, K.E. and Diggans, L.T. 1990. Increased selection response in large populations. II selection for ethanol vapour resistance in *Drosophila melanogaster* at two population sizes. *Genetics* **125**: 585–597.

Wray, N.R. and Goddard, M.E. (1994). Increasing long term response to selection. *Genet. Sel. Evol.* **26**: 431–451.

Yoo, B.H. (1980). Long term selection for a quantitative character in large replicate populations of *Drosophila melanogaster*. II lethals and visible mutants with large effects. *Genet. Res.* **35**: 19–31.

Zhang, X.-S. and Hill, W.G. (2005). Predictions of the patterns of response to artificial selection in lines derived from natural populations. *Genetics* **169**: 411–425.

9 Appendix Effect of Selection for Milk Yield on Fertility in Dairy Cows

Mean 6 week pregnancy rate = 0.8
Phenotypic standard deviation = 0.4
Heritability of 6 week pregnancy rate = 0.04
Genetic standard deviation of 6 week pregnancy rate = 0.2*0.4 = 0.08
Genetic correlation between 6 week pregnancy rate and milk yield = −0.35
Genetic improvement in milk yield = 0.125 *genetic standard deviation per year*
Predicted decline in 6 week pregnancy rate = 0.35*8*0.125 = 0.35% per year

Discussion

John W. James

The major theme of the discussion during Session 1 arose from the suggestion by Goddard that, although fitness is an important concept in population and evolutionary genetics, animal breeders should ignore it. Fitness could determine the arrays of gene and haplotype frequencies at different loci in natural populations but, as others had pointed out, domestic animal populations are usually very far removed from natural conditions, though to a lesser extent in developing countries. What is important in animal breeding is the productivity and profitability of livestock, and therefore breeders need to consider how components of fitness such as reproductive rate and viability contribute to the economic value of the population. Provided this is done properly all relevant aspects of fitness will be taken into account.

It was objected that market models are often fallible and, for instance, often ignore externalities, so that reliance on economic models was therefore risky. In reply, it was said that the best models should be used. Index selection has been shown repeatedly to be remarkably insensitive to moderate errors in the economic weights given to traits, provided that all important traits are included in the breeding objective (with the right signs). However, if improvement of important traits is based on selection for correlated traits this comforting conclusion may not apply, as genetic correlations are notoriously difficult to estimate accurately, and have often been found to be unstable.

Another suggestion was made that multi-trait selection would spread selection pressure over a larger number of QTLs than would selection for one or two traits, which would tend to concentrate selection on a small number of loci. On the other hand, it was argued that this would not happen, because pleiotropy would mean that some QTLs would have positive effects on many desirable characters, while others would have positive effects on some characters and negative effects on others. The result would be that only those QTLs with generally positive effects would be strongly selected. Those with a mixture of positive and negative effects would be

J.W. James (✉)
Centre for Advanced Technologies in Animal Genetics and Reproduction (Reprogen), Faculty of Veterinary Science, University of Sydney, NSW 2006, Australia
e-mail: jwj@vetsci.usyd.edu.au

J. van der Werf et al. (eds.), *Adaptation and Fitness in Animal Populations*,

only weakly selected and be subject to random genetic drift. The main difference would be to put major selection pressure on a different set of QTLs. No empirical evidence was offered in support of either viewpoint, and it might well be the case that both assertions are correct in some situations. However, it does seem that neither assertion is necessarily correct.

Management can hide genetic differences by altering the environment, so that losses in fitness could occur but not show up because environments were changed to overcome these losses. But it can be argued that breeding decisions should be made for the commercial environment, and if this involves mitigating fitness problems this has to be accepted. Presumably the necessary alterations have an economic cost, and this would be factored into the breeding objective, and loss of fitness would be acceptable only if the costs of managing the fitness decline were outweighed by consequent improvements in other traits. A different problem arises if studs provide special conditions which are not commercially relevant and breeding decisions are made on the basis of performance in such special conditions.

Genomic selection may offer a new approach to this problem. Samples of DNA can be taken at any point in the production chain in order to identify QTLs. As a perhaps extreme, but not unreasonable example, somebody wishing to identify favourable markers for meat tenderness and flavour could buy steaks from butchers' shops, take DNA samples from, cook and evaluate the steaks, and then search for SNPs associated with superior eating quality. Of course, this would raise the problem that effects of SNPs identified in this way on other important traits would be unknown, and on the assumption that pleiotropy is widespread these effects would need to be estimated by further studies before introducing selection for these QTLs. Of course, current work on gene discovery does not always satisfy this criterion and may be of restricted generality.

Another point made was that the importance of fitness might be very different in breeding stock and in production animals. For example, in terminal sire breeds reproductive rate has little importance other than being sufficiently high to produce the required number of sires for sale, whereas dams of commercial stock should be highly fertile. The correct way to deal with this is to optimise the entire system. However, this can be very difficult since in many cases the decisions are in different hands and market signals are not clear. The poultry and pig industries are more integrated than the sheep and cattle industries and they therefore have greater opportunities to optimise the complete production system, including fitness components.

The idea that global fitness can be ignored provided components with economic value are included in the breeding objective can be dangerous if fitness components are selected for through correlated characters, because the results of this approach have sometimes been found disappointing. Broilers have responded to selection for many generations, but some of the correlated responses have not been in agreement with the originally estimated correlations. So, even if fitness is dealt with by including relevant components in the breeding objective, there is a non-negligible chance that the response will not be as expected. It is therefore very important that fitness components, like other economically important traits, should be regularly monitored.

A related point is that Comstock showed, 30 years ago, that the only way to guarantee that the favourable allele at any locus is subjected to positive selection pressure is to select directly for the breeding objective. A conventional selection index, even if it included all traits to be improved, could in some situations place negative selection pressure on a desirable allele in order to speed response at other loci. This would obviously be a greater problem when not all important traits were included in the index.

One important problem is that of correlation between the sexes for fitness traits. If the allele at any locus which is favourable in males is deleterious in females this will maintain segregation in a natural population and thus such loci might contribute significantly to variation in fitness. It may be best to resort to structured breeding systems in order to deal with this.

It has generally been recommended that marker assisted selection (MAS) should be used preferably for traits which are hard to measure, and fitness could be regarded therefore as a good candidate for MAS. But MAS might not work so well for fitness because there is a great deal of non-additivity in gene action. There is reason to believe that a good deal of variation in many fitness traits is due to recessives at low frequency, and MAS is not well suited to dealing with such cases. If there is no additive effect, heritability is zero, but even with extreme non-additivity on the physiological level there will often be a significant amount of additive genetic variance on the statistical level, and this can be made use of in the normal way.

There should be a possibility with the development of genomics to select for specific multi-locus genotypes, but it may not be easy to identify the desired genotypes. If the desired genotypes are highly heterozygous, this could be dealt with by reciprocal recurrent selection, and genomic selection could be helpful in pursuing this, and may be valuable for fitness traits.

Present applications of genomic selection have used highly accurate EBVs to identify the genes for which to select, and by definition EBVs deal with additive genetic effects, so that so far genomic selection for non-additive effects has not been attempted. EBVs could not be used to develop the selection criteria, so individual animal phenotypes would be used. But to have worthwhile power large samples are needed and with current technology it would be very expensive to attempt such work.

With markers being used in selection there can be problems of linkage phase. For example, suppose there are two bulls, both of which carry two favourable rare alleles at closely linked loci, both being double heterozygotes. If one has the genes in coupling and one has them in repulsion, we would prefer the bull with the coupling phase, since some of his sons would carry both plus alleles, while sons of the other bull would carry only one plus allele each, and thus be less favoured as breeding animals. While this is true, the example shows that the problem is sorted out in the next generation. Nevertheless, if we know phase it is an advantage.

Part II
Maintaining Fitness

Maintaining Genetic Variation in Fitness

William G. Hill and Xu-Sheng Zhang

Abstract Although natural selection is expected to remove additive genetic variation in fitness, the recent evidence is that in natural populations considerable amounts of such variation remain in fitness and related traits, just as for other quantitative traits. In assessing this it is important to distinguish between heritability which may be low and the genetic coefficient of variation which may be high. Here we concentrate mainly in the artificial selection context, where selection criteria are known and experiments have been conducted, and review rather than present new results. In an artificial selection programme, the selection index can be regarded as the measure of fitness. In this case, analyses of rates of loss of variation can be used that have been derived for quantitative traits in general. These indicate that effective population size is a critical factor, in particular when mutation to useful variation is taken into account. In many long term selection experiments and breeding programmes variation has been maintained for very many generations, without a plateau in response. If fitness is regarded as the set of health, viability and reproductive traits not directly related to the selection objective, for example in a breeding programme for meat animals, then maintenance of variation in fitness is more highly dependent on effective population size than is variation in the directly selected traits. Examples of improvement programmes show that, while such measures of fitness have declined in the past, there has remained genetic variation such that subsequent selection to reduce or reverse the trends in some fitness associated traits has been effective.

Keywords Genetic variation · fitness · selection · selection response · animal breeding · evolution

---

W.G. Hill (✉)
Institute of Evolutionary Biology, School of Biological Sciences,
University of Edinburgh, West Mains Road, Edinburgh, EH9 3JT, UK
e-mail: w.g.hill@ed.ac.uk

Dedicated to Stuart Barker in honour of his many contributions at the interface of animal improvement and evolution.

J. van der Werf et al. (eds.), *Adaptation and Fitness in Animal Populations*,

1 Introduction

The fitness of an individual can be defined in terms of the number of its descendents or of genes transmitted to the next generation, and is a function of its survival and reproductive performance. Fitness can be regarded as a quantitative trait, but in natural populations it may be very difficult to determine how other traits influence fitness. Although it is clear that in *Drosophila*, for example, there is an overt relationship between female egg production and fitness, the relationship between bristle number and fitness is less obvious. For some traits there is likely to be a monotonic relationship with fitness, for others it may be quadratic with intermediate optimum (Falconer and Mackay 1996, ch. 21). Fisher's (1958) fundamental theorem shows that the rate of change in fitness is equal to the additive genetic variance in fitness, while Robertson's (1966) secondary theorem shows that the change in another trait equals its additive genetic covariance with fitness. Thus natural or artificial selection would be expected to use up the useful variance, but even so variation remains, as we review in the next section. Genetic variation in fitness can be maintained by mutation and other factors such as heterogeneous environments; genetic variation in other traits can remain through these factors and also because the covariance with fitness may be small (for discussion see, for example, Falconer and Mackay 1996; Lynch and Walsh 1998; Bürger 2000; Zhang and Hill 2005a).

In the animal or plant improvement context, fitness can be defined simply as the index on which selection is practiced. This may be, say for a broiler chicken population, just a function of juvenile traits such as growth rate and body composition. At its simplest, therefore, we can discuss the evolution of fitness and its variation solely in the context of selection index theory. An important difference from the natural selection context is that the components of fitness are, at least in principle, completely known. Also fitness under artificial selection may be a function not only of the individual's but also its relatives performance, which is equivalent to altruism in a natural population. Other more overtly 'fitness' traits, such as viability or leg strength of broilers, may be recorded and included in the index. In any practical situation, however, there are likely to be other components of fitness that are unrecorded or unknown that may influence the reproductive performance of selected individuals, such that overall fitness is not exactly the same as that defined by the index. Deteriorations in such traits may be of concern.

In this paper we shall consider the extent to which genetic variation is maintained in fitness, basically as a fuel for future change, and do so in different contexts: where natural selection is on fitness per se; in artificial selection in which the defined index is regarded as exactly the same as fitness; and in artificial selection where it is recognised that components of fitness are not incorporated in the selection criterion per se. We shall concentrate mainly in the artificial selection context, where selection criteria are known and experiments have been conducted, and review rather than present new results.

When quantifying genetic variation it is important to consider the scale on which the measurement is made. The standard measure, heritability (h^2) describes the correlation among relatives and selection response, but is a ratio of variances not a measure of their magnitude. For many descriptive and predictive purposes it is better to scale variation in relation to the mean (μ), such as the coefficient of variation (*CV*). Thus in evolutionary genetic analysis widespread use is made of the genetic coefficient of variation or 'coefficient of evolvability', $CV_A = \sigma_A/\mu = hCV$ (Houle 1992). In the animal breeding context CV_A is a useful parameter for comparing possible rates of improvement among traits and species because $i\rho(CV_A)$ predicts the rate of response in a trait as a percent of its mean, with selection intensity (i) and accuracy (ρ) depending on the programme's design. We shall concentrate mainly on levels of genetic variance. It must also be noted that the magnitude of the residual phenotypic variation of individuals of specified genotypic value must itself be influenced by the genotype – i.e. there is genetic variation in the level of environmental variance. The implications of this are discussed in a later section, but first we consider genetic variance as it influences population means.

2 Genetic Variance in Fitness in Natural Populations

As shown by Fisher's fundamental theorem, the rate of increase in fitness in natural populations is equal to its additive genetic variance, so a priori additive variance is expected to become very small or indeed exhausted. That is not the case.

There have been numerous estimates of heritability for many traits in species in the laboratory, showing that estimates of h^2 are generally much higher for morphological and growth than fitness associated traits. See for example the review by Roff and Mousseau (1987) of data on *Drosophila*, showing mean h^2 of 0.26 for life history traits and 0.46 for morphological traits. Genetic coefficients of variation, though, are typically higher for fitness associated traits (Price and Schluter 1991; Houle 1992), indicating that life history traits do not have less genetic variance but more environmental variance. Similar results are of course found for traits of livestock: a simple illustration of the contrast in magnitudes of h^2 and CV_A is for growth rate and litter size in pigs, the former having a heritability of typically 25% and CV 10%, the latter 10% and 30%, respectively, giving CV_A of 5% and almost 10%, respectively.

Estimation of genetic variance in total fitness rather than its components is necessarily difficult. It is likely to be a composite trait of those that can be observed and recorded, and some components may be missed. In a natural population there may be limited or no pedigree information. One solution is to study populations which have adapted to the laboratory environment, for example *Drosophila melanogaster* caught from the wild but maintained in closed populations in cages for many years, equivalent to hundreds of generations. Thus Fowler et al. (1997) allowed replicated competition of a sample of wild type third chromosomes against balancer chromosomes in such a population cage and recorded changes in chromosome frequency to estimate fitness. There were highly significant differences between chromosomes in relative fitness, and effects appeared to be largely additive. The estimate of the genetic variance in $\log_e$ fitness, albeit based on a sample of only twelve third chromosomes, was equal to 0.175. Assuming contribution according to chromosome length, this would be equivalent to 0.45 for the whole genome, a remarkably high figure.

Genetic variance in fitness can be estimated directly in a natural population in which pedigrees have been obtained either by observation and marking animals over many generations in long term field studies or now by using DNA markers to predict relationships. The many different relationships present require the use of sophisticated analysis, and now animal model REML or similar methods developed initially for use in animal breeding are used (Kruuk 2004). In these studies estimates of heritability of life history traits have typically been low compared to those on morphological traits, but show clear evidence of genetic variance in fitness associated traits being present (e.g. Kruuk et al. 2000). A problem with such an analysis is actually to record fitness per se and not a component which may, for example, be showing additive genetic variability but be strongly negatively correlated with other unrecorded component traits, such that the overall 'index' of fitness actually has a low genetic variance.

There appears to be proportionality more dominance variance (e.g. expressed as $V_D/(V_A + V_D)$ for life history than growth or morphological traits (Crnokrak and Roff 1995). The life history traits also show more inbreeding depression, a likely but not invariable associate as it requires directional dominance. In view of the problems of estimation there is little information on epistatic variance. For some traits there is direct evidence that epistatic variance is small or almost non-existent. For example, the heritability of finger ridge count in humans is almost 100%; and, the repeatability across segments for abdominal bristle number in *Drosophila*, lactations for milk yield in cattle and parities for litter size in pigs, are not much greater than the respective heritabilities, implying either little non-additive genetic variance or very low genetic correlations across records (Falconer and Mackay 1996).

There has been extensive discussion about the magnitude of additive genetic variance in fitness and closely associated traits in natural populations (e.g. Falconer and Mackay 1996; Lynch and Walsh 1998; Merila and Sheldon 1999), with a major problem being to obtain estimates free of bias from other genetic or environmental components. Although there have been clear cases where there appears to be no genetic variation in fitness or a closely associated trait (e.g. Bradshaw 1991; Hoffmann et al. 2003), overall there is strong evidence for the presence of substantial amounts of additive genetic variance in fitness in natural populations (for further review see, for example, Merila and Sheldon 1999). In livestock, fitness per se has multiple definitions as noted, but the presence of substantial and utilisable additive variation for the component traits such as viability and litter size seems unequivocal.

We now consider the extent to which variation is maintained and how it is influenced in populations undergoing artificial selection, where it is known (to a good approximation) on what set of traits selection is actually practiced. In subsequent sections we shall return to mechanisms for maintenance of variation in fitness in natural populations, concentrating particularly on the role of mutation.

3 Artificial Selection on Fitness

Here we consider the maintenance of genetic variance in the trait or index of traits under direct selection in a breeding programme. We use a simple example to predict formulae for artificial selection in the context of fitness and the formulae of Fisher (1958) and Robertson (1966). With truncation selection of a proportion p of individuals on a quantitative trait X, the expected fitness of is $1/p$ for selected individuals and 0 for unselected individuals. Therefore the phenotypic variance in fitness is $V_{\mathrm{PF}} = p(1-p)(1/p)^2 = (1-p)/p$. Using the formulae relating all-or-none traits (in this case selection or not) to the underlying distribution (Falconer and Mackay 1996), the heritability of fitness is $h_{\mathrm{F}}^2 = i^2 h_{\mathrm{X}}^2 p/(1-p)$. Hence the additive genetic variance in fitness is $V_{\mathrm{AF}} = i^2 h_{\mathrm{X}}^2$, where i is the selection intensity. To complete the circle, the response in the trait is the additive genetic covariance with fitness, $cov_{\mathrm{AXF}} = \sigma_{\mathrm{AF}}\sigma_{\mathrm{AX}} r_{\mathrm{A}} = ih_{\mathrm{X}}^2\sigma_{\mathrm{PX}}$, because the genetic correlation between X and fitness is $r_{\mathrm{A}} = 1$.

The extensive research over several decades into the rate of loss of variation in artificial selection programmes and how this can be minimised has been limited mainly to assumptions of the infinitesimal model, and we shall not review it in detail. Quantitative genetic variation is lost mainly through decline in heterozygosity at loci affecting the trait(s) under selection. This is likely to be greatest for genes that have intermediate frequency and large effect on the trait, as these contribute most of the variation and change frequency most rapidly. There is also loss through gametic disequilibrium (Bulmer effect), but for unlinked loci most of this takes place very quickly and essentially a steady state of gain and loss of disequilibrium is reached. With linked loci, the magnitude of the gametic (linkage) disequilibrium may be larger and the time to reach the asymptote longer. We consider the magnitude of standing variation and that arising from mutation under different model assumptions.

3.1 Infinitesimal Model

Under the infinitesimal model, all the genetic variation (V_{G}) is additive ($V_{\mathrm{A}} = \sigma_{\mathrm{A}}^2$) and its rate of loss in a closed population is simply inversely proportional to the effective population size, $V_{\mathrm{A}}/(2N_{\mathrm{e}})$. With truncation selection the rate of improvement is equal to $i\rho_{\mathrm{IA}}\sigma_{\mathrm{A}}$, where i is the selection intensity and ρ_{IA} is the accuracy of selection. If the breeding objective remains the same, the variance after t generations is given by $V_{\mathrm{A}t} = V_{\mathrm{A0}}[1 - 1/(2N_{\mathrm{e}})]^t \sim V_{\mathrm{A0}}[1 - t/(2N_{\mathrm{e}})]$ and the response by $R_t = t[1 - \frac{1}{2}(t-1)/(2N_{\mathrm{e}})]R_1$ approximately assuming t/N_{e} is small, where V_{A0} and R_1 are the initial variance and response in the first generation. The prediction of response assumes no reduction in accuracy, and a somewhat lower figure applies if accuracy also changes because of the reduction in heritability (see Wei et al. 1996 for formulae).

It then becomes a matter of judgment as to how long a time scale is used for predictions and thus what is an acceptable rate of inbreeding. More intense selection and more accurate selection using relatives' information both reduce effective population size, so there is a trade-off between rate of improvement over a finite time period and rate of loss of variation (Robertson 1960), and the breeder has to judge their relative importance. Under the infinitesimal model, however, the changes in the genetic parameters can be explicitly predicted from observable quantities such as the genetic variances and inbreeding effects. These formulae can be extended to take account of the discount rate applying to a breeding programme, in which case early response has more value and gets more weight than later responses. This provides an objective way to weight early vs. later responses but, in practice, if discount rates of a few percent per year are used, for large animal species these equate to very large rates per generation (e.g. 4% per year may be almost 20% per generation for cattle), and the optimum population sizes to maximise investment are very small indeed (James 1972). In practice, therefore, much lower rates of inbreeding are therefore recommended.

A limitation of the above simplistic model is the assumption of truncation selection (Bijma and Woolliams 2000). The relative size of the response to the rate of loss of variation can be increased by taking more offspring from the very best parents and fewer from those less close to the top, and by choice of individuals to minimise relationships among them and thus future inbreeding. There is now an elegant and extensive literature on this, including focusing the changes over generations in terms of Mendelian sampling terms, and algorithms available to facilitate selection decisions. See, for example, Villanueva et al. (2006) for references and a recent analysis and discussion by one of the active groups. This is clearly the main way the breeder has to influence maintenance of variability. Furthermore, the formulation is relevant to conservation of genetic resources (see Oldenbroek 2007 for a series of relevant papers), but we shall not review it in detail here.

3.2 Relaxation of Infinitesimal Model – Utilising Existing Variation

The expected change in frequency of an additive gene with genotypic values of homozygotes differing by a/σ_P phenotypic standard deviations on a single trait is $i(a/\sigma_P)(\rho_{IA}/h)q(1-q) = sq(1-q)$. For genes with effects sufficiently large that the change in gene frequency is not negligible, the relationship between response, variance maintained and N_e is less straightforward (Robertson 1960). (Introducing a selection index of multiple traits complicates the formulae to the detriment of reducing clarity, so we take the simple case.) The probability that the favourable allele is ultimately fixed is a function of the product $N_e sq$, with the probability exceeding 0.86 if $N_e sq > 1$. Examples of the population size needed to be confident to fix genes of different effects and initial frequency are given by Hill (2000). For example, for a gene at frequency 0.2, $N_e s > 5$ is required, and with the further assumption that $i\rho\sigma_A > 1.67$ (e.g. selection intensity 1.5, accuracy a little greater than h) it follows that $N_e a/\sigma_P > 3$ is needed for > 86% fixation probability. So for a gene of quite large effect $a/\sigma_P = 0.3$, an effective population size of only 10 would be required. Such a gene contributes a heritability of ~0.2%. For each halving of gene effect, the required population size would double and its contribution to heritability reduce four-fold. For each halving of q the required N_e also doubles, but the contribution to heritability is roughly only halved.

Hence only those favourable genes that are at low frequency or of very small effect, which therefore contribute a very small proportion of the genetic variability, have an appreciable probability of being lost by chance. The relation between long term response and population size from *extant* variation is thus likely to be much less sensitive than predicted from the infinitesimal model. (Robertson's (1960) figure of $2N_e\times$ the response in the first generation is a simplistic infinitesimal model assumption.) It is therefore reasonable to conclude that the infinitesimal model calculations on progress and variance maintained are unnecessarily conservative when considering the utilisation of increasing variation. New mutants, however, are much more easily lost.

3.3 De novo (Mutational) Variation

Let V_M be the increment in the additive genetic variance of the quantitative trait from one generation of mutation, which is usually expressed in terms of V_M/V_E, the mutational heritability, where V_E is the environmental variance. There have been many estimates of the level of mutational variance in quantitative traits. The typical value, expressed relative to the environmental variance as estimates are usually made in inbred or isogenic lines, is $V_M/V_E = 0.001$ or a mutational heritability of 0.1%, quite consistent across traits and species (Houle et al. 1996; Keightley 2004). If expressed as the mutational coefficient of variation, σ_M/μ, however, generally higher values are obtained for fitness associated than morphological traits (Houle et al. 1996).

Under the infinitesimal model, the steady state variance from mutation, when the gain equals the loss by genetic drift, is $2N_e V_M$. This quantity is not influenced directly by selection because selective values are so small under the infinitesimal model that the probability of fixation is close to $1/2N_e$. If $Ns>1$ the fixation probability of the mutant is given by $s = i(a/\sigma_P)(\rho_{IA}/h)$ (Hill 1982). For genes with positive s, the expected response if the mutant is fixed is a and hence it turns out that the asymptotic variance is approximately $4N_e V_M[E(a^2|a>0)]/[E(a^2)]$, where $E(a^2)$ is the mean square effect of new mutations and $[E(a^2|a>0)]$ is that quantity summed only over those increasing the trait. Hence it follows that the variance equals $2N_e V_M$ as in the infinitesimal case if the distribution of effects is symmetric (i.e. mutants are equally likely to increase or decrease the trait). If mutations are predominately disadvantageous, it will be correspondingly less (Hill and Keightley 1988). The asymptotic response from mutations in the case of symmetry of effects of additive genes or the neutral model is then $2N_e i(\rho/h)V_M/\sigma_P$. For genes with partial or complete dominance, the fixation probability depends on the magnitude of the effect in the heterozygote, but the prediction formulae are not greatly affected. Recessive mutants have a very small probability of fixation, even if their effects are large.

In the medium term, therefore, the critical factor in maintaining variation is population size through its impact on maintaining segregation of genes of very small effect, while in the long term the major impact is likely to be through the utilisation of mutation. Calculations based on amounts of inbreeding depression per se, when all fitness associated traits such as reproductive rate are included in the index are also therefore inadequate, because selection will itself lead to purging of deleterious recessives and influence their rate of fixation.

3.4 Information from Selection Experiments

The role of population size and selection intensity in long term response was demonstrated in an important series of experiments by Stuart Barker and colleagues in the 1960s (Frankham et al. 1968a,b; Jones et al. 1968), who showed results generally in line with the theory of Robertson (1960). More recently Weber has conducted experiments with very much larger population sizes and generally shown increased responses as N_e has been increased. His and other results on long term responses in *Drosophila* selection experiments are summarised by Weber (2004). He plotted R_{50}/R_1, the response up to 50 generations relative to that in the first generation, as a function of population size, and showed a monotonic but non-linear increase to about $N = 300$ with little further increase to $N = 1000$, the size of his largest experiments. The apparent asymptote is at about $R_{50}/R_1 = 40$, which, if the Bulmer effect is taken into account, does not appear far from that expected with the infinitesimal model.

Several selection experiments have been continued well beyond 50 generations, however. In the classic Illinois corn experiment, for example, there are clear and continuing responses to generation 100 in both oil and protein content of the kernel in the high lines (Dudley and Lambert 2004). In the low lines asymptotes have been reached, at essentially 0% oil and at a protein level probably minimal for seedling growth. Further experiments are reviewed elsewhere (Hill and Bünger 2004). More recently, Holt et al. (2005) have reported on 122 generations of selection for litter size in mice, which included crossing of sublines and also periods of reversed selection. Responses were not consistent, but with a general trend of increasing litter size in the high selected line to the end. The genetic variance and heritability from REML analyses increased in the high selected line, but decreased in the low line, averaging little change overall. Some additive genetic variation was clearly maintained.

The role of mutation in selection response, and thus estimates of the mutational rate for quantitative traits has been investigated mainly by starting with isogenic or highly inbred populations (see Keightley 2004 for a summary). These include a study by López and López-Fanjul (1993a) in which the impact of population size was tested using *Drosophila* populations of population size 10 and 50. They obtained substantially larger responses, roughly five-fold as predicted, with the five-fold larger population. These experiments also demonstrate that mutant genes may be of large effect (López and López-Fanjul 1993b), and the infinitesimal model certainly does not fit much of the data.

The general observation is that variation is maintained in the traits for which selection is practised, i.e. for 'fitness' as defined here. For the lines started from an inbred base, mutation must have been the source of variation. For those started from an outbred base, the situation is less clear cut in the medium term because predicted responses using the infinitesimal model ignoring mutation may not differ greatly from those assuming there is contribution from mutation but that gene effects on the trait differ in effects and frequency. The longer the experiment continues (in terms of t/N) the more important the role of mutation is likely to become, such that Walsh (2004) could demonstrate that mutation must have had a role in the responses in the Illinois maize lines. We undertook an analysis (Zhang and Hill 2005b) using a model which incorporated the possible distribution of gene effects on the trait allowing for previous natural selection, the Bulmer effect allowing for the restricted recombination in *Drosophila*, and mutation. We did not, however, achieve a much better fit to the data tabulated by Weber (2004) than with the infinitesimal model. Experiments started from an outbred base population do not discriminate well between models, but all indicate long term maintenance of genetic variability.

3.5 Results of Livestock Improvement Programmes

Let us now consider what long term responses have been made in practical breeding programmes in livestock and the extent that, in so far as we can tell, variation has been maintained in the main traits under selection. Subsequently, we shall also consider those in fitness traits. The most appropriate data would seem to be from broiler poultry, which have been under intensive selection with roughly annual generations for over 50 years since specialised broiler and layer populations diverged from dual purpose breeds. We also have good information on genetic changes over this period because unselected stocks have been maintained for comparisons as controls. In particular Havenstein et al. (1994, 2003a,b) compared commercial broilers of 1991 with those of controls started from the commercial birds of 1957, and then compared 2001 birds with the same controls. Thus we obtain estimates of genetic change over a total of 44 years and also over the 10 most recent of these years, which enables a check on whether response continues.

The comparisons between the modern and old stock were made on both a modern diet and one formulated to represent that used at the time the old stock dated from. For example, at 56 days of age the weights of the 2001 and 1957 birds on the 2001 diet were 4.40 and 1.01 kg, respectively, and on the 1957 diet were 3.17 and 0.92 kg respectively, i.e. the modern birds were about 4 fold heavier on the modern

Table 1 Comparison of broiler chickens in two trials reared on own diet (i.e. diet used at year of stock): D1 estimates response from 1957 to 1991, D2 estimates response from 1957 to 2001, and D2–D1 estimates response from 1991 to 2001

	1991 trial			2001 trial			Both
	1991	1957	D*1*	2001	1957	D2	**D2–D1**
Body weight (kg)	3.11	0.79	*2.32*	3.95	0.81	*3.14*	**0.82**
Carcass weight (kg)	2.07	0.50	*1.51*	2.81	0.48	*2.33*	**0.82**
Carcass yield (%)	69.7	61.2	*8.5*	74.4	60.8	*13.6*	**5.1**
Breast yield (%)	15.7	11.8	*3.9*	21.3	11.4	*9.9*	**6.0**
Carcass fat (%)	15.3	9.4	*5.9*	15.9	10.6	*5.3*	**−0.6**

(Source: Havenstein et al. 1994, 2003a,b)

diet and about 3.5 fold heavier on the old diet (Havenstein et al. 2003a). A similar pattern of weights applied at other ages ranging from 21 to 84 days. Thus enormous changes in rate of body weight gain to market age or weight have been made, and of that total change, most has been genetic. Further, the estimates of changes between 1991 and 2001 are also very large (Table 1), a rate of about $2\frac{1}{2}$%/year in growth rate. There is thus no evidence of a plateau in response due to loss in variability. There were relatively larger changes in carcass and meat yield, as proportions of these components also increased Havenstein et al. 2003b). Although breast yield continued to rise, fatness increased to 1991 but then was reduced to 2001 (Table 1). Thus variation has also been maintained in other body components.

Commercial breeding populations are unlikely to be completely closed as breeders may wish to bring in novel material if they see opportunities to increase performance or introduce genes of large effect (such as happened with the halothane gene in pigs). As populations become more extreme, however, such as those of the modern broiler, opportunities to bring in material likely to enhance performance diminish. Further, there has been a steady reduction in the number of breeders, reducing opportunities for immigration; and even if the population is regarded as the total of the nucleus populations held by the major breeders, it is increasingly limited in size.

There have similarly been large and continuing, albeit less spectacular, responses in egg production in layer strains of poultry (notwithstanding longstanding concerns of an impending plateau), in growth rate and leanness of pigs and in milk yield of dairy cattle (see Hill and Bünger 2004 for review). In trials in which seed from older and modern hybrids are grown together, there has also been shown to be continuing genetic improvements in grain yield of maize in commercial practice, with a nearly linear annual change in yield over a 60 years period (Duvick et al. 2004). The regression of yield on year of introduction was about 0.071t/ha/year, without obvious sign of plateau (albeit there may be a declining response in % change in mean/year).

In some of these cases, such as egg production in poultry, milk production in cattle and yield in maize, the commercial trait would seem to be a major component of fitness by measures of the latter other than as a selection objective. Recent data on commercial poultry show heritability of egg production of the order of 20%

(Preisinger and Flock 2000), little different from those of long ago, so genetic variation seems to be retained.

The most notable exception to this pattern of continuing change is in Thoroughbred horses and Greyhounds in terms of winning times in classic races. In both cases there were substantial gains until about 1950, followed by little subsequent response (Hill and Bünger 2004). For example the record time in the Kentucky Derby was set in 1973. The reasons for the plateau in performance are not clear, for there is evidence of substantial genetic variation in performance (Gaffney and Cunningham 1988). We discuss possible explanations later.

4 Fitness as a Correlated Trait in Artificial Selection

We turn now to consider the consequences on mean and variation in fitness and in associated traits when selection is practised on other traits.

4.1 Some Theory

The change in fitness, regarded most simply as a single correlated trait (*Y*), as a consequence of selection on an index of production traits (*X*) equals the change in $X\times$ the genetic regression of *Y* on *X*, or $i\rho_{\mathrm{IAX}}r_{\mathrm{A}}\sigma_{\mathrm{AY}}$, where r_{A} is the genetic correlation between the 'traits'. Under the infinitesimal model the genetic variance in the fitness trait *Y* simply reduces as $(1-1/(2N_{\mathrm{e}})]V_{\mathrm{A}Y}$. Selection itself does not reduce variation in fitness except initially through the Bulmer effect, just as it does not reduce variation in the trait under selection. Mean fitness may of course deteriorate as a correlated response to the direct selection.

The correlated response when the infinitesimal model is relaxed depends on the effects at individual loci and the relative rates of gene frequency change (Bohren et al. 1966). Thus if a_{X} and a_{Y} denote gene effects on the traits, the correlated response due to this locus depends on $a_{\mathrm{X}}a_{\mathrm{Y}}$. If, for example, a small positive covariance derives from rather few loci with large effects on both traits and of the same sign, and from many loci with smaller effect on both traits, but tending to be of opposite sign, then with selection on *X*, the genes of large effect will tend to become fixed earlier. Thus the genetic variance in *Y* will increasingly be due to the genes of opposite effect, and the covariance between the trait and fitness will gradually become more unfavourable. If all the initial variance in the directly selected trait is exhausted, the remaining variation in a trait on which no section was practiced would be $(1-r_{\mathrm{A}}^{2})V_{\mathrm{AY}}$, but that would assume no variation lost by drift or new variation arising by mutation. The steady state genetic variance maintained by mutation in the correlated trait is simply $2N_{\mathrm{e}}V_{\mathrm{MY}}$, but if the infinitesimal model does not apply, the variance maintained will depend on the joint distribution of effects of mutants on the selected and correlated trait. It becomes hard to draw general conclusions! Even so, it seems more likely that variance is lost proportionately more in the trait under selection than in other traits including fitness. At best the correlation with the index is small such that loss of variance in fitness is only a function of N and gain of NV_M.

4.2 Information from Laboratory Experiments

Deleterious changes in traits associated with fitness have frequently been found in selection experiments for quantitative traits, as discussed for example by Falconer and Mackay (1996). For example Yoo (1980) found a rapid reduction in bristle number in *Drosophila* when artificial selection to increase the trait was relaxed, associated with segregating lethals and other genes deleterious to fitness. López and López-Fanjul (1993b) for example showed that genes with substantially deleterious effect on fitness, including a homozygous lethal, were segregating in the lines derived from an inbred base, indicating the selection of mutants which would, in the absence of the selection, have been quickly eliminated. To look on the bright side, these experiments also illustrate, of course, the maintenance of variation in fitness despite the long term continued selection!

4.3 Consequences on Fitness of Livestock Improvement Programmes

There are substantial data and practical experience of declines in fitness associated traits, such as fertility, viability and leg strength in livestock, reviewed and documented for example by Rauw et al. (1998). These are inevitable consequences of selection on performance traits when fitness traits are unfavourably correlated, for example on leg weakness and abnormalities in broiler chickens and pigs. In the absence of genetic variation in these fitness associated traits it would not be possible to recover fitness, but there is substantial evidence that unfavourable trends can be reversed. We again mainly consider broiler poultry, where changes have been largest.

In the first trial of Havenstein et al. (1994) there was considerably higher mortality in the 1991 commercial and 1957 control populations. In the subsequent trial (Havenstein et al. 2003a), overall mortality rates were lower, but the 2001 commercials were little different from the 1957 controls, and in terms of mortality to market weight rates were lower.

McKay et al. (2000) report that legs abnormalities increased in incidence in Ross nucleus broiler and in commercial populations as a consequence of selection on growth, with % leg lesions in male lines around 7% in 1989 and 1990. Following selection on growth plate quality, incidence fell into the range 1–3% during 1991 and 1996, and below 1% in 1997 and 1998. Changes were also produced in the female lines and in the commercial population (McKay et al. 2000). Aviagen (formerly Ross Breeders) has maintained unselected controls also, and in a recently published study showed comparisons of 1972 and 2005 birds. There had been enor-

Table 2 Live weight at 42 days, Feed conversion ratio to 2 kg live weight, Mortality to 42 days, and proportion of birds with good leg health (GLH) in three modern (2005) lines and unselected control (1972) lines from which they were derived (e.g. MML came from CML in 1972)

	Line	42d LWT (g)	2 kg FCR	42d Mort (%)	2 kg GLH (%)
Modern	MML	2880	1.57	5.0	73
	MFL	2449	1.72	3.3	66
	MIML	2455	1.71	4.3	87
Control	CML	1204	2.26	5.3	82
	CFL	1216	2.19	4.7	55
	CIML	1229	2.24	1.8	66

Source: Fleming et al. 2007.

mous changes in rates of body weight growth, but leg abnormalities were little more frequent (Table 2, from Fleming et al. 2007). Results in Table 2 also show there has been little overall reduction in viability (or a worsening followed by improvement as Havenstein et al.'s data indicate). Comparisons for reproductive rate are complicated by the major change in environment, breeder females now being heavily feed restricted, such that what would otherwise be an unfavourable correlated response is masked by change in management.

There are relevant data on other species, showing changing trends in fitness associated traits. Thus Merks (2000) shows a decline in genetic merit for total number born in Dutch Landrace and Yorkshire pig populations up to the early 1990s and a subsequent improvement and also strongly positive genetic trends for piglet viability in the late 1990s. There is now much concern about unfavourable changes in fitness related traits in dairy cattle, in particular in fertility, with evidence that conception rates fell by about 1% per year over 23 years to 1998 (Royal et al. 2000). Consequently, selection pressure is now being put on predictors of fertility, such as non-return rate and calving interval (Brotherstone and Goddard 2005). There are not reliable estimates of such parameters prior to the era of intensive section on yield, but estimates have been obtained recently. These show low heritabilities but quite sizable estimates of additive genetic variance, indicating that genetic variation in fitness has been retained, and there are opportunities for improvement in fitness related traits (Wall et al. 2003).

The conclusion is that, while fitness has worsened when selection has not been practiced on it, trends can be reversed by changing selection goals because variation remains in the populations. It is also instructive to note that these reversals of performance in fitness associated traits have been made by changing selection goals within the extreme commercial populations, not by going back to old unimproved stock, for which a major loss in performance would be the cost of any gain in fitness associated traits. In view of the unfavourable correlations between production traits, such as growth rate, and fitness associated traits such as litter size, breeders have to continue to devote selection effort to preventing the latter worsening. It does not seem, however, that they need to take any additional action to maintain variability in fitness than what is done to maintain variability in the production traits, with effective population size playing the major role.

5 Maintenance of Variation in Fitness: Role of Mutation

Natural selection removes variation in fitness, so while such factors as heterozygote superiority or multiple niches which can lead to stable polymorphism may be important in some circumstances, mutation clearly has a very important role in maintaining variation in fitness. Hence there has been extensive theoretical analysis of its role and of experimental work aimed at determining the relevant parameters of mutation effects on fitness. Although we restrict the discussion to this area, we do not imply that other forces do not operate (see for example Falconer and Mackay 1996, ch. 21; Charlesworth and Hughes 2000; Bürger 2000). For example in further work using the chromosome competition method of Fowler et al. (1997, see above), Gardner et al. (2005) found persistent maintenance of fitness differences among the chromosomes over many generations. As mutation-selection balance leads to a dynamic equilibrium, stable adaptive mechanisms must also have been operating.

At its simplest, if a mutant has a fitness effect $-s$ in the heterozygote, its contribution to variation in fitness is $2x(1-x)s^2$ where x is the frequency of the mutant. In a population of effective size N_e sufficiently large that $N_e s >> 1$, i.e. the mutant is never fixed, the heterozygosity of the mutant is approximately $2u/s$ where u is the mutation rate per locus (Kimura 1969). Therefore the variance maintained under mutant-selection balance (MSB) is $\Sigma_i(2u_i s_i) = 2\lambda E(s)$ where λ is the genome-wide mutation rate; thus the variance is proportional to the mean decline in fitness per generation from mutation. For new mutants which are nearly neutral, the variance maintained is $4N_e us^2 = 2N_e V_M$, where V_M is the mutational variance in fitness. Houle et al. (1996) give estimates of the levels of standing variation in natural populations relative to the mutational variance, which would equal $2N_e$ in the absence of selection. The ratios of standing to mutational variance are distributed around about 50 for life history traits, and around 100 for morphological traits. These values are smaller than the likely sizes of natural populations, indicating that mutation is a substantial source of standing variation in fitness associated traits and presumably in fitness itself. Further, the relatively small differences between the life history and other traits imply that the effects of the natural selection, however acting, are not having a great impact on the variance. In the absence of selection favouring heterozygosity, the extent to which the upper or lower of these bounds on the variance in fitness depends on the distribution of mutational effects and on additional factors such as their degree of dominance and any association of mutation rate with size of effects.

Much effort has therefore been extended to determine the distribution of fitness effects of mutants, both directly using mutation accumulation (MA) experiments (Baer et al. 2005) and indirectly using genome-wide sequence data by examining differences in fixation rates at potentially silent (e.g. third base, intron) compared to non-silent sites of the DNA (Keightley and Eyre-Walker 1999; Eyre-Walker et al. 2006). The empirical data suggest an L-shaped distribution of mutation effects on fitness, with most mutants having very small effects and a few having a large effect, described by a gamma distribution substantially more leptokurtic than the exponential distribution. Both mean fitness effect and rate of mutations vary greatly among species. The value of V_M also varies somewhat among species. Experimental data also show that the action of genes on fitness is not additive and mutants of large effect appear to be recessive or partially recessive. To find whether MSB is a possible mechanism for the maintenance of genetic variance in fitness, detailed theoretical analyses of mutation-selection balance are needed to take into account all the empirical knowledge of mutation properties.

6 Genetic Variation in Phenotypic and Environmental Variance

The magnitude of the genetic variance defines the opportunities for genetic change, but the rate also depends on what proportion of the phenotypic variance this component comprises. Thus we have also to consider the magnitude of the environmental or residual variance, V_E. It has many components: developmental noise, perhaps also including, in modern terms, epigenetic variation; response to environmental knocks to which the animal is exposed; and, in many analyses, epistatic and dominance genetic variance components which are not fitted in the model. Indeed, we actually have a poor understanding of what does comprise the 'environmental' variance. If, for example, calves of intermediate birth weight have a higher probability of survival than heavy or light calves such that less variable genotypes are on average fitter, genetic variation in V_E implies genetic variation in fitness.

There has been extensive study of canalisation, not least in Australia by Rendel (1967). It is basically a reduction in phenotypic variance at particular levels of the trait. More recently we have given attention to what maintains levels of V_E more generally in populations, and have assumed that the level, formally of phenotypic variance given genotypic value, $V(P|E)$, is itself under genetic control. This is not an unreasonable assumption in that the response to micro-environmental change (plasticity) has a genetic component (De Jong and Gavrilets 2000) and presumably the level of developmental noise is under some genetic control. An illustration of the latter is that the variation within individuals, or fluctuating asymmetry, of *Drosophila* in wing length is much less relative to its mean than in sternopleural bristle number. There is other direct evidence from *Drosophila melanogaster*, provided by Mackay and Lyman (2005) on variation in V_E among chromosome substitution lines and by Whitlock and Fowler (1999) on its variation among inbred lines. Mackay and Lyman reared each line in replicate vials to provide an error term, and found highly significant variation among substitution lines, both as homozygotes and heterozygotes, in the variance within lines for abdominal and sternopleural bristle number. Some mutant genes, such as *bobbed* (Frankham 1980) and $smooth^{\mathrm{lab}}$ (Mackay 1985) have higher environmental variance than the wild type for abdominal bristle number in *Drosophila*. There have been a number of studies on livestock using structural models in which a genetic variance in residual variance is fitted (SanCristobal-Gaudy et al. 1998; Sorensen and Waagepetersen 2003) or simple comparisons made between families in within sire variance. In most of these, evidence of such genetic variance has been shown (see Mulder et al. 2007 for a summary).

Natural selection would be expected to reduce levels of V_E under any circumstances in which the plot of fitness against phenotype for the trait is concave downward, such that the mean fitness is less than the fitness at the mean of the trait. Nevertheless much V_E remains, so the problem is to establish what forces contribute to its maintenance. We furthermore need to understand what determines levels of CV and of heritability and why they differ substantially among traits, including why, for example, the genetic CV is higher for fitness than morphological traits.

Genetic variation is maintained by segregation, such that mutation selection balance is a plausible model for maintenance of V_G. However segregation per se does not generate V_E; it is a parameter more like the mean of a trait. We have found that if there is very substantial mutation bias, such that mutants are far more likely to increase than decrease the residual variance (Baer 2008), then V_E could be maintained, particularly in smaller populations where selection against mutants is less effective; but overall it does not seem an entirely plausible model (Zhang and Hill 2008). It is, however, the only model that we have constructed or seen which leads to any prediction of the magnitude of the heritability, which becomes a function of the mutational effects on mean and variance. Perhaps more surprisingly, while competition for resources among individuals of similar phenotype may lead to maintenance of polymorphism of genes affecting mean performance, and thus V_G, it does not maintain stable levels of V_E (Zhang and Hill 2007).

Environmental variance may reflect responses to micro-environmental differences, and thus selection for phenotypic plasticity which defines changes of mean phenotype as a function of environmental quality and therefore may be highly adaptive, is a potential route to maintenance of V_E. Analyses show that, if a positive association between mean phenotype and environment quality builds up, variation is maintained (Zhang 2006).

In view of the ubiquitous presence of V_E and the consistency of its magnitude and of CV and h^2 for different traits, it seems reasonable to assume that there must be some fundamentally simple but quite general force acting to maintain V_E. We have therefore considered what we believe to be the reasonable, but unsubstantiated possibility that there is an intrinsic cost of homogeneity, an 'engineering' cost (C) in producing a consistent product. A simple model is if overall fitness is reduced by a factor $\exp(-C/V^2)$ for a genotype of residual variance $V(P|G) = V$ (Zhang and Hill 2005c). The steady state environmental variance is given by $V_E = (2C\omega^2)^{1/2}$, where ω^2 denotes the strength of stabilizing selection (the 'variance' of the fitness profile). Furthermore, the overall reduction in fitness is small. The model does not, however, lead to predictions of the magnitude of the heritability and thus directly also to joint quantification of the magnitude of V_A and V_E maintained.

Analyses of predicted selection response (SanCristobal et al. 1998; Hill and Zhang 2004; Mulder et al. 2008) show that changes in V_E should be achievable, for example to improve product homogeneity or, perhaps, reduce variation in litter size which has a cost in terms of piglet or lamb survival. Whilst some progress can be made by individual selection, in practice useful change is likely to require evaluation using large families (Mulder et al. 2008).

We do not yet have, however, much information on how genetic variance in the residual component has changed as a consequence of selection, in part because it is fairly straightforward to monitor changes in phenotypic variance in a selection experiment, but harder to partition it into components. On the one hand there are clear examples where it has risen with high and low bristle number selected lines showing more variance than the base population (e.g. Clayton and Robertson 1957, likely in the low line due mainly to bobbed mutants, Frankham 1980). On the other hand Falconer and Robertson (1956) failed to obtain changes in V_E for body size in mice; and while phenotypic variance has changed in the same direction as selection in most populations under selection, typically the CV has changed much less, as if the determinant was only one of scale (Hill and Bünger 2004). For example, the CV for body weight has remained almost unchanged over 100 generations of selection in the Dummerstorf high body weight selected line (Hill and Bünger 2004). In view of the changes in genetic variance that likely have occurred, there must also have been some compensating genetic change in the residual variation. Further analysis and understanding of these results of selection are needed.

7 Discussion

7.1 Much Variation

There appear to be substantial amounts of genetic variance in fitness in natural populations, not only in nature where environmental shifts may change fitness profiles and promote heterozygosity, but also in long established population cages kept at constant environment as evidenced by the study of Fowler et al. (1997) on fitness directly and on many life history components. There is therefore no reason to assume that there is or that there is likely to be little genetic variance in fitness associated traits in livestock populations simply due to selection. The main force reducing variation is likely to be reduced population size, which increases the rate of loss of both existing variation and that arising by mutation.

What is also quite clear is that there are many unfavourable correlations between production traits such as growth rate or milk yield and fitness associated traits such as fertility. There is no reason to assume these will become less important. Indeed, there is some evidence that correlations become less favourable when animals become more extreme, as those between weight gain and egg number in turkeys show (Kranis et al. 2006). Hence selection effort will have to be maintained simultaneously on all the relevant traits. For such overall indices, however, there is no reason to assume that genetic variance will not be found: fitness in natural populations or in a population cage is itself a multi-trait compromise. The animal breeder should keep selecting!

7.2 Genetic Conservation

Much is made of the value of conserved populations as a hedge against loss of variation or environmental change. In fact little if any use is made of them in mainstream commercial production of livestock, apart from some recent niche marketing of, for example, meats of local breeds. In practice, modern populations, notably of chickens, are so far ahead of conserved strains in production traits that adaptation of these offers far more opportunity than crossing back to far out of date stocks. Whilst it is possible that important genes are present in such conserved populations, the information to identify them is unlikely to be obtained. Further, with moves towards genomic selection rather than marker-assisted selection for particular loci, the emphasis moves more towards best utilisation of the large amounts of variation present in the currently most extreme commercial populations. Livestock breed conservation, at least in the developed world, is important for our cultural heritage, rather than as a potential resource for food production for the masses.

7.3 Slow Horses and Dogs

There is little evidence of selection limits, except where there are real physical barriers (the low Illinois line of maize has almost 0% oil, Dudley and Lambert 2004), but there has been little improvement of race times in Thoroughbred horses or Greyhound dogs in the last 50 years. One hypothesis is just that the base population was very small, limiting the amount of variation present, but that discounts the impact of mutations which contribute to variation as analyses of other traits in other species have shown. Further, there is evidence of substantial genetic variance within the Thoroughbred horse population (Gaffney and Cunningham 1988), as shown by heritability estimates. So how can the results be explained? The variance estimates may be substantially biased by positive environmental covariances among relatives, but even so it is hard to believe there can be no genetic variance. There is evidence of heterozygote advantage for myostatin induced double muscling for racing performance in whippet dogs (Mosher et al. 2007). Whilst such a locus can maintain only non-additive genetic variation, unless it also has epistatic effects over other loci associated with speed it would not preclude further response to selection. One possible hypothesis is that indeed mutations occur for speed, but almost all of them are deleterious to performance and to components such as strength of leg (essentially multivariate constraints as discussed by Walsh and Blows in this volume). Further, as horses have also to satisfy the conformation required by the breeders and horse lovers, oddballs, even fast ones, are less likely to receive training.

7.4 Molecular Genetic Applications

We have not considered the impact of knowledge and application of molecular genetics in this review, which has been entirely at the quantitative level. The application of marker assisted selection clearly has a large potential impact not only on rates of progress but also on the effective population sizes of nucleus breeding populations. By concentrating simply on a single locus, especially one at low frequency, effective population size can of course be greatly reduced. More generally, however, use of marker data allows individuals to be selected more on the basis of their own characteristics, their genotype as well as phenotype, and less on their family (e.g. sibs') performance for sex limited or lowly heritable traits. This thereby reduces co-selection of relatives with an increase in N_e for a given number of potential candidates (Villanueva et al. 2004). This positive impact would seem to increase further with the use of genomic selection (Meuwissen et al. 2001) where the evaluation is of each region of the genome separately and reduces concentration on one or a few QTL. Hence there is the potential to have less impact on variability and inbreeding depression in fitness traits.

If there is adequate information, selection indices can in principle be constructed using marker data, taking into account the pleiotropic effects of QTL on production and fitness traits. Just how much such an ideal will be realised is more moot, however, as it is much harder to identify QTL for lowly heritable traits and get good estimates of their effects.

More controversially, transgenic technologies offer the opportunity to introduce new variation into the population by modification of the sequence of genes already present or by introducing new ones. It is really hard to see limits to long term response in production or fitness associated traits if transgenic technology can be used. The limited successes to date for production traits in animals merely reflect lack of knowledge, work and opportunity.

7.5 Adaptation

The presence of genetic variation in fitness implies the ability of the species or population to adapt. In the animal breeding context, where fitness is defined solely in terms of a single trait, even if animals are managed the same way in successive generations, adaptation implies an increasingly extreme phenotype. If the management or climate changes, adaptation implies there should be both continued increase in performance and the ability to do so under the new conditions. The best phenotype is defined by a new index of traits, and what may have been neutral (cryptic) or deleterious genes become an advantage. As there is always room for improvement of livestock, artificial selection is equivalent to adaptation to a changing economic environment. The substantial variation arising from past and new mutations provides the fuel for adaptation, even if all mutants are deleterious in the current environment. Hence large amounts of variation in fitness or, equivalently, adaptive fitness and in traits associated with it are found in livestock and natural populations. The magnitude can be influenced by many factors, but clearly population size is the most tangible. Even so, we have much yet to learn about the forces that maintain variability in populations at all levels, from the level of bases in the DNA to the broadest of all, fitness.

Acknowledgments We are grateful to Dick Frankham for many constructive comments on an earlier draft.

References

Baer, C.F. 2008. Quantifying the decanalizing effects of spontaneous mutations in rhabditid nematodes. *American Naturalist* **172**: 272–281.

Baer, C.F., Shaw, F., Steding, C., Baurngartner, M., Hawkins, A., Houppert, A., Mason, N., Reed, M., Sinnonetic, K., Woodard, W. and Lynch, M. 2005. Comparative evolutionary genetics of spontaneous mutations affecting fitness in rhabditid nematodes. *Proc. Natl. Acad. Sci. USA* **102**: 5785–5790.

Bijma, P. and Woolliams, J.A. 2000. Prediction of rates of inbreeding in populations selected on best linear unbiased prediction of breeding values. *Genetics* **156**: 361–373.

Bohren, B.B., Hill, W.G. and Robertson, A. 1966. Some observations on asymmetrical correlated responses to selection. *Genet. Res.* 7: 44–57.

Bradshaw, A.D. 1991. Genostasis and the limits to evolution. The Croonian lecture, 1991. *Phil. Trans. R. Soc. B* **333**: 289–305.

Brotherstone, S. and Goddard, M.G. 2005. Artificial selection and maintenance of variance in the global cow population. *Phil. Trans. R. Soc. B* **360**: 1479–1488.

Bürger, R. 2000. *The Mathematical Theory of Selection, Recombination, and Mutation*, Wiley, Chichester, UK.

Charlesworth, B. and Hughes, K.A. 2000. The maintenance of genetic variation in life history traits. In *Evolutionary Genetics: From Molecules to Morphology*, (R.S. Singh and C.B. Krimbas, eds.), Cambridge University Press, UK, pp. 369–392.

Clayton, G.A. and Robertson, A. 1957. An experimental check on quantitative genetical theory. II. The long-term effects of selection. *J. Genet.* **55**: 152–170.

Crnokrak, P. and Roff, D.A. 1995. Dominance variance – associations with selection and fitness. *Heredity* **75**: 530–540.

De Jong, G. and Gavrilets, S. 2000. Maintenance of genetic variation in phenotypic plasticity: the role of environmental variation. *Genet. Res.* **76**: 295–304.

Dudley, J.W. and Lambert, R.J. 2004. 100 generations of selection for oil and protein content in corn. *Plant Breed. Rev.* **24**: part 1, 79–110.

Duvick, D.N., Smith, J.S.C. and Cooper, M. 2004. Long-term selection in a commercial hybrid maize breeding program. *Plant Breed. Rev.* **24**: part 2, 109–151.

Eyre-Walker, A., Woolfit, M., and Phelps, T. 2006. The distribution of fitness of new deleterious amino acid mutations in humans. *Genetics* **173**: 891–900.

Falconer, D.S. and Mackay, T.F.C. 1996. *Introduction to Quantitative Genetics*, 4th ed. Longman, Harlow, UK.

Falconer, D.S. and Robertson, A. 1956. Selection for environmental variability of body size in mice. *Z.tschrift für induct. Abstammungs- und Vererbungslehre* **87**: 385–391.

Fisher, R.A. 1958. *The Genetical Theory of Natural Selection*, Dover, New York.

Fleming, E.C., Fisher, C. and McAdam, J. 2007. Genetic progress in broiler traits – implications for welfare. *Proc. Brit. Soc. Anim. Sci. 2007* p. 50.

Fowler, K., Semple, C., Barton, N.H. and Partridge, L. 1997. Genetic variation for total fitness in *Drosophila melanogaster. Proc R. Soc. B* **264**: 191–199.

Frankham, R. 1980. Origin of genetic variation in selected lines. In *Selection Experiments in Laboratory and Domestic Animals* (A. Robertson, ed.), Commonwealth Agricultural Bureaux, Slough, England, pp. 56–58.

Frankham, R., Jones, L.P. and Barker, J.S.F. 1968a. The effects of population size and selection intensity in selection for a quantitative character in *Drosophila*. I. Short-term response to selection. *Genet. Res.* **12**: 237–248.

Frankham, R., Jones, L.P. and Barker, J.S.F. 1968b. The effects of population size and selection intensity in selection for a quantitative character in *Drosophila*. III. Analyses of the lines. *Genet. Res*. **12**: 237–248.

Gaffney, B. and Cunningham, E.P. 1988. Estimation of genetic trend in racing performance of thoroughbred horses. *Nature* **332**: 772–724.

Gardner, M.P., Fowler, K., Barton, N.H. and Partridge, L. 2005. Genetic variation for total fitness in *Drosophila melanogaster*: complex yet replicable patterns. *Genetics* **169**: 1553–1571.

Havenstein, G.B., Ferket, P.R. and Qureshi, M.A. 2003a. Growth, livability, and feed conversion of 1957 versus 2001 broilers when fed representative 1957 and 2001 broiler diets. *Poult. Sci.* **82**: 1500–1508.

Havenstein, G.B., Ferket, P.R. and Qureshi, M.A. 2003b. Carcass composition and yield of 1957 versus 2001 broilers when fed representative 1957 and 2001 broiler diets. *Poult. Sci.* **82**: 1509–1518.

Havenstein, G.B., Ferket, P.R., Scheideler, S.E. and Larson, B.T. 1994. Growth, livability, and feed conversion of 1957 vs 1991 broilers when fed typical 1957 and 1991 broiler diets. *Poult. Sci.* **73**: 1785–1794.

Hill, W.G. 1982. Rates of change in quantitative traits from fixation of new mutations. *Proc. Natl. Acad. Sci.* **79**: 142–145.

Hill, W.G. 2000. Maintenance of quantitative genetic variation in animal breeding programmes. *Livest. Prod. Sci.* **63**: 99–109.

Hill, W.G. and Bünger, L. 2004. Inferences on the genetics of quantitative traits from long-term selection in laboratory and farm animals. *Plant Breed. Rev*. **24**: part2, 169–210.

Hill, W.G. and Keightley, P.D. 1988. Interrelations of mutation, population size, artificial and natural selection. In *Proc. 2nd Int. Conf. Quantitative Genetics* (B.S. Weir, E.J. Eisen, M.M. Goodman and G. Namkoong, eds.), Sinauer, Sunderland, Massachusetts, pp. 57–70.

Hill, W.G. and Zhang, X.-S. 2004. Effects on phenotypic variability of directional selection arising through genetic differences in residual variability. *Genet. Res.* **83**: 121–132.

Hoffmann, A.A., Hallas, R.J., Dean, J.A. and Schiffer, M. 2003. Low potential for climatic stress adaptation in a rainforest *Drosophila* species. *Science* **301**: 100–102.

Holt, M., Meuwissen, T. and Vangen, O. 2005. Long-term responses, changes in genetic variances and inbreeding depression from 122 generations of selection on increased litter size in mice. *J. Anim. Breed. Genet.* **122**: 199–209.

Houle, D. 1992. Comparing evolvability of quantitative traits. *Genetics* **130**: 195–204.

Houle, D., Morikawa, B. and Lynch, M. 1996. Comparing mutational variabilities. *Genetics* **143**: 1467–1483.

James, J.W. 1972. Optimum selection intensity in breeding programs. *Anim. Prod.* **14**: 1–8.

Jones, L.P., Frankham, R. and Barker, J.S.F. 1968. The effects of population size and selection intensity in selection for a quantitative character in *Drosophila*. II. Long-term response to selection. *Genet. Res.* **12**: 237–248.

Keightley, P.D. 2004. Mutational variation and long-term selection response. *Plant Breed. Rev.* **24**: part 1, 227–247.

Keightley, P.D. and Eyre-Walker, A. 1999. Terumi Mukai and the riddle of deleterious mutation rates. *Genetics* **153**: 515–523.

Kimura, M. 1969. The number of heterozygous nucleotide sites maintained in a finite population due to steady flux of mutations. *Genetics* **61**: 893–903.

Kranis, A., Hocking, P.M., Hill, W.G. and Woolliams, J.A. 2006. Genetic parameters for a heavy female turkey line: impact of simultaneous selection for body weight and total egg number. *Brit. Poult. Sci.* **47**: 685–693.

Kruuk, L.E.B. 2004. Estimating genetic parameters in natural populations using the 'animal model'. *Phil. Trans R. Soc. B* **359**: 873–890.

Kruuk, L.E.B., Clutton-Brock, T.H., Slate, J., Pemberton, J.M., Brotherstone, S. and Guiness, F.E. 2000. Heritability of fitness in a wild mammal population. *Proc. Natl. Acad. Sci. USA* **97**: 698–703.

López, M.A. and López-Fanjul, C. 1993a. Spontaneous mutation for a quantitative trait in *Drosophila melanogaster*. I. Response to artificial selection. *Genet. Res.* **61**: 107–116.

López, M.A. and López-Fanjul, C. 1993b. Spontaneous mutation for a quantitative trait in *Drosophila melanogaster*. I. Distribution of mutant effects on the trait and on fitness. *Genet. Res.* **61**: 117–126.

Lynch, M. and Walsh, B. 1998. *Genetics and Analysis of Quantitative Traits*, Sinauer, Sunderland, MA, USA.

Mackay, T.F.C. 1985. Transposable element-induced response to artificial selection in *Drosophila melanogaster. Genetics* **111**: 351–374.

Mackay, T.F.C. and Lyman, R.F. 2005. *Drosophila* bristles and the nature of quantitative genetic variation. *Phil. Trans. R. Soc. B* **360**: 1513–1527.

McKay, J.C., Barton, N.F., Koerhuis, A.N.M. and McAdam, J. 2000. The challenge of genetic change in the broiler chicken. In *The Challenge of Genetic Change in Animal Production* (W.G. Hill, S.C. Bishop, B.J. McGuirk, J.C. McKay, G. Simm and A.J. Webb, eds.), Occ. Publ. *Brit. Soc. Anim. Sci.* no. 27, pp. 1–7.

Merila, J. and Sheldon, B.C. 1999. Genetic architecture of fitness and non-fitness traits: empirical patterns and development of ideas. *Heredity* **83**: 103–109.

Merks, J.W.M. 2000. One century of genetic changes in pigs and future needs. In *The Challenge of Genetic Change in Animal Production* (W.G. Hill, S.C. Bishop, B.J. McGuirk, J.C. McKay, G. Simm and A.J. Webb, eds.), Occ. Publ. *Brit. Soc. Anim. Sci.* no. 27, pp. 8–19.

Meuwissen, T.H.E., Hayes, B.J. and Goddard, M.E. 2001. Prediction of total genetic value using genome-wide dense marker maps. *Genetics* **157**: 1819–1829.

Mosher, D.S., Quignon, P., Bustamante, C.D., Sutter, N.B., Mellersh, C.S., Parker, H.G. and Ostrander, E.A. 2007. A mutation in the myostatin gene increases muscle mass and enhances racing performance in heterozygote dogs. *PLoS Genetics* **3**(5): e79.

Mulder, H.A., Bijma, P. and Hill, W.G. 2007. Prediction of breeding values and selection responses with genetic heterogeneity of environmental variance. *Genetics* **175**: 1895–1910.

Mulder, H.A., Bijma, P. and Hill, W.G. 2008. Selection for uniformity in livestock by exploiting genetic heterogeneity of environmental variance. *Gen. Sel. Evol.* **40**: 37–59.

Oldenbroek, K. (Ed.). 2007. *Utilisation and Conservation of Farm Animal Genetic Resources*, Wageningen Academic Publishers, The Netherlands.

Preisinger, R. and Flock, D.K. 2000. Genetic changes in layer breeding: historic trends and future prospects. In *The Challenge of Genetic Change in Animal Production* (W.G. Hill, S.C. Bishop, B.J. McGuirk, J.C. McKay, G. Simm and A.J. Webb, eds.), Occ. Publ. *Brit. Soc. Anim. Sci.* no. **27**, pp. 20–28.

Price, T. and Schluter, D. 1991. On the low heritability of life history traits. *Evolution* **45**: 853–861.

Rauw, W.M., Kanis, E., Noordhuizen-Stassen, E.N. and Grommers, F.J. 1998. Undesirable side effects of selection for high performance efficiency in farm animals: a review. *Livest. Prod. Sci.* **56**: 15–33.

Rendel, J.M. 1967. *Canalization and Gene Control*, Logo Press, Academic Press, New York.

Robertson, A. 1960. A theory of limits in artificial selection. *Proc. R. Soc. B* **153**: 234–249.

Robertson, A. 1966. A mathematical theory of the culling process in dairy cattle. *Anim. Prod.* **8**: 95–108.

Roff, D.A. and Mousseau, T.A. 1987. Quantitative genetics and fitness: lessons from *Drosophila. Heredity* **58**: 103–118.

Royal, M.D., Darwash, A.O., Flint, A.P.E., Webb, R., Woolliams, J.A. and Lamming, G.E. 2000. Declining fertility in dairy cattle: changes in traditional and endocrine measures of fertility. *Anim. Sci.* **70**: 487–501.

SanCristobal-Gaudy, M., Elsen, J.M., Bodin, L. and Chevalet, C. 1998. Prediction of the response to a selection for canalisation of a continuous trait in animal breeding. *Genet. Sel. Evol.* **30**: 423–451.

Sorensen, D. and Waagepetersen, R. 2003. Normal linear models with genetically structured residual variance heterogeneity: a case study. *Genet. Res.* **82**: 207–222.

Villanueva, B., Avendaño, S. and Woolliams, J.A. 2006. Prediction of genetic gain from quadratic optimization with constrained rates of inbreeding. *Genet. Sel. Evol.* **38**: 127–146.

Villanueva, B., Dekkers, J.C.M., Woolliams, J.A. and Settar, P. 2004. Maximising genetic gain over multiple generations with quantitative trait locus selection and control of inbreeding. *J. Anim. Sci.* **82**: 1305–1314.

Wall, E., Brotherstone, S., Woolliams, J.A., Banos, G. and Coffee, M.P. 2003. Genetic evaluation of fertility using direct and correlated traits. *J. Dairy Sci.* **86**: 4093–4102.

Walsh, B. 2004. Population- and quantitative-genetic models of selection limits. *Plant Breed. Rev.* **24**: part 1, 177–225.

Weber, K. 2004. Population size and long-term selection. *Plant Breed. Rev.* **24**: part 1, 249–268.

Wei, M., Caballero, A. and Hill, W.G. 1996. Selection response in finite populations. *Genetics* **144**: 1961–1974.

Whitlock, M.C. and Fowler, K. 1999. The changes in genetic and environmental variance with inbreeding in *Drosophila melanogaster. Genetics* **152**: 345–353.

Yoo, B.H. 1980. Long-term selection for a quantitative character in large replicate populations of *Drosophila melanogaster*. II. Lethals and visible mutants with large effects. *Genet. Res.* **35**: 19–31.

Zhang, X.-S. 2006. The phenotypic variance within plastic traits under migration-mutation-selection balance. *Evolution* **60**: 1125–1136.

Zhang, X.-S. and Hill, W.G. 2005a. Genetic variability under mutation-selection balance. *Trends Ecol. Evol.* **20**: 468–470.

Zhang, X.-S. and Hill, W.G. 2005b. Predictions of patterns of response to artificial selection in lines derived from natural populations. *Genetics* **169**: 411–425.

Zhang, X.-S. and Hill, W.G. 2005c. Evolution of the environmental component of the phenotypic variance: stabilizing selection in changing environments and the cost of homogeneity. *Evolution* **59**: 1237–1244.

Zhang, X.-S. and Hill, W.G. 2007. Competition can maintain genetic but not environmental variance in the presence of stabilizing selection. *Evolution* **61**: 1532–1545.

Zhang, X.-S. and Hill, W.G. 2008. Mutation selection balance for environmental variance. *Am. Nat.* **171**: 394–399.

Spherical Cows Grazing in Flatland: Constraints to Selection and Adaptation

Mark Blows and Bruce Walsh

Abstract The vector of traits that a breeder is trying to improve and/or natural selection is acting upon has a distribution of phenotypic and breeding values that lives in a complex space. This space is not simply a sphere with equal variation in all dimensions, but rather a much more constrained structure and these constraints have critical implications for selection responses. Here we emphasise the importance of the resurgence of interest in Fisher's geometric model of adaptation, and the necessity of taking a multivariate view of selection. We review basic matrix tools, such as the angle between two vectors, the projection of a vector into a matrix subspace, and more recent advances such as the estimation of the dimensionality of a covariance matrix, that provide different ways to quantify potential constraints on evolutionary change. A key goal of quantitative genetics is to now understand the geometry of the genetic covariance matrix **G** from both the point of view of mutation generating genetic variance and from selection depleting it. Initial studies using *Drosophila* have suggested that **G** may be very ill-conditioned, with a number of phenotypic dimensions displaying very little genetic variance. Such multivariate constraints may be quite important, potentially resulting in very little usable genetic variation in the direction of persistent directional selection, despite significant heritabilities in each of the component traits.

Keywords Selection · adaptation · G matrix · genetic constraints · matrix subspace projection

1 Adaptation and the Multivariate Phenotype

Organisms are complex structures that are also highly variable. Couple this with the fact that scientists like to measure things such as heights, weights, bristle numbers, length of horns, size of genitalia, wing spans, and mRNA levels (to name

M. Blows (✉)
School of Integrative Biology, University of Queensland, St. Lucia Queensland 4072, Australia
e-mail: m.blows@uq.edu.au

J. van der Werf et al. (eds.), *Adaptation and Fitness in Animal Populations*,

infinitesimally few of the possibilities). Finally, throw in measures of the same character over different times and/or environments and the list of potential traits is literally endless. Thus, organisms are an essentially unmeasurable complex of traits, some of which are targets for selection while others are effectively neutral. This is the essence of the task facing evolutionary biologists when trying to understand adaptation. Even when one chooses just the smallest subset of possible traits as candidates for targets of selection, a surprisingly complex multivariate phenotype still emerges.

An additional complication is that organisms are not the mythical "spherical cow". Traits have constraints that may limit the likelihood of particular combinations of phenotypic values. Even worse, the same holds for breeding values, potentially closing off (or at least rendering extremely inefficient) certain paths for improved fitness. Thus, any view of adaptation has a strong geometric component. This view dates back to D'Arcy Thompson (1917) and the founders of the modern synthesis, R. A. Fisher (1930), Sewall Wright (1932), and G. G. Simpson (1944). In the modern analysis of adaptation, we also take a geometric view, considering the distribution of values in both phenotypic and genetic (breeding value) space. Here we examine some recent theoretical developments in this geometric view (both genetical and statistical), and some recent empirical work that suggests a central role for the geometric representation of constraints to selection and adaptation.

2 Spherical Cows: Geometric Models for the Adapativeness of New Mutations

Geometric descriptions of organismal morphology began with the classic treatment of Thompson (1917), wherein different trait values can be thought of as axes in some morphological space. We can superimpose a fitness surface (Wright 1932; Simpson 1944; Lande 1979) upon this space, adding an axis that represents the mean fitness of any particular combination of trait values. On such a fitness space, Fisher (1930) suggested that the actual number of independent traits under selection has important consequences for adaptation. His model has subsequently been investigated by a number of other workers (e.g. Kimura 1983; Leigh 1987; Rice 1990; Hartl and Taubes 1996; Orr 1988, 2000). Fisher's model, in a classic spherical cow approximation, assumes an overly simplistic geometry for the resulting fitness surface, with a single optimal fitness peak (at the optimal point θ) with fitness uniformly decreasing as we increase the Euclidean distance from θ (Fig. 1). Fisher's interest was how likely a new mutation was to be adaptive (increasing its fitness relative to the current value). In particular, given a phenotype at point z, which is at distance d from the optimal value θ, what is the probability that a new mutation with effect r (the distance between the wild-type and mutant phenotypes) increases fitness? For a two-dimensional fitness surface, Fisher's model gives the contour of equal fitness of the current phenotype as a circle, allowing a simple geometric argument to be used to determine how likely it is that a mutation ends up with higher fitness

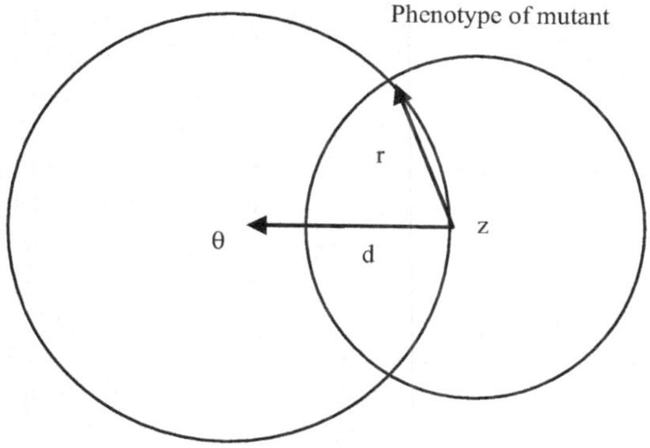

Fig. 1 Fisher's model of the chance that a new a mutation is adaptive (shown here for the two dimensional case). The current phenotype is at point z, while the optimum of this fitness surface is at point θ, which is (Euclidean) distance d away from the current phenotype. A random mutation creates a new phenotype at distance r from the wildtype. The circle centred at θ, and passing thorough z, represents the equal-fitness contour of the current phenotype. The circle with diameter r centred at z represents the potential range of new mutation given a move of r in phenotypic space. The probability of adaptation is simply that part of this circle that is inside (i.e. closer to θ) of the current fitness contour of the wildtype. In this case, roughly (by eyeball) 30%

(Fig. 1). Fisher's key insight was that, for phenotypes far away from the optimum value, large moves can have a higher probability of adaptation than small moves. However, as we approach the optimum, large moves almost always have reduced fitness. In particular, it is the scaled ratio r/d, the fraction of the move relative to the distance from the optimum, that characterizes the probability of adaptation.

For n traits, Fisher assumed that the contours of equal fitness are hyper-spheres around the optimum value, in which case the probability of adaptation is determined by the single quality

$$x = \frac{r\sqrt{n}}{2d} \tag{1}$$

which now scales the relative move r/d by the dimensionality n of the fitness space. For large n, Fisher (1930) showed that the probability of such a mutation being favourable (increasing fitness) is

$$p_{fav} = \frac{1}{\sqrt{2\pi}} \int_x^\infty \exp(-y^2/2)dy = 1 - \text{erf}(x) \tag{2}$$

where **erf** is the error function (Rice 1990 gives exact expressions for any n). Thus, as noted by Fisher, increasing the number of independent dimensions of selection (increasing n) makes a new mutation increasingly unlikely to be adaptive. The direct (but often unstated) implication of this result is that organisms must be somewhat constrained in the number of independent dimensions that fitness can act upon.

Kimura (1983) and Orr (1988, 2000) developed an important extension of Fisher's model. Fisher simply considered the probability of a favorable mutant. However, if a favorable new mutation is introduced as a single copy in an infinite population, then its probability of fixation is just $2s$, twice its selective advantage as a heterozygote, so that fixation of most favorable new mutations is unlikely. Thus, while mutations of larger effect (x larger) are increasingly unlikely to be favorable, when they are, they may have a significantly larger effect, and hence a higher probability of fixation. The quantity of interest is the fixation rate, the product of the average selective advantage (when favorable) times the probability of being favorable (Equation (2)). For example, as the mutational effect r becomes infinitesimally small, the probability of it being advantageous approaches 50%, but its average advantage (when favorable) also approaches 0 (as $s \to 0$). Orr (2000) showed that the optimal "size" for a mutation (to maximize this product) is roughly $x \cong 0{:}925$, so that during an adaptive walk trying to move from z to θ through a successive series of mutations, the optimal mutation size gets smaller with each successive fixation (i.e. as d gets closer to zero). Specifically, from Equation (1), the optimal mutation size is approximately

$$r_{opt} \cong \frac{0.925.2.d}{\sqrt{n}} \cong 1.85\frac{d}{\sqrt{n}} \tag{3}$$

Orr further showed that there is a considerable cost to complexity (dimensions of selection n), with the rate of adaptation (favourable mutation rate times fixation probability) declining significantly faster than $1/n$. Thus, the constraint on dimensionality may be much more severe than originally suggested by Fisher.

Much evolutionary angst has been generated over the "cost of complexity" given these results. However, Fisher's model makes the simplifying assumption that all traits have the same fitness curvature, which is highly unlikely. More likely is a fitness surface with a few dimensions under strong selection (large curvature) and a very large number of dimensions under very weak selection (low curvature) (Barton 1990; Johnson and Barton 2005). Rice (1990) has examined such cases where the curvature of the fitness surface is not constant across all dimensions. The probability of adaptation on these surfaces depends upon their "effective curvature", roughly the harmonic mean of the individual curvatures. Recalling that the harmonic mean is dominated by small values, it follows that the probability of adaptation is likewise dominated by those fitness surfaces with low curvature (weak selection). However, on such surfaces, s is small, and hence the fixation probability small. Clearly, adaptive walks in such variable-curvatures surfaces is an interesting area for much further work. The key, however, is that while it is sophisticated to think about the geometric space generated by a fitness surface, we may obtain very misleading results if our assumed geometry is oversimplified.

3 Multivariate Phenotypes and Selection Response

As we have stressed, all selection is multivariate. Even in those (rare) cases where artificial selection is focused on just a single trait, natural selection on other (potentially) correlated traits is occurring as well, and this additional selection can help or (more likely) hinder the artificial selection objectives. In natural populations, wherein organisms are attempting to adapt to ever-changing environments, the adaptive target may be constantly shifting. Indeed, as organisms evolve, they necessarily change the environment around them, and hence organisms must evolve just to keep pace with other organisms with which they compete for resources. This is van Valen's (1973) red-queen hypothesis, wherein (after Lewis Carroll's Red Queen in *Through the Looking Glass*) one must run just to stay in place.

Attempts to model multivariate selection response have two components. The first, prediction of response given the nature of selection, is a concern shared by both breeders and evolutionary biologists. For example, Smith (1936), Hazel (1943), and others have developed selection indices to maximize some linear combination of traits given a desired set of weights. Evolutionary biologists face a second task, estimating the actual nature of selection. Even in an apparently simple univariate setting, this approach is fraught with peril. Particularly distressing is that there are numerous examples of well-studied natural populations (most monitored for over a decade) that show significant heritability for a trait, a (relatively) constant selection differential on that trait, and yet while h^2 and S are both significant, the simple breeders equation ($R = h^2 S$) is not followed, as no response is seen (Merilä et al. 2001). One (but not the only) explanation is that a third trait, influencing both fitness and our focal trait, can generate a focal trait-fitness correlation. If the causative trait is not heritable, there is no response. A classic example is antler size in Red Deer, where males with larger antlers gather more mates, and antler size is highly heritable. However, analysis using animal model BLUP shows that there is no direct selection in breeding values for antler size (Kruuk et al. 2002). Rather, the suggestion is that an animal's nutritional state, if high, allows them to grow larger antlers. Further, if nutritional state is high, males are also better fighters, getting more mates. This creates an antler-fitness correlation entirely due to the effect of a third trait, nutritional state, on both.

A formal theory for multivariate selection was proposed by Lande (1979), ironically at roughly the same time that multivariate BLUP was proposed (Henderson and Quaas 1976). The multivariate version of the breeders' equation relates the vector **R** of responses and the vector **S** of selection differentials through the phenotype **P** and genetic (breeding value) **G** covariance matrices,

$$\mathbf{R} = \mathbf{G}\mathbf{P}^{-1}\mathbf{S} \qquad (4)$$

Note that in the multivariate version $\mathbf{GP}^{-1}$ replaces $h^2 = \sigma_A^2(1/\sigma_z^2)$. The multivariate breeders' equation follows under the assumption that the joint distribution of phenotypes and breeding values is multivariate normal.

Robertson (1966) and Price (1970) showed that the (univariate) selection differential is mathematically equivalent to the covariance between trait value and fitness. Thus, **S** is a vector of covariances, the i-th element being the covariance between trait i and fitness, $S_i = \sigma(z_i, w)$, and from the theory of multiple regression it follows that

$$\boldsymbol{\beta} = \mathbf{P}^{-1}\mathbf{S} \tag{4a}$$

is the vector of weights of the multiple regression of fitness w on the vector $(z_1, z_2, \ldots, z_n)$ of phenotypes,

$$w = a + \sum_{i=1}^{n} \beta_i z_i + e_i \tag{4b}$$

β_i is called the directional selection gradient on the ith trait, and is a measure of the amount of direct selection acting on that trait, after (phenotypic) correlations with all of the other traits in the sample having been removed (Pearson 1903; Lande and Arnold 1983). An important caveat is that unmeasured traits under direct selection that are phenotypically correlated with trait i can create a false picture of the nature of selection on i.

Putting all of this together, we have the Lande Equation, relating a measure of phenotypic selection $\boldsymbol{\beta}$ (removing the effects of phenotypic correlations) with the usable space of breeding values (given by **G**),

$$\mathbf{R} = \mathbf{G}\boldsymbol{\beta} \tag{5}$$

If phenotypes are multivariate-normally distributed, then Lande (1979) shows that the direction the current mean must move to maximize the local change in mean population fitness is given by the vector $\boldsymbol{\beta}$. Formally, $\boldsymbol{\beta}$ is a gradient with respect to (log) mean population fitness as a function of the trait means. Thus fitness change is maximized by moving the population in the direction $\boldsymbol{\beta}$, but genetics *constrains* this response through the variance-covariance matrix **G**. When a vector is multiplied by a matrix, it results in a rotation and scaling. Any rotation results in the direction of change not being along the optimal direction $\boldsymbol{\beta}$ to increase fitness, while scaling changes the relative rate of response along the final direction. This is also a well-known lesson for index selection: specification of the breeding objective (the vector of economic weights on a selection index) does not fully specify the direction of genetic change.

Equations (4a) and (5) show how constraints arise in multivariate selection and how (under multivariate normality) covariance matrices quantify these constraints. The first potential constraint is on phenotypes. Suppose selection is directly acting on k traits, but the resulting phenotypic covariance matrix **P** for these traits is singular (or nearly so). In such cases, the effects of phenotypic selection will not fully translate into direct selection on individual traits. If **P** is ill-conditioned (nearly singular), then small random fluctuations may result in large changes in individual $\boldsymbol{\beta}$ values, decreasing the efficiency of selection response. When **P** is singular, then so is **G** (Pease and Bull 1988), as $\mathbf{P} = \mathbf{G} + \mathbf{E}$, the sum of the genetic and environmental covariance matrices. However, the converse is not true, and **P** can be of full rank even when **G** is singular. Indeed, most discussion of constraints in the response to selection focus on constraints imposed by **G**, but this may be only part of the story. Biologically, singularities in a covariance matrix can arise when a trait has essentially no (phenotypic or genetic) variation and/or (more likely) when traits highly covary with each other.

4 Living in Flatland: The Misleading Univariate World

Selection and response are intrinsically high-dimensional phenomena. However, the focus of breeders, and even evolutionary biologists, is on a much restricted subset of this space, often one or just a few traits. The problems of mapping from a high dimensional space into a lower dimensional one were first noted in Edwin Abbott Abbott's (1884) wonderful *Flatland*. In Abbott's world, geometric shapes (such as triangles) living in two dimensions, take on different appearances to their fellow Flatlandians as their orientation changes. Visitors from 3-D space sojourning through Flatland have even more dramatic changes, starting out as a circle of increasing size, which reaches maximal size and then shrinks back to zero as the sphere passes through its 2-D projection in Flatland. As biologists (and breeders) we tend to work in flatland, ignoring (indeed, we are almost always ignorant of) the full geometry underlying the traits we are intending to follow and/or modify, and instead work with a projection of the full space onto a lower dimensional one.

As early as Dickerson (1955), some of the consequences of a univariate view of multivariate selection were noted. Dickerson (in the context of a selection index) remarked that each trait in an index may have a significant heritability, but the resulting heritability of the index itself may be zero (or nearly so). Thus, with multivariate selection there can be genetic variation (i.e. nonzero heritabilities) in all traits under selection, but little or no genetic variation along the direction that selection is trying to move the population. To see this point, consider the following simple example. Suppose the additive variance of traits 1 and 2 are 20 and 40, respectively. Further, suppose that there is direct selection on both traits, with selection favoring a two unit change in trait 1 for each one-unit decrease in trait 2, e.g.,

$$\boldsymbol{\beta} = \begin{pmatrix} 2 \\ -1 \end{pmatrix}$$

If we consider each trait separately, in a fully univariate framework ignoring all other traits, the selection response for trait i is just $R_i = \sigma_A^2(i)\beta_i$, which would be $10.2 = 20$ for trait one and $40.(-1) = -40$ for trait two. When $\sigma(A_I, A_2) = 0$, these are indeed the responses predicted from the multivariate breeders' equation,

$$R = G\beta = \begin{bmatrix} 10 & 0 \\ 0 & 40 \end{bmatrix} \begin{bmatrix} 2 \\ -1 \end{bmatrix} = \begin{bmatrix} 20 \\ -40 \end{bmatrix}$$

While there is response in both traits, there is actually a (minor) constraint here (as **R** does not point in the same direction as $\boldsymbol{\beta}$), a point we return to shortly.

Now suppose that the genetic covariance between traits 1 and 2 is $\sigma(A_1, A_2) = 20$, giving

$$\boldsymbol{R} = \boldsymbol{G\beta} = \begin{bmatrix} 10 & 20 \\ 20 & 40 \end{bmatrix} \begin{bmatrix} 2 \\ -1 \end{bmatrix} = \begin{bmatrix} 0 \\ 0 \end{bmatrix}$$

Thus, even though there is considerable additive genetic variation in both traits, there is no response. This occurs because **G** has a zero eigenvalue, whose associated eigenvector exactly corresponds to our $\boldsymbol{\beta}$. There is no additive genetic variance along this particular direction, and hence no response. This is an example of an absolute constraint (Pease and Bull 1988). Quantitative constraints can arise when $\boldsymbol{\beta}$ is only a few degrees away from an eigenvector associated with a zero eigenvalue, in which case the resulting response would be very small. In theory, especially in a rapidly changing environment, a quantitative constraint could result in population extinction, as although the population can respond in the desired direction, it may not be able to do so quickly enough to keep pace with environmental change.

One straightforward way to quantity the constraints on the response imposed by **G** is to consider the angle θ between the optimal $\boldsymbol{\beta}$ and actual response **R**. Recall that the angle θ between two vectors is given by

$$\cos(\theta) = \frac{\boldsymbol{x}^T \boldsymbol{y}}{\|\boldsymbol{x}\| \,.\, \|\boldsymbol{y}\|}$$

Hence the constraint angle (between the desired direction $\boldsymbol{\beta}$ and the actual direction **R**) is

$$\theta = \cos^{-1}\left(\frac{\boldsymbol{R}^T \boldsymbol{\beta}}{\|\boldsymbol{R}\| \,.\, \|\boldsymbol{\beta}\|}\right) \tag{6}$$

There is no constraint if $\theta = 0$ (**R** and $\boldsymbol{\beta}$ have the same orientation), while $\theta = 90$ degrees corresponds to an absolute constraint. Now let's reconsider the response above with no genetic covariance between the traits. Selection ($\boldsymbol{\beta}$) is maximized by a two unit response in trait one for every (negative) unit of response in trait two. However, the actual response in this case was in a (slightly) different direction, doubling the response in trait two for every unit of response in trait one (Fig. 2). Equation (6) gives $\theta = \cos^{-1}(0.8)$, for $\theta = 37$ degrees. Thus, the actual response is rotated 37 degrees from the optimal direction for selection response by **G**. Notice that this (mild) constraint occurs even though the two traits are genetically uncorrelated. Trait two has four times the genetic variance of trait one, and hence gives a larger response per unit of selection than trait one. Thus, even a diagonal (and

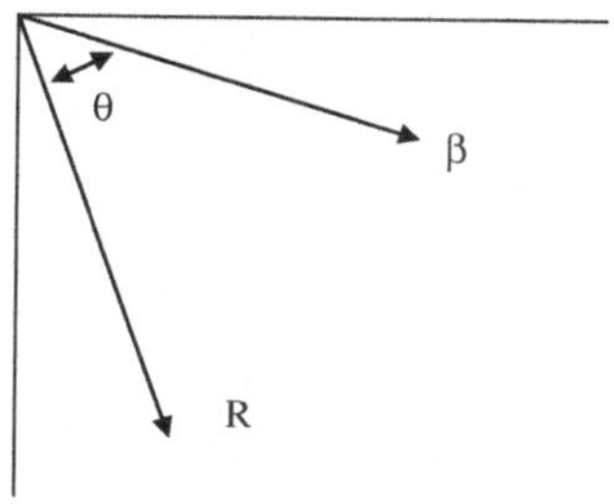

Fig. 2 Mild constraints can be imposed even for a diagonal G matrix (i.e. no genetic covariances between traits). Unless the pattern of genetic variances corresponds to the strength of selection ($\beta_i = c.G_{ii}$), then $\boldsymbol{\beta}$ is not an eigenvector of G, and the actual response R will be rotated somewhat away from the optimal response $\boldsymbol{\beta}$. These two vectors are plotted for our example where $\beta_1 = 2$, $\beta_2 = -1$, $G_{11} = 1$, $G_{22} = 40$, $G_{12} = 0$. The resulting angle θ between R and $\boldsymbol{\beta}$ is 37 degrees

non-singular) **G** matrix can impose constraints to response, although these are expected to be mild unless the largest values of β_i are associated with the smallest G_{ii}.

5 Constraints and Consequences

While the inherent genetic constraints imposed by **G** have long been appreciated by evolutionary biologists (e.g. Lande 1979; Pease and Bull 1988), the general view of many was that the conditions for $\boldsymbol{\beta}$ to align with a zero-eigenvalue were rather pathologic and hence unlikely (by chance) to happen. However, additive genetic variation tends to erode under constant directional selection. Hence, if a trait has been under long-term selection to move in a particular direction, it is possible that any (initially) usable genetic variation along this direction has been largely eroded, with any residual variation likely generated since selection started (such as by mutation and/or recombination). Thus eigenvectors with near-zero eigenvalues along directions of response may be an evolved phenomena from persistence selection, rather than rare chance events.

How common are genetic constraints? Unfortunately, there are not a sufficient number of studies to offer any comprehensive view, for two reasons. First, if one is focusing on two or three traits, and yet the actual constraint is in a higher dimensional space, we might not observe any restrictions on the lower dimensional projection we are considering (Pease and Bull 1988). Second, as we have seen above it is important to know both the genetic basis of the traits under consideration and the direction of selection, and there are not many studies that estimate both $\boldsymbol{\beta}$ and **G** for a vector of traits under natural selection.

Recently, Blows et al. (2004) examined 8 cuticular hydrocarbons (CHC) in the Australian fruit fly *Drosophila serrata*, which are important cues in mate choice. Both **G** for these traits, as well as their associated gradients $\boldsymbol{\beta}$ under sexual selection (mate choice), were estimated. In a large half-sib design, **G** was found to be ill-conditioned, with its first two eigenvalues accounting for 78% of the total additive genetic variation. The resulting two leading eigenvectors ($\mathbf{g}_1$ and $\mathbf{g}_2$) of $\mathbf{G}$ and $\boldsymbol{\beta}$ were as follows:

$$g_1 = \begin{bmatrix} 0.232 \\ 0.132 \\ 0.255 \\ 0.536 \\ 0.449 \\ 0.363 \\ 0.430 \\ 0.239 \end{bmatrix}, \quad g_2 = \begin{bmatrix} 0.319 \\ 0.182 \\ 0.213 \\ -0.436 \\ 0.642 \\ -0.362 \\ -0.014 \\ -0.293 \end{bmatrix} \quad \boldsymbol{\beta} = \begin{bmatrix} -0.099 \\ -0.055 \\ 0.133 \\ -0.186 \\ -0.133 \\ 0.779 \\ 0.306 \\ -0.465 \end{bmatrix}$$

As (one) measure of constraints on the evolution of this vector of traits, consider the angle θ between the direction of maximal genetic variation ($\mathbf{g}_1$) and the optimal direction favored by selection ($\boldsymbol{\beta}$). From Equation (6),

$$\theta = \cos^{-1}\left(\frac{g_1^T \boldsymbol{\beta}}{\|g_1\| \cdot \|\boldsymbol{\beta}\|}\right) = \cos^{-1}\left(\frac{0.147496}{\sqrt{(0.99896) \cdot (0.999502)}}\right) = \cos^{-1}(0.1476)$$

Giving $\theta = 81{:}5$ degrees. Thus, the vector of maximal genetic variation and the vector of optimal response are almost at right angles. Likewise, the angle between $\mathbf{g}_2$ and $\boldsymbol{\beta}$ is $\theta = 99{:}65$ degrees. For this population, there is very little standing additive genetic variation in the direction of the optimal selection response. While all of the CHC traits showed significant genetic variation, and indeed have responded to selection under a number of environmental conditions in different experiments, there is very little usable genetic variation in the direction that sexual selection is trying to move the population.

Assuming $\mathbf{G}$ remains (relatively) constant, can we relate population divergence to any feature of $\mathbf{G}$? Schluter (1996) suggested that we can, as he observed that populations tend to diverge along what he called genetic lines of least resistance, $\mathbf{g}_{max}$ (the first principal component of $\mathbf{G}$). Schluter looked at morphological divergence data in a small set of vertebrates (stickleback fish, mice, and three species of birds), and observed that populations tend to diverge in the direction of $\mathbf{g}_{max}$, specifically the angle between the vector of between-population divergence in means and $\mathbf{g}_{max}$ was small.

Thus, there is at least some empirical evidence that populations tend to evolve along lines of least genetic resistance (i.e. lines of maximal genetic variance). There are two ways to interpret this observation. The first is that such lines constrain selection, with departures away from such directions being difficult. The second is that such lines are also the directions on which maximal genetic drift is expected to occur (as the between-mean variance is proportion to the total amount of genetic variation). To see this last point, consider a trait entirely under random genetic drift. Using a Brownian motion model as an approximation (Lande 1976, 1979), the expected vector of divergences in the vector of traits from their starting value is multivariate-normally distributed, so that if the population starts at mean vector μ in generation 0, the distribution in the vector of means at generation t is

$$\mu(t) = MVN(\boldsymbol{\mu}, \frac{t}{2N_e} \cdot \boldsymbol{G}) \quad (7)$$

Maximal divergence is along the lines of greatest variation in **G**, i.e. along the first few eigenvectors of **G**.

Both of these explanations involve drift, but their distinction is a lack of adaptive response (despite pressure for such a change) versus random drift in phenotypic space. An interesting perspective on this problem is offered from considerations of the divergence in two species of Australian rainbow fish (genus *Melanotaenia*) that each have populations differentially adapted to lake vs. stream hydrodynamic environments (McGuigan et al. 2005). Several morphological landmarks for hydrodynamic morphology were followed as a vector of traits. Divergence between species, as well as divergence among replicate hydrodynamic populations within each species, followed Schluter's results (small angular departures from the vector **d** of divergent means and $\mathbf{g}_{max}$). However, hydrodynamic divergence between lake versus stream populations within each species were along directions that were quite removed from $\mathbf{g}_{max}$ (as well as the other eigenvectors of **G** that described most of the genetic variation). Thus, the between- and within-species divergence within the same hydrodynamic environment are consistent with drift, while hydrodynamic divergence within each species had to occur against a gradient of very little genetic variation. Of course, one cannot rule out that the adaptation to these environments resulted in a depletion of genetic variation along these directions. Indeed, this may indeed be the case, and is a point we return to below.

It is worth noting here that empirical investigations have yet to incorporate further elaborations that may help determine how **G** may influence population divergence. The first of these involves taking full advantage of the partitioning of the additive genetic variance within and among populations. Prout and Barker (1993; Spitze 1993) developed the approach of comparing the level of among-population genetic variance in quantitative traits (Q_{st}) with genetic variation measured by neutral markers (F_{st}) to test for a signal of population divergence as a consequence of selection. This approach has the potential to be expanded to the estimation of the among population **G** matrix, that in turn should enable a much closer examination of how the within-population **G** matrix influences population divergence. The second involves the issue of variation among populations in selection. Although among-population variation in selection has received little empirical attention (Arnold et al. 2001), Zeng (1988) has developed a framework for determining how such variation will influence population divergence that relies upon the estimation of among-population variance-covariance matrices for selection gradients. Clearly, we are just at the beginning of investigations of how multiple traits diverge among populations.

6 Using Matrix Subspace Projection to Measure Constraints

Schluter's approach only considers the leading eigenvector $\mathbf{g}_1$ of **G**. If its associated eigenvalue λ_1 dominates all of the others (and hence accounts for most of the variance), this single vector may be an adequate description of the space of usable genetic variation. More generally, we might expect the first few eigenvalues together may account for most of the variation, so that focusing only on the largest may miss a significant fraction of the variation. In such cases, matrix subspace projection provides a more realistic description of the usable subspace contained within **G** (Blows et al. 2004).

A second advantage of using a subspace projection deals with the common problem that the **G** matrix is often ill-conditioned, in that $\lambda_{max}/\lambda_{min}$ (the ratio of the largest to smallest eigenvalues) is large. In such cases (as well as others!) estimation of the **G** matrix may result in estimates of eigenvalues that are very close to zero or even negative (Hill and Thompson 1978). Negative estimates arise due to sampling (Hill and Thompson 1978), but values near zero may reflect the true biology in that there is very little variation in certain dimensions. One can extract (estimate) a subspace of **G** that accounts for the vast majority of useable genetic variation by, for example, taking the leading *k* eigenvectors. It is often the case that **G** contains several eigenvalues whose associated eigenvectors account for almost no variation (i.e. $\lambda_l/\text{tr}(\mathbf{G}) \cong 0$, recalling that the trace tr of a matrix [the sum of its diagonal elements] equals the sum of its eigenvalues, $\text{tr}(\mathbf{G}) = \sum_i G_{ii} = \sum_i \lambda_i$). In such cases, most of the genetic variation resides on a lower-dimensional subspace.

Quantifying genetic constraints within this subspace follows using the projection of the full space of **G** into this subspace (this is just the matrix extension to the projection of one vector onto another). Suppose we have included the first *k* eigenvectors in our analysis. These form a projection matrix by first defining the matrix **A**, where

$$\mathbf{A} = (\mathbf{g}_1, \ \mathbf{g}_2, \ldots\ldots, \ \mathbf{g}_k) \tag{8}$$

so that the **A** matrix consists of the first *k* eigenvectors of **G**. The resulting projection matrix becomes

$$\mathbf{P}_{roj} = \mathbf{A}(\mathbf{A}^T\mathbf{A})^{-1}\,\mathbf{A}^T \tag{9a}$$

and in particular, the projection of the optimal vector of selection response $\boldsymbol{\beta}$ onto this subspace of **G** (the subspace that essentially contains all of the useable additive variation) is given by

$$\mathbf{p} = \mathbf{P}_{roj}\,\boldsymbol{\beta} = \mathbf{A}(\mathbf{A}^T\mathbf{A})^{-1}\mathbf{A}^T\beta \tag{9b}$$

Let's reconsider the *Drosophila serrata* data on the eight CHC traits involved in mate choice. The first two eigenvalues account for roughly 80% of the total variation in **G**, i.e. $(\lambda_l + \lambda_2)/\sum \lambda_i = 0.78$. The resulting **A** matrix becomes

$$A = (g_1, g_2) = \begin{bmatrix} 0.232 & 0.319 \\ 0.132 & 0.182 \\ 0.255 & 0.213 \\ 0.536 & -0.436 \\ 0.449 & 0.642 \\ 0.363 & -0.362 \\ 0.430 & -0.014 \\ 0.239 & -0.293 \end{bmatrix}$$

Applying Equation (9a) gives an 8×8 projection matrix (not show here), and Equation (9b) gives the projection vector **p** of $\boldsymbol{\beta}$ onto the subspace given by **A** as

$$p = p_{roj}\boldsymbol{\beta} = \begin{bmatrix} -0.0192 \\ -0.0110 \\ 0.0019 \\ 0.1522 \\ -0.0413 \\ 0.1142 \\ 0.0658 \\ 0.0844 \end{bmatrix}$$

The angle θ between $\boldsymbol{\beta}$ and the projection **p** of $\boldsymbol{\beta}$ into the subspace of the genetic variance is given by

$$\theta = \cos^{-1}\left(\frac{p^T\boldsymbol{\beta}}{\|p\| \cdot \|\boldsymbol{\beta}\|}\right) = \cos^{-1}(0.223) = 77.1 \text{ degrees}$$

Thus the direction of optimal response is 77 degrees away from the genetic variation described by this subspace (which spans 78% of the total variance).

A second recent study examined a vector of CHC traits (again involved in mate choice) in *Drosophila bunnanda*, a recently described northeastern Australian sibling species of *D. serrata* (Van Homrigh et al. 2007). A total of 15 potential candidate CHC were used, and in mating trails, 9 of these appear to be involved in mate choice. Further, an estimate of the amounts of selection on these nine traits ($\boldsymbol{\beta}$) can be directly estimated from these mate choice experiments. The resulting estimated **G** for these traits had 98% of the total genetic variation in the first five principal components (the first four had 95% of the total variance). The angle between $\boldsymbol{\beta}$ and its projection into the 5-dimensional subspace of **G** was 88.2 degrees. Likewise, if the first four PCs were considered for the subspace, the projection is even more constrained, being 89.1 degrees away for $\boldsymbol{\beta}$. Hence, in this second example, the constraints were even stronger. When the entire space of **G** is considered, the resulting angle θ between **R** and $\boldsymbol{\beta}$ is 67 degrees.

7 The Dimensionality of the G Matrix

An obvious question relating to constraints is the actual dimensionality of the genetic covariance matrix **G**. If the rank of **G** is much less than the number of measured traits n, then there are considerable absolute constraints. Further, even if **G** is of full rank, if several of the eigenvalues are near zero, this also imposes a significant constraint on some paths of response. Recalling that a singular phenotypic covariance matrix **P** also imposes constraints on adaptation, comparison of the ranks of both **P** and **G** should formally be done to examine the degree to which the constraints are genetic (McGuigan and Blows 2007). For example, if both **P** and **G** are singular, but of similar rank, then genetic constraints may simply be reflections of the phenotypic constraints. In contrast, if **P** is of full rank, but **G** is not, then genetic constraints are important. For example, McGuigan and Blows (2007) examined ten wing traits in *Drosophila bunnanda*. The phenotypic space had ten full supported dimensions, while the supported dimensionality of **G** for this sexually-dimorphic trait was two in males and five in females. Thus, there appears to be significant genetic constraints on these traits that are not the result of a constrained phenotypic space.

For obvious reasons of power, accurately estimating the dimension of **G** is not trivial. For example, even if the true matrix is of full rank, it is not unusual (especially if the dimensionality is large) for random samples drawn from this **G** to have not just zero estimates of eigenvalues, but negative ones (Hill and Thompson 1978; Amemiya 1985). Thus it can be very difficult to estimate the actual number of positive eigenvalues of the true **G** given a sample covariance matrix. Three different approaches to estimating dimensionality have been applied to genetic covariance matrices. First, Amemiya (1985) develops a procedure based on estimating the characteristic roots λ of the equation

$$\det(\mathbf{M}_b - \lambda \mathbf{M}_\mathrm{w}) = 0$$

where $\mathbf{M}_b$ and $\mathbf{M}_w$ are the between- and within-mean square matrices. If all of the λ_i roots for Equation (7) are ≥ 1, then our estimate $\mathbf{M}_b$ of the genetic covariance matrix will be non-negative definite, otherwise, it will be indefinite. Thus, Amemiya's procedure successively tests for how many of the λ_I are indeed significantly greater (or equal) to one. If the first k are, but the $k+1$st is not, then the estimate of the dimensionality of **G** is given by k. When this approach is applied to the eight *Drosophila serrata* CHC traits mentioned previously, the dimensionality of **G** is not significantly greater than two (Hine and Blows 2006). Likewise, the dimensionality of **G** for the nine *D. bunnanda* CHCs was examined by Van Homrigh et al. (2007), finding strong support for 4 dimensions, and marginal support ($p = 0.081$) for a fifth.

A second approach, based on bootstrap resampling (at the sire family level for a half-sib design) was used by Mezey and Houle (2005), who examined 20 *Drosophila melanogaster* wing traits. These authors found that the resulting **G** matrix could not be shown to be significantly less than full rank. Finally, Kirkpatrick and Meyer (2004) and Meyer and Kirkpatrick (2005) have suggested estimating the principal components (PCs) directly and in successive fashion. In this factor-analytic approach, the dimensionality of **G** is the number of PCs that are added until the next does not significantly improve model fit. There are several powerful features of this approach. First, the resulting covariance matrix is guaranteed to be non-negative-definite (by construction). Further, the actual number of parameters that need to be estimated can be greatly reduced (on the order kn for k PCs versus roughly order n^2 for estimation of the full matrix). Preliminary simulation studies (Hine and Blows 2006) find that the bootstrap method almost always overestimates the true dimensionality, while Amemiya's method and factor-analytic modeling do a reasonable job of estimation when heritability is high. When traits have low heritability, Amemiya's method has low power, resulting in underestimation of the true dimensionality.

8 Evolution Under Constraints or Evolution of Constraints?

G both constrains selection and also evolves under selection. Over short time scales, if most alleles have modest effects, **G** changes due to selection generating linkage disequilibrium. It is important to note that under the infinitesimal model (and hence no change in allele frequencies), selection still generates considerable disequilibrium, even among unlinked loci (Bulmer 1971). Indeed, any selection (such as stabilizing or any form of directional selection) that generates a decrease in the variance following selection generates negative disequilibrium, meaning that gametes containing favourable alleles are less frequent than expected simply from the product of their allele frequencies.

In particular, assuming only directional selection (no curvature in the fitness surface), the within generation change in the additive-genetic covariance matrix under the infinitesimal model is given by

$$\Delta \mathbf{G} = -\mathbf{G}\boldsymbol{\beta}\beta^T\mathbf{G} = -\mathbf{R}\mathbf{R}^T \tag{11a}$$

(Tallis and Leppard 1988; Lande 1979; Phillips and Arnold 1989). Thus, the (within-generation) change in the additive genetic covariance between traits i and j is

$$\Delta G_{ij} = \Delta\sigma(A_i A_j) = -R_i R_j \tag{11b}$$

Equation (11b) shows that selection decreases the additive genetic variance at traits responding to selection (Bulmer 1971) and decreases the covariances if traits i and j respond in the same direction, while increasing the covariance if i and j respond in different directions. The net result is that linkage disequilibrium increases any initial constraints. A simple way to see this is to consider selection on the index given by Equation (4b), $I = \sum_i z_i \beta_i$. Selection on this index (which is the predicted fitness) results in decreased additive variance in this composite trait (Bulmer 1971). Thus, as pointed out by Shaw et al. (1995), if one estimates **G** by first having several generations of random mating in the laboratory under little selection, existing linkage disequilibrium decays, and the resulting estimated **G** matrix may show less of a constraint than the actual **G** operating in nature (with its inherent linkage disequilibrium).

The second avenue for change in **G**, allele frequency change, occurs over longer time scales (unless there are major genes). Unlike the relatively simple dynamics under linkage disequilibrium (Equation (11)), effects on **G** from allele frequency change are extremely unpredictable (Bohren et al. 1966). A general qualitative trend was suggested by Lush (1948) and Lerner (1950). They predicted that selecting two traits in the same direction results in a negative change in their genetic covariance as alleles with desirable (++) and undesirable (−−) pleiotropic effects on the two traits are fixed (and lost) by selection leaving only alleles segregating +− pleiotropic effects. While compelling, there are counterexamples to this simple prediction, e.g. Sheridan and Barker (1974). Thus, during the course of selection, **G** itself changes, often so as to increase constraints (as usable variation continues to be removed).

Hence, it is certainly not surprising that little usable genetic variation may remain along a direction of persistent directional selection. What is surprising, however, is that considerable genetic variation may exist along other directions. The quandary is not why is there so little usable variation but rather why is there so much. Indeed, quantitative genetics is in the embarrassing position as a field of having no models that adequately explain one of its central observations – genetic variation (measured by single-trait heritabilities) is common and typically in the range of 0.2–0.4 for a wide variety of traits. As nicely reviewed by Johnson and Barton (2005), neither of two general classes of theoretical models (mutation-selection and balancing selection) proposed to account for standing variation adequately account for the observed levels of variation with realistic estimates of mutation rates. As Johnson and Barton point out, the resolution of these issues likely resides in more detailed considerations of pleiotropy, wherein new mutations influence a number of traits. Once again, it is likely we need to move to a higher dimensional space to reasonably account for observations based on a projection into one dimension (i.e. standing heritability levels for a trait).

The final consideration then with pleiotropy is not just the higher-dimensional fitness surface for the vector of traits they influence but also the distributional space of pleiotropic mutations themselves. Is the covariance structure **G** itself some optimal configuration for certain sets of highly-correlated traits? Indeed, has there been selection on developmental processes to facilitate morphological integration (the various units of a complex trait functioning smoothly together), which in turn would result in constraints on the pattern of accessible mutations under pleiotropy (Olson and Miller 1958; Lande 1980)? Our spherical cow may in reality have a very non-spherical distribution of new mutation phenotypes around a current phenotype.

9 Closing Comments: G is Dead, Long Live G

In a recent evaluation of the evolutionary quantitative genetics research program, Pigliucci (2006) claimed that the field needed "a fundamental reconsideration of its goals", particularly with regard to the utility of **G** as an informative construct. Pigliucci reminds us of three classic limitations of quantitative genetics: the environment-dependent nature of its parameters, the summative nature of those same parameters, and the temptation to use such parameters in making evolutionary inferences based solely on corelational evidence. Rather than following Pigliucci in advocating the abandonment of unresolved questions, we would argue that a major focus of future efforts is escaping the Flatlandian prison in which much of biology finds itself captive. It is perhaps ironic given Pigliucci's dismal of the usefulness of **G**, that it is an understanding of the geometry of **G** from both the point of view of mutation generating genetic variance and from selection depleting it, which may result in a more complete understanding of those basic observations concerning genetic variation and selection that our field has strived to explain for so long.

It would be easy for empiricists to take a depressing message away from our discussion; without being able to measure all the traits that are under selection, the tyranny of Flatland and its potential for generating incorrect interpretations is ever present. To some extent this is likely always to be true, but are there any mitigating circumstances that make the task more manageable than this message might suggest? It's difficult to say, and a lack of theory that models sufficient dimensions is a hindrance here. Although the investigation of a modest number of dimensions in recent experiments has made some progress in finding associations between selection and the distribution of genetic variance, Pigliucci's (2006) admonishment concerning a reliance on correlational evidence is well made; empirical research programs need to incorporate either manipulations to test the associations found, or sampling designs from natural populations which allow clear tests of the evolutionary inferences that are being made.

Finally, there are very deep issues at the core of whether depletion of genetic variation along certain orientations, while still maintaining significant variance in all component directions, is a common phenomenon. The heart of the problem is the nature of the maintenance of genetic variation. We see an explosion of variance in just about every (univariate) trait we examine, values too high for standard models to account for. However, Fisher's original model (spherical cow or not) may provide a key bit of insight. Genetic variation may indeed be ruthlessly exploited by persistent selection along a particular direction, with further progress (adaptation) dependent upon adaptive new mutations. As Fisher showed, improving fitness in a high dimensional space is not trivial, because while a mutant could have enhanced fitness along several dimensions, it may have reduced fitness along others. The net result may be equivalent to the nearly-neutral models from molecular population genetics (Kimura 1983), wherein mutation is pumping into the population new variants that are close to effectively neutral, but not too close. Thus, their variation is higher than expected under standard mutation-selection models (as they have weak selection) but far less than expected under pure drift-mutation models (as they have some selection). Perhaps this is what we are observing with most quantitative variation. In essence, we may have a misleading view; due to projections of variation on single traits, there appears to be considerable genetic variation to exploit. There may well be, *provided* we are trying to move in a different direction relative to persistent selection. If we attempt to improve along lines of historically persistent selection, very little gain may be the expectation.

Acknowledgments We are extremely pleased to dedicate this paper to Stuart Barker. For someone learning the trade of quantitative genetics in the late 1980s, Stuart's work was like a beacon of interest in a sea of allozymes; incisive reviews, classic experimental designs (even with allozymes!), and above all the innovative application of quantitative genetics to important and interesting questions in evolutionary biology. MWB would like to acknowledge his close collaborators on many aspects of this work, Steve Chenoweth, Emma Hine and Katrina McGuigan. Finally, we thank both reviewers for very helpful comments.

References

Abbott, E.A. 1884. *Flatland. A romance of many dimensions*, Seeley and Co. Ltd., London. (Reprinted in 1922 by Dover Publications).

Amemiya, Y. 1985. What should be done when an estimated between-group covariance-matrix is not nonnegative definite. *Am. Stat.* **39**: 112–117.

Arnold, S.J., Pfrender, M.E. and Jones, A.G. 2001. The adaptive landscape as a conceptual bridge between micro- and macro-evolution. *Genetica* **112**: 9–32.

Barton, N.H. 1990. Pleiotropic models of quantitative variation. *Genetics* **124**: 777–782.

Blows, M.W., Chenoweth, S.F. and Hine, E. 2004. Orientation of the genetic variance-covariance matrix and fitness surface for multiple male sexually selected traits. *Amer. Natl.* **163**: 329–340.

Bohren, B.B., Hill, W.G. and Robertson, A. 1966. Some observations on asymmetrical correlated responses to selection. *Genet. Res. Camb.* **7**: 44–57.

Bulmer, M.G. 1971. The effect of selection on genetic variability. *Amer. Nat.* **105**: 201–211.

Dickerson, G.E. 1955. Genetic slippage in response to selection for multiple objectives. *Cold Spring Harbor Symp. Quant. Biol.* **20**: 213–224.

Fisher, R.A. 1930. *The Genetical Theory of Natural Selection*, Oxford University Press, Oxford, U. K.

Hartl, D. and Taubes, C.H. 1996. Compensatory nearly neutral mutations: selection without adaptation. *J. Theor. Biol.* **182**: 303–309.

Hazel, L.N. 1943. The genetic basis for constructing selection indexes. *Genetics* **28**: 476–490.

Henderson, C.R. and Quaas, R.L. 1976. Multiple trait evaluation using relatives' records. *J. Anim. Sci.* **43**: 1188–1197.

Hill, W.G. and Thompson, R. 1978. Probabilities of non-positive definite between-group or genetic covariance matrices. *Biometrics* **34**: 429–439.

Hine, E. and Blows, M.W. 2006. Determining the effective dimensionality of the genetic variance-covariance matrix. *Genetics* **173**: 1135–1144.

Johnson, T. and Barton, N. 2005. Theoretical models of selection and mutation on quantitative traits. *Phil. Trans. R. Soc. B* **360**: 1411–1425.

Kimura, M. 1983. *The Neutral Theory of Molecular Evolution*, Cambridge University Press, Cambridge, U.K.

Kirkpatrick, M. and Meyer, K. 2004. Direct estimation of genetic principal components: simplified analysis of complex phenotypes. *Genetics* **168**: 2295–2306.

Kruuk, L.E.B., Slate, J., Pemberton, J.M., Brotherstone, S., Guinness, F. and Clutton-Brock, T. 2002. Antler size in red deer: heritability and selection but no evolution. *Evolution* **56**: 1683–1695.

Lande, R. 1976. Natural selection and random genetic drift in phenotypic evolution. *Evolution* **30**: 314–334.

Lande, R. 1979. Quantitative genetic analysis of multivariate evolution, applied to brain:body size allometry. *Evolution* **33**: 402–416.

Lande, R. 1980. The genetic covariance between characters maintained by pleiotropic mutations. *Genetics* **94**: 203–215.

Lande, R. and Arnold, S.J. 1983. The measurement of selection on correlated characters. *Evolution* **37**: 1210–1226.

Leigh, E.G. 1987. Ronald Fisher and the development of evolutionary theory. II. Influences of new variation on the evolutionary process. In *Oxford Surveys in Evolutionary Biology 4* (P.H. Harvey and L. Partridge, eds.), pp. 213–263.

Lerner, I.M. 1950. *Population Genetics and Animal Improvement*, Cambridge University Press.

Lush, J.L. 1948. *The Genetics of Populations*, Iowa State University, Ames.

McGuigan, K. and Blows, M.W. 2007. The phenotypic and genetic covariance structure of Drosphilid wings. *Evolution* **61**: 902–911.

McGuigan, K., Chenoweth, S.F. and Blows, M.W. 2005. Phenotypic divergence along lines of genetic variance. *Amer. Nat.* **165**: 32–43.

Merilä, J., Sheldon, B.C. and Kruuk, L.E.B. 2001. Explaining stasis: microevolutionary studies in natural populations. *Genetica* **112–113**: 199–222.

Meyer, K. and Kirkpatrick, M. 2005. Restricted maximum likelihood estimation of genetic principal components and smoothed covariance matrices. *Genet. Sel. Evol.* **37**: 1–30.

Mezey, J.G. and Houle, D. 2005. The dimensionality of genetic variation for wing shape in *Drosophila melanogaster. Evolution* **59**: 1027–1038.

Olson, E. and Miller, R. 1958. *Morphological Integration*, University of Chicago Press, Chicago.

Orr, H.A. 1988. The population genetics of adaptation: the distribution of factors fixed during adaptive evolution. *Evolution* **52**: 935–949.

Orr, H.A. 2000. Adaptation and the cost of complexity. *Evolution* **54**: 13–20.

Pearson, K. 1903. Mathematical contributions to the theory of evolution. XI. On the influence of natural selection on the variability and correlation of organs. *Phil. Trans. Roy. Soc. Lond. A* **200**: 1–66.

Pease, C.M. and Bull, J.J. 1988. A critique of methods for measuring life history trade-offs. *J. Evol. Biol.* **1**: 293–303.

Phillips, P.C. and Arnold, S.J. 1989. Visualizing multivariate selection. *Evolution* **43**: 1209–1222.

Pigliucci, M. 2006. Genetic variance-covariance matrices: a critique of the evolutionary quantitative genetics research program. *Biol. Philos.* **21**: 1–23.

Price, G.R. 1970. Selection and covariance. *Nature* **227**: 520–521.

Prout, T. and Barker, J.S.F. 1993. F statistics in *Drosophila buzzatii*: Selection, population size and inbreeding. *Genetics* **134**: 369–375.

Rice, S.H. 1990. A geometric model for the evolution of development. *J. Theor. Biol.* **143**: 319–342.

Robertson, A. 1966. A mathematical model of the culling process in dairy cattle. *Anim. Prod.* **8**: 95–108.

Schluter, D. 1996. Adaptive radiation along genetic lines of least resistance. *Evolution* **50**: 1766–1774.

Shaw, F.H., Shaw, R.G., Wilkinson, G.S. and Turelli, M. 1995. Changes in genetic variances and covariances: G whiz! *Evolution* **49**: 1260–1267.

Sheridan, A.K. and Barker, J.S.F. 1974. Two-trait selection and the genetic correlation. II. Changes in the genetic correlation during two-trait selection. *Aust. J. Biol. Sci.* **27**: 89–101.

Simpson, G.G. 1944. *Temp and Mode in Evolution*, Columbia University Press, New York.

Smith, H.F. 1936. A discriminant function for plant selection. *Ann. Eugen.* **7**: 240–250.

Spitze, K. 1993. Population structure in *Daphnia obtusa*: Quantitative genetic and allozymic variation. *Genetics* **135**: 367–374.

Tallis, G.M. and Leppard, P. 1988. The joint effects of selection and assortative mating on multiple polygenic characters. *Theor. Appl. Genet.* **75**: 278–281.

Thompson, D.W. 1917. *On Growth and Form*. Cambridge University Press, Cambridge.

Van Homrigh, A., Higgie, M., McGuigan, K. and Blows, M.W. 2007. The depletion of genetic variance by sexual selection. *Curr. Biol.* **17**: 528–532.

Van Valen, L. 1973. A new evolutionary law. *Evol. Theory* **1**: 1–30.

Wright, S. 1932. The roles of mutation, inbreeding, crossbreeding, and selection in evolution. *Proc. VI Int. Congr. Genet.* **1**: 356–366.

Zeng, Z.B. 1988. Long-term correlated response, interpopulation covariation, and interspecific allometry. *Evolution* **42**: 363–374.

Maintaining Fitness by Within Breed Selection

Piter Bijma

Abstract Genetic improvement of productivity in livestock has been very successful in the past. Improvement of traits related to fitness, such as health, fertility and disease resistance, however, has proven more difficult. Reasons are (i) low heritabilities, (ii) difficulty to record traits related to fitness, (iii) genotype by environment interaction, e.g. between commercial and breeding environment, (iv) presence of heritable effects other than the direct genetic effect of the individual itself. This paper argues that improvement of fitness related traits is not fundamentally different from improvement of other traits. Important factors are (i) the identification of all heritable components affecting trait values, (ii) the collection of phenotypic data from commercial environments combined with marker assisted estimation of breeding values (so-called "genomic selection"). Marker assisted breeding value estimation does not threaten genetic diversity, and is operationally much easier to use than methods based on QTL mapping. The paper concludes with a section on heritable social interactions among individuals. Results show that heritable social interactions represent an extra level of variation which may increase the heritable variation available to breeders to values considerably higher than the usual additive genetic variance. First results in pigs and poultry are promising, but more data analysis is needed to substantiate the relevance of heritable social interactions in other species.

Keywords Fitness · social interactions · maternal effects · selection

1 Introduction

Probably the first issue to address given the subject is what animal breeders mean by "fitness". In its formal population genetic definition, fitness refers to the reproductive output of an individual. It may be interpreted as the contribution of an individual

P. Bijma (✉)
Animal Breeding and Genomics Centre, Wageningen University, Marijkeweg 40,
6709 PG Wageningen, The Netherlands
e-mail: piter.bijma@wur.nl

J. van der Werf et al. (eds.), *Adaptation and Fitness in Animal Populations*,

to population size. From that perspective, fitness in livestock breeding populations would be largely determined by traits in the breeding goal, since the individuals with the highest (estimated) breeding values for those traits contribute most offspring to the next generation. Among animal breeders, however, the term "fitness" is often interpreted as an overall measure of health, survival and reproduction of an individual. That interpretation will also be the focus of this paper.

Fitness is a key trait in the study of natural populations. Animal breeders can therefore learn from evolutionary biologists. In livestock genetic improvement, breeders often have access to large amounts of data, whereas in evolutionary biology the amount of data is often insufficient for quantitative genetic analyses. Animal breeders have therefore focused primarily on the statistical genetic analyses of large data sets and on the development of tools for this purpose. Less attention has been given to biological, ecological, environmental or social factors underlying the observed trait values. Evolutionary biologists, in contrast, have developed extensive theory related to, for example, optimal strategies for allocation of resources, parental investment in offspring, social interactions among individuals, fluctuating environments and the consequences for life history and behavioural traits (e.g. West-Eberhard 2003; Rice 2004). Such knowledge can help animal breeders to understand the mechanisms underlying trait values in livestock, and to decide whether knowledge of those mechanisms can be utilised to improve breeding strategies. Knowledge of those mechanisms may, for example, lead to analysing livestock data from a different perspective, which may identify new factors that can contribute to genetic improvement. For example, the consequences of social interactions among individuals on evolution by natural selection have received a lot of attention in evolutionary biology (Hamilton 1964; Frank 1998; Michod 1999). Recent work shows that social interactions may also contribute substantially to the heritable variation in livestock breeding populations (Muir 2005; Bijma et al. 2007a,b; see below).

Though breeders can learn from knowledge of natural populations, this does not mean that they should simply copy strategies observed in nature. Animal breeders aim to improve livestock populations so that they better serve human purposes. Natural selection, however, does not target human interest. Natural selection is largely directed at self interest of the individual and of the gene. (With the exception of social interactions among relatives, in which case natural selection is aimed at inclusive fitness; Hamilton 1964). Consequently, there is no reason to expect the structure of natural populations or ecosystems to be a reliable blueprint for agricultural systems (Denison et al. 2003). Natural selection may result in selfish individuals that are successful at the expense of others, rather than in maximisation of output on the population, flock or herd level. A clear example comes from the difference between wild wheat and rice versus modern cultivars. In natural populations, individual plants compete for sunlight. Natural varieties and traditional cultivars, therefore, have long stems making them sensitive to lodging (falling over). Modern wheat and rice cultivars, however, owe their high yield potential largely to their short stature, relative to traditional cultivars. Short stature reduces lodging and increases growth of grain at the expense of stems, but artificial selection for these traits has resulted in decreased individual competitiveness. Forced to compete with traditional, low yielding cultivars, modern high-yield rice cultivars completely disappear within as few as three generations (Denison et al. 2003). This result clearly illustrates that natural selection may produce a phenotype that is suboptimal for human purposes. Nevertheless, this example illustrates the relevance of competition among individuals in natural populations, which may also be relevant for livestock genetic improvement.

In its basics, livestock genetic improvement is straightforward. The breeding goal should refer to the commercial environment and include all traits of importance, either as a sum of traits weighted by economic values, or expressed as desired gains. Breeding value estimation should utilise all available information, ideally in multi-trait animal model evaluations, and selection decisions should maximise response while restricting the rate of inbreeding by using optimal contribution selection (Meuwissen 1997). Selection index theory and stochastic simulations can be used to optimise the structure of breeding schemes. Further improvement can be obtained by mating least related individuals, applying factorial mating when possible (Sonesson and Meuwissen 2000), and by using molecular information in particular for traits not recordable on selection candidates (Meuwissen and Goddard 1996).

The above strategy has been very successful. (Not so much yet in the case of genetic markers.) In many species, yields have increased dramatically over the past ~25 years, particularly in species with short generation intervals in which the focus has been on traits of moderate to high heritability recordable on the selection candidate, such as growth rate in broiler chickens. In many cases, however, there has been a trend for aspects of fitness to decline. A key question for breeders is whether this decline is due to (i) a lack of selection emphasis on fitness related traits in the past, or (ii) insufficient understanding of fitness related traits to effectively improve those traits, leading to unexpected and unfavourable correlated responses. In the first case, there is no fundamental difficulty inherent to improvement of fitness related traits; redirecting genetic improvement by shifting selection emphasis and data recording towards fitness related traits will be sufficient to overcome existing problems. In this case, limited improvement or even a decline of fitness related traits that has occurred in the past is a price that has been paid in order to attain larger progress in productivity so as to achieve higher profit or market share. In the second case, however, we have to address improvement of fitness related traits in a manner different from the past. In that case, we will need to focus on a better understanding of fitness related traits and on how such traits differ from conventional production traits. In this context, it is important to think about what we mean with sufficient understanding. Since genetic improvement is generated by selecting parents for the next generation, our understanding has to be sufficient to identify the genetically superior individuals for the traits of interest. Identifying genetically superior individuals usually does not require a complete understanding of the biological mechanisms underlying the observed phenotypes. Domestication and classical breeding has been very successful without breeders understanding the biological basis of observed traits. We don't need to know the functions of all the genes to make successful genetic improvement.

(Though such an understanding may enable more efficient and better understood selection strategies, and thus accelerate response to selection).

Key issues for animal breeders are (i) identifying all the genetic components of interest, and (ii) estimating the breeding values of the selection candidates for those genetic components, aiming to maximise the accuracy. This topic is addressed in Section 2 with an emphasis on traits related to fitness. Finally, I discuss the consequences of heritable social interactions on the prospects for livestock genetic improvement (Section 3).

2 Response to Selection

For any trait, response to selection can be described by

$$\Delta\bar{G} = i\rho\sigma_{\text{TBV}}, \tag{1}$$

in which i is selection intensity, ρ the correlation between the selection criterion and the total breeding value (TBV) of an individual, and σ_{TBV} the standard deviation in TBVs of individuals. The TBV summarises the quality of an individual judged by its value for genetic improvement, i.e. its "total value for breeding" (see below). Equation (1) applies to both single and multitrait selection because ρ is the correlation between the overall selection criterion and the TBV. (With single trait selection, ρ corresponds to the accuracy of the EBV for the single trait.) Because Equation (1) applies to any trait, we can study its components to investigate how to increase response in traits related to fitness.

2.1 The Heritable Standard Deviation

In classical quantitative genetic theory, the heritable standard deviation is simply equal to the additive genetic standard deviation, $\sigma_{\text{TBV}} = \sigma_A$ and TBV $= A$. In this case, A is the usual breeding value, as in the model $P = A + E$. However, G may be defined more generally as the value of an individual for response to selection. I illustrate this with a maternally effected trait (Wilham 1963). For a maternally affected trait, the trait value is given by $P_i = A_{D,i} + A_{M,dam(i)} + E_i$, in which $A_{D,i}$ is the direct breeding value of individual i, $A_{M,dam(i)}$ the maternal breeding value of its dam, and E_i summarises all non-heritable components in P_i. It follows from taking means in the current and next generation, that response to selection equals $\Delta\overline{\text{TBV}} = \Delta\bar{A}_D + \Delta\bar{A}_M$. In other words, response to selection is the sum of responses in direct and maternal effects. Therefore, we may define a total breeding value for each individual, $\text{TBV}_i = A_{D,i} + A_{M,i}$, which represents the value of individual i measured by its heritable impact on the population mean trait value. Note that TBV_i contains $A_{M,i}$, not $A_{M,dam(i)}$, because i passes on its own alleles, not those of its dam. Thus TBV_i represents the value of individual i for breeding purposes;

response to selection equals $\Delta\bar{G} = \Delta\overline{\text{TBV}} = \overline{\text{TBV}}_{t+1} - \overline{\text{TBV}}_t$; and the heritable variance in the trait equals $\sigma^2_{\text{TBV}} = \sigma^2_{A_D} + 2\sigma_{A_{DM}} + \sigma^2_{A_M}$. The σ^2_{TBV} represents the heritable variance in trait value that can be used to generate response to selection. Note that σ^2_{TBV} differs from the genetic component of the phenotypic variance. The phenotypic variance equals $\text{Var}(P) = \sigma^2_{A_D} + \sigma_{A_{DM}} + \sigma^2_{A_M} + \sigma^2_E$, with the coefficient of $\sigma_{A_{DM}}$ being $\frac{1}{2}$ times 2, reflecting the $\frac{1}{2}$ relationship between individuals and their mothers; thus the genetic component of the phenotypic variance equals $\sigma^2_{A_D} + \sigma_{A_{DM}} + \sigma^2_{A_M} \neq \sigma^2_{\text{TBV}}$. It is the σ^2_{TBV} that determines the potential response to selection, not the $\sigma^2_{A_D} + \sigma_{A_{DM}} + \sigma^2_{A_M}$. The response to selection equals $i\rho\sigma_{\text{TBV}}$, where ρ is the correlation between the selection criterion and the TBV.

This example illustrates that the total heritable variance in trait value, σ^2_{TBV}, is not always a component of the phenotypic variance in trait value; $\sigma^2_P \neq \sigma^2_{\text{TBV}} + \sigma^2_E$. As a consequence, there exists no longer a heritability that represents the *proportion* of the phenotypic variance which is heritable. In other words, the heritable variance expressed relative to the phenotypic variance, $\sigma^2_{\text{TBV}}/\sigma^2_P$, may exceed the usual limit of 1. This will be discussed further in the section on social interactions below.

Compared to production traits, traits related to fitness may be affected more by factors other than direct genetic effects. Such factors may be, e.g. maternal effects, social environment and competition among individuals, genotype by environment interaction, epigenetic effects such as inheritance of environmental experience (e.g. learning), etc. Because those factors may have a heritable component, they will affect the prospects for genetic improvement. In that case, those effects are not merely a nuisance that needs to be corrected for in statistical analyses, but they may contribute to the heritable variance in trait values and to response to selection, as illustrated for maternal effects above. An important step in genetic improvement of fitness related traits, therefore, is the identification of all heritable components affecting trait values. Identifying those components provides a measure for the total breeding value of an individual and for the heritable variance in trait value, and guides selection decisions. Consider, for example, a maternally affected trait in groups composed of four individuals, and assume that competition among group members affects the trait value. In this case, the phenotype of individual i may be expressed as $P_i = A_{D,i} + A_{M,dam(i)} + \sum_{j=1}^{3} A_{S,j} + E_i$, in which j denotes a group member and $A_{S,j}$ the heritable social effect of this group member on the trait value of individual i. Response to selection is $\Delta\bar{G} = \Delta\text{TBV} = \Delta\bar{A}_D + \Delta\bar{A}_M + 3\Delta\bar{A}_S$. The total breeding value of an individual is given by $\text{TBV}_i = A_{D,i} + A_{M,i} + 3A_{S,i}$, in which all components refer to individual i. The TBV_i summarises the value of individual i for genetic improvement. Selection decisions should, therefore, maximise the mean TBV_i of the selected parents. The heritable variance in trait value equals $\text{Var(TBV)} = \sigma^2_{A_D} + 2\sigma_{A_{DM}} + 6\sigma_{A_{DS}} + \sigma^2_{A_M} + 6\sigma_{A_{MS}} + 9\sigma^2_{A_S}$. The standard deviation of the TBV is a measure of the amount of response that can be obtained (Equation 1). For example, when interest is in a single trait, an efficient breeding scheme can generate a response of approximately one standard deviation, σ_{TBV}, per generation. Response to selection equals $\Delta\bar{G} = i\rho\sigma_{\text{TBV}}$, in which ρ is the correlation between the selection criterion and the TBV of individuals.

In conclusion, identification of all heritable components underlying phenotypic values of individuals provides a goal value, the TBV, for each trait, and a measure of the total heritable variation in trait value, Var(TBV), which is a measure of the response to selection that can be obtained. The next step is to estimate TBVs for selection candidates with the highest possible accuracy.

2.2 Accuracy

With multitrait selection, the ρ in the expression $\Delta\bar{G} = i\rho\sigma_{\mathrm{TBV}}$ is not the accuracy of EBV, but the correlation between the selection criterion and the TBV for the trait of interest. The ρ depends on two underlying parameters, (i) the relative emphasis on the trait in the breeding goal, and (ii) the accuracy of the EBV for this trait. In other words, the ρ depends on the emphasis given to a trait and on the accuracy with which we can estimate the breeding values for the trait. The emphasis given to a trait is a breeder's decision and will always involve trade-offs; increasing response in a specific trait by increasing its weight in the breeding goal will be accompanied by lower weight and response for other traits. This is primarily a matter of decision making based on market-share or economic considerations, not a biological or genetic issue. Therefore, I will focus on the second parameter, the accuracy of the EBV for the trait of interest.

2.2.1 Phenotypic Data

Traits related to fitness are characterised by low heritability, often more GxE-interaction than observed for production traits (e.g. between-country correlations for conception rate in dairy cattle are lower than correlations for yield traits, Interbull http://www-interbull.slu.se), difficulty of recording, recording requiring exposure to poor environments (e.g. commercial environment), low incidence of expression (disease), *etc*. In many cases, therefore, selection candidates do not have own performance records for traits related to fitness. In other cases, records are available but do not accurately represent the trait of interest because they have been collected in good breeding environments rather than the commercial environment. Thus, compared to production traits, improvement of traits related to fitness will often rely to a larger degree on records of other individuals, such as sibs or progeny. With efficient breeding schemes, however, sib information is of relatively little value. First, sib information suffers from the reduction of the genetic variance due to selection (the "Bulmer effect"). Second, a large number of sib records are required for traits of low heritability, which is usually only possible for half sibs. Half sibs, however, provide information only on the breeding value of the sire of the candidate, the value of which is most reduced by the Bulmer effect. Third, sib information generates response by selection between families. Between family selection contributes little to response in schemes that maximise response while constraining inbreeding using optimal contributions. For those reasons, the value of sib information is substantially smaller than indicated by simple selection index calculations. Progeny testing is a robust selection strategy that does not suffer from the above mentioned phenomena. However, progeny testing extends generation intervals, in particular for traits related to fitness such as reproduction and longevity. Moreover, due to the limited reproductive capacity of females, progeny testing is usually limited to selection among males. When using classical breeding strategies, therefore, efficient improvement of traits related to fitness is difficult.

2.2.2 Molecular Data

The use of molecular information has long been advocated as a promising tool to improve difficult traits. Up to now, however, molecular information has contributed little to genetic progress in livestock breeding schemes. Most molecular genetic studies in the past have focused on detecting QTL. With the exception of some QTL of major effect, such as the halothane gene, however, QTL have been difficult to use in applied breeding. Reasons are that QTL have often been detected in experimental crosses, the number of QTL soon becomes impractically large, linkage phase between markers and QTL may change over time, and QTL effects may change over time.

A major step forward will come from the implementation of marker assisted breeding value estimation (MA-BVE) using dense maps covering the entire genome (Meuwissen et al. 2001). In that approach, interest is solely in the breeding value of the candidates; the objective is to estimate a breeding value with the highest possible accuracy using all phenotypic and genomic information. There is no interest in the location or effect of individual QTL. Using a mixed model, the method gives an estimate for the effect of each marker or haplotype. The breeding value of an individual is simply the sum of the effects of its markers or haplotypes. The most appealing property of this approach is its operational power, which makes it very suitable for practical breeding, as opposed to QTL based approaches. There is no need to keep track of individual QTL or markers, there is no need to consider linkage phase, and a single set of genome-wide markers can be used for all traits. Essentially, this approach replaces individuals, which are the subject of the conventional mixed model equations, by markers or haplotypes; breeding values are estimated for each marker or haplotype instead of for each individual. First results in dairy cattle indicate that accuracies of ~0.6 to ~0.8 are feasible in practical breeding schemes. Compared to sib information, these are very high values.

Improvement of traits related to fitness will be accelerated by combining MA-BVE with the phenotypic information that best represents the trait of interest, but which is not available on the selection candidates. Hence, data recording should shift from the breeding environment to commercial environments, and selection on correlated traits should be replaced by direct selection on MA-EBV for the trait of interest. For example, data on crossbred individuals kept in commercial environments can be used to estimate markers effects. Subsequently, estimated marker effects can used to calculate EBVs of selection candidates kept in clean breeding environments. When a dense map is used, one can probably skip one or two generations, leading to short generation intervals combined with high accuracy. Thus, the traditional trade-off between accuracy and generation interval can be removed. When the current trend in genotyping costs continues, it may become feasible to test large numbers of juvenile candidates, which would increase intensity of selection. The key factor will be to collect a sufficient amount of data of good quality on the traits of interest and in the relevant environments. Compared to current breeding strategies, this will require closer collaboration between breeders and producers. The rapid developments in molecular technology and informatics may provide opportunities that seem rather far fetched at present, such as the use of nanobots to record health status on large numbers of individuals. The impacts on response to selection can be quite dramatic. Selection index calculations show, for example, that response in longevity in dairy cattle can easily be doubled (results not shown).

2.2.3 Maintaining Diversity When Using Molecular Information

Predicted responses to selection when using MA-BVE are so large that one may fear that we will rapidly deplete within-breed genetic diversity. The key issue here would seem to be whether selection for an EBV based on molecular information differs fundamentally from classical selection strategies. Classical selection strategies have yielded extraordinary responses without depleting genetic diversity, both experimentally and in livestock populations, in particular when effective population size has been large (e.g. Weber and Diggins 1990). To answer the question whether the use of MA-BVE will accelerate the loss of genetic diversity required to investigate the impact of MA-BVE on the mechanisms causing loss of diversity.

Any successful selection strategy will affect genetic diversity, simply because response to selection requires changing allele frequencies. Thus a change of allele frequency is desired, but should be restricted as much as possible to favourable genes. Factors affecting allele frequency change are drift, mutation and selection (assuming absence of migration). Loss of diversity due to drift can be restricted by restricting the rate of increase in kinship, i.e. by maintaining a sufficient kinship effective population size, $N_E = 1/(2\Delta f)$. The equilibrium variance that would be obtained under mutation-drift equilibrium depends on the mutational variance, V_m, a biological parameter outside the breeder's control, and effective population size, $\sigma^2_{Aeq.} = 2N_E V_m$. Hence, when considering mutation and drift, the key parameter is effective size. Selection may cause both desired and non-desired changes in allele frequency. Non-desired changes in allele frequency may occur because regions neighbouring selected loci may hitchhike to fixation due to linkage disequilibrium (LD). LD between selected and neutral loci is generated by drift; smaller effective size yields higher LD. Thus, reducing undesired loss of diversity due to hitchhiking requires maintainance of sufficient effective size. In addition, it will be important to frequently re-estimate marker effects, e.g. each generation, so that decay of LD can be utilised. This will both reduce the degree of hitchhiking and increase proportion of the heritable variance that is explained by markers (i.e. the squared accuracy). Thus, the use of MA-BVE does not fundamentally alter the criterion for maintenance of genetic diversity; kinship effective size remains the relevant parameter.

At present, it is unclear whether effective sizes required to avoid problems due to inbreeding ($N_E >\sim 100$), are sufficient to prevent substantial loss of diversity due to LD leading to hitchhiking. Hitchhiking will, however, occur also with conventional selection strategies, and probably to a similar degree. MA-EBVs of individuals are the sum of estimated average effects of markers (haplotypes) spread across the entire genome; they are linear combinations of observed phenotypes. This seems to be similar to conventional EBVs based on progeny testing. When considering a limited number of QTL, selection may rapidly deplete a part of the genetic variance, i.e. the part explained by those QTL. However, when applying MA-EBV with a dense map covering the entire genome resulting in a high accuracy, selection must be directed at a large proportion of the true breeding value, similar to conventional progeny testing. Moreover, the use of MA-BVE to replace selection based on sib information will increase effective population size, because selection will rely less on family information. In conclusion, therefore, despite the large responses that seem to be feasible, MA-BVE does not pose a great threat for genetic diversity. Compared to current breeding schemes, the use of MA-BVE may even maintain more diversity, or at least yield higher response at the same rate of increase in pedigree kinship.

Though MA-EBV relate directly to genomic regions, their use may not substantially affect the increase of kinship observed on the molecular level. This is because the number of genes underlying traits of interest seems to be fairly large, and breeding programs are targeting multiple traits at the same time. Hence, a total merit index will be the based on hundreds of genes, resulting in selection pressure spread across the entire genome. This situation seems to resemble selection on EBV based on progeny testing, or mass selection for a trait of high heritability.

2.2.4 Direct Versus Correlated Response

In many cases, there is a trade-off between intensity and accuracy of selection. The expression for response to selection, $\Delta\bar{G} = i\rho\sigma_G$, suggests that the term of interest is the product of intensity and accuracy, rather than intensity or accuracy in itself. In the short term, this is likely to be true. In the long term, however, it may matter how the accuracy is obtained. Selection for a correlated trait may lead to an inefficient organism spending resources on traits that are not of direct interest. Consider, for example, fertility, which is the most problematic trait in dairy cattle breeding. Selection for milk yield has deteriorated negative energy balance during early lactation, which has compromised fertility. A solution is to use body condition score (BCS) as a predictor of negative energy balance, and hence, fertility. (BCS is favourably correlated to fertility; it essentially provides indirect information on feed intake.) With long-term selection, however, BCS may change from a predictor of energy balance into a trait that competes for resources with yield and fertility. In other words, selection for BCS may lead to animals that value BCS as a goal in itself, rather than as a means to improve fertility. In more technical terms, the relationship between BCS, fertility and yield will be non-linear, leading to changing genetic correlations over time. This does not invalidate the use of common animal breeding tools such as BLUP and selection index, but indicates that we have to be aware of changing genetic parameters. Though this example is probably a little extreme, selection for BCS will lead to individuals spending more energy on BCS than they would have done had they been selected directly for fertility. In the long term, therefore, selection strategies aiming directly at the trait of interest are likely to outperform strategies based on correlated traits. In other words, accuracy may be more important than intensity of selection in the long term.

2.3 Discussion

In the above, I have addressed response in traits related to fitness from a quantitative genetic perspective, which is largely statistical in nature. Little attention has been given to biological considerations such as resource allocation models or the physiological basis of observed phenotypes. Resource allocation models are often intuitively appealing, because they confirm the intuitive notion that selection solely for productivity may cause health related problems in the longer term (de Jong 1993; Rauw et al. 2002).

The question is how animal breeders can benefit from such models. In the short term, resource allocation models have to comply with the genetic parameters of traits. Both approaches are merely a different way of looking at the same issue. In the short term, therefore, breeders do not need resource allocation models; the genetic parameters of the current populations are sufficient to predict response to selection. In the short term, therefore, it is probably better to give priority to collecting sufficient amounts of data of good quality on the traits of interest in the relevant environments, so that the relevant genetic correlations can be estimated with high accuracy. The utility of resource allocation models to support operational selection decisions seems to be limited.

In the longer term, resource allocation models may show non-linear behaviour, which corresponds to a change in the genetic relationships between traits. As long as we do not know all the genes underlying observed trait values, their effects and their interactions, quantitative genetics does not provide a prediction of the change in the genetic relationships between traits (apart from the short-term Bulmer effect, which is not of interest in this context). Hence, resource allocation models may offer an approach to predict changes in the genetic relationships between traits, which is required to predict response to long term selection. The problem is that we do not know which traits are related linearly (if any), and which traits are a result of non-linear processes (Meuwissen and Goddard 1997). For example, one may model the relationship between resources allocated to fitness and resources allocated to production as the result of two linear traits, the first representing resource intake and the second representing the proportion of resources allocated to fitness (the remainder is allocated to production). With such a model, long term selection will result in a changing relationship between fitness and production, which is intuitively appealing (De Jong 1993; Rauw et al. 2002; van der Waaij 2004). However, this result is a direct consequence of the *a priori* decision to treat resource intake and allocation as linear causal variables, and fitness and production as resulting non-linear variables. This decision has not been sufficiently justified, neither theoretically nor experimentally. An important objective, therefore, is the empirical justification of model choice. Nevertheless, resource allocation models offer a means to understand changing relationships between traits, for which there are few alternative solutions. Their utility is not primarily to support operational selection decisions, but to help breeders foresee the trade-offs that may occur, either in future generations that will have a higher genetic level, or in environments where the breed has not yet been used. When breeders are aware of such trade-offs, measures may be taken in time so as to avoid surprises.

Section 2.1 has stressed the need to identify all the heritable components affecting the traits of interest. In some cases, those heritable components may not be restricted to the population or species considered, but may also come from other species. When considering disease incidence, for example, co-evolution of the pathogen may play an important role in (lack of) response to selection. Breeding strategies to reduce disease incidence may affect co-evolution of the pathogen, and it may be possible to adapt them so as to reduce or even benefit from pathogen co-evolution. In this respect, breeders can probably learn from evolutionary biologists, who have extensively studied the evolutionary arms race between hosts and pathogens (e.g. Thrall and Burdon 1997).

2.4 Conclusions

Genetic improvement of traits related to fitness is not fundamentally different from genetic improvement of other traits of interest to livestock breeders. In principle, quantitative genetic theory applies to any trait. For traits related to fitness, however, the models may need to be more extensive. An important element of successful genetic improvement is to identify all the heritable components that affect the trait of interest. Moreover, traits related to fitness are often more difficult to record, in particular on the candidate itself, and are often sensitive to genotype by environment interaction. This problem can be partly overcome by the use of marker assisted breeding value estimation (MA-BVE) combined with large scale recording of phenotypic data in commercial environments. Given the acceleration of response to selection that is expected with MA-BVE, monitoring the consequences of genetic selection will become increasingly important. With MA-BVE, selection decisions can be made in a juvenile stage. For certain traits, such as longevity, this is long before the phenotype of the candidate itself can be recorded, so that selection decisions may be taken a full generation earlier than the observation of phenotypes, which will increase risk in breeding schemes. Timely collection of data on the relevant traits in the commercial environment will become increasingly important to observe and correct undesired consequences of selection as soon as possible. When applied in populations of sufficient effective size, the use of MA-BVE has high potential to contribute significantly to the genetic improvement of traits related to fitness while maintaining genetic diversity.

3 Social Interactions

Livestock breeds are usually kept in groups, ranging from groups of four laying hens kept in cages, individuals kept in pens of moderate size, such as in fattening pigs and beef cattle in feeding lots, to large flocks of sheep or beef cattle. Performance of individuals may, therefore, depend not only on properties the individual itself, but also on (social) characteristics of other individuals present in the same group. A well-known example is mortality due to cannibalistic behaviours in non-beak trimmed laying hens (Muir and Aggrey 2003). It is unlikely, however, that effects of social interactions are restricted to cannibalism in laying hens. Our first results in pigs, for example, indicate that growth rate is substantially affected by social interactions among individuals, even in *ad libitum* feeding (unpublished results). In general, other individuals kept in the same group are an important component of the environment experienced by an individual. Social interactions may, therefore, affect many traits of interest in livestock genetic improvement.

From a breeder's perspective, a social environment is fundamentally different from a physical environment because it may have a heritable component (Wolf et al. 1998). In contrast to the physical environment, the social environment experienced by individuals can be improved by means of selection. Thus, when social effects have a heritable component, performance of individuals can be improved by breeding for a better social environment. With the exception of maternal effect models, however, classical breeding theory does not account for heritable social effects. Thus generalisation of quantitative genetic theory is required to understand how interactions among individuals affect performance, and which selection methods are suited to improve traits affected by social interactions. This involves both the development of basic theory of inheritance and response to selection, and the development of methodology to estimate the genetic parameters of social effects in real populations.

Social interactions create a discrepancy between the fitness of an individual judged from the performance of the individual itself, versus judged from the impact of the individual on fitness of its entire group. As a consequence, selection for individual fitness may lead to an increase in selfish or competitive behaviour, creating a decrease of population mean fitness. For example, in non-beak trimmed laying hens, selecting individuals that best survive may yield a decrease of survival in the next generation (Muir and Aggrey 2003). This is because the survivors may have negative social effects on their pen mates. Thus selection for individual fitness may correspond to selection for anti-social individuals. In livestock production, interest is in performance of populations. Thus, rather than selecting for individual fitness or performance, selection should be aimed at the total effect of individuals on group performance, and thus include both direct and social effects.

Here I summarise the basic theory presented in Muir (2005), Bijma et al. (2007a,b), and Ellen et al. (2007), and discuss its relevance for livestock genetic improvement. The focus will be on translating the basic results into terms familiar to animal breeders, such as accuracies of selection.

3.1 Model

Consider a population consisting of groups of n individuals each. Social interactions occur among individuals within a group and affect phenotypic values of individuals. The phenotypic value of individuals will be the sum of a direct effect due to the individual itself, and the social effects of the $n - 1$ other individuals kept in the same group. Both the direct effect and the social effect may consist of a heritable component, A, and a non-heritable component, E. The observed phenotypic value of each individual, therefore, is the sum of the direct breeding value and direct environmental effect of the individual itself, and the summed social breeding values and social environmental effects of its $n - 1$ group members (Griffing 1967),

$$P_i = A_{D,\,i} + E_{D,\,i} + \sum_{i \neq j}^{n} A_{S,j} + \sum_{i \neq j}^{n} E_{S,j}, \tag{2}$$

In Equation (2), $A_{D,i}$ is the direct breeding value (DBV) of individual i, and $A_{S,j}$ the social breeding value (SBV) of its group member j. The DBV corresponds to the classical (direct) breeding value, whereas the SBV is a generalisation of a breeding value for maternal effect. The $E_{D,i}$ and $E_{S,j}$ are the corresponding non-heritable effects ("environment"). The magnitude of social effects will depend on group size (n). In larger groups, the per individual social effect is probably smaller than in small groups. Thus $\mathrm{Var}(A_S)$ will be a decreasing function of group size (see below and discussion of Bijma et al. 2007b).

3.2 Response to Selection

The DBV and SBV represent the heritable components underlying phenotypic trait values. Thus response to selection will depend on improvement in both DBVs and SBVs. The relative impact of DBVs and SBVs on response to selection follows from decomposition of the overall population mean into mean direct and social effects, $\bar{P} = \frac{1}{m}\sum_m P_i = \frac{1}{m}\left[\sum_m P_{D,i} + (n-1)\sum_m P_{S,i}\right] = \bar{P}_D + (n-1)\bar{P}_S$, where averages are taken over m individuals, m denoting population size. Total genetic response per generation, therefore, equals the increase of the DBV (ΔA_D) plus $(n - 1)$ times the increase of the SBV (ΔA_S),

$$\Delta\bar{P} = \Delta A_D + (n-1)\Delta A_S \tag{3}$$

This result shows that social effects should be weighted by group size minus one. The $(n - 1)\Delta A_S$ in Equation (3) represents response to selection in the social environment that individuals experience.

In classical breeding theory, response to selection equals the change of the (direct) breeding value per generation (Lynch and Walsh 1998). With interaction among individuals, however, response contains a component due to SBV Equation (3). We therefore generalise the definition of breeding value to incorporate social interactions, and define a total breeding value,

$$\text{TBV}_i = \text{DBV}_i + (n-1)\text{SBV}_i = A_{D,i} + (n-1)A_{S,i} \tag{4}$$

Analogous to classical theory, response equals the change of the TBV per generation, and the TBV replaces the usual breeding value. This result shows that, with social interactions among individuals, the total breeding value of an individual will depend on group size (n). Note that the social component of the TBV of an individual is expressed not in the individual itself, nor in its offspring, but in the phenotypes of their group members.

In addition to a total breeding value, we can also identify the total heritable variation in the trait. Due to social interactions, the total heritable variation that is available to generate response to selection will differ from the usual (direct) additive genetic variance. Analogous to classical theory, the total heritable variation is the variance of TBVs among individuals, $\sigma^2_{\text{TBV}} = \text{Var}[A_{D,i} + (n-1)A_{S,i}] =$

$$\sigma^2_{\text{TBV}} = \sigma^2_{A_D} + 2(n-1)\sigma_{A_{DS}} + (n-1)^2\sigma^2_{A_S} \tag{5}$$

Analogous to the usual σ^2_A, the σ^2_{TBV} determines the potential of a trait to respond to selection. The term $\sigma^2_{A_D}$ in Equation (5) corresponds to the additive genetic variance in classical theory. The term $(n-1)^2\,\sigma^2_{A_S}$ shows that presence of heritable social interactions can substantially increase the total heritable variance. This may explain the rapid responses observed with group selection (e.g. Muir 1996). Essentially, the term $(n-1)^2\,\sigma^2_{A_S}$ represents the heritable variance present in the social environment. The term $2(n-1)\sigma_{A_{DS}}$ in Equation (5) shows that a negative genetic covariance between DBVs and SBVs ($\sigma_{A_{DS}}$) reduces the total heritable variance. With negative $\sigma_{A_{DS}}$, individuals with positive breeding values for their own phenotype (DBV) have on average negative social effects on the phenotypes of their group members (SBV). Thus negative $\sigma_{A_{DS}}$ may be interpreted as "heritable competition". Heritable competition, therefore, reduces the total heritable variance and thus the potential of the trait to respond to selection. In contrast, heritable cooperation ($\sigma_{A_{DS}} > 0$) increases the total heritable variance.

In stark contrast to classical theory, the total heritable variance in a trait can be larger than the phenotypic variance in the trait, Var(TBV) > Var(P). [When groups consist of unrelated individuals, the phenotypic variance equals Var(P) = $\sigma^2_{A_D} + \sigma^2_{E_D} + (n-1)(\sigma^2_{A_S} + \sigma^2_{E_S})$]. In that case, a population contains a heritable variance which is greater than the observed phenotypic variance. Social interactions, therefore, result in hidden heritable variation. The reason that the total heritable variance is not a component of the phenotypic variance is because the TBV of an individual is distributed over multiple phenotypes; the total social effect, $(n-1)$AS is not expressed in a single individual, but each of the $n-1$ group members receives an effect AS,i. The total effect, therefore, does not surface in the phenotypic variance, but is hidden to direct observation. Our preliminary results in pigs indicate that, in a real population, the heritable variance may indeed exceed the phenotypic variance. As a consequence, response in traits affected by social interactions can be very large when compared to the phenotypic standard deviation of the trait. This agrees with the very large responses that have been observed in some selection experiments (Muir 1996). The potential for genetic improvement of traits affected by social interactions may, therefore, be substantially larger than currently perceived.

3.3 Selection Methods

The above has shown that social interactions create additional heritable variation. However, this additional heritable variation may not be utilised to the full by classical selection strategies, such as mass selection or selection based on sib or progeny information. Improvement of traits affected by social interactions, therefore, may require modification of our selection methods. Analogous to classical theory, response for each selection strategy may be expressed as the product of intensity, accuracy and genetic standard deviation,

$$\Delta G = i\rho\sigma_{\mathrm{TBV}}, \tag{6}$$

in which ρ is the accuracy of selection, which is the correlation between the selection criterion and the TBV of the selection candidate. (Equation (6) is equivalent to Equation (1); with multitrait selection, ρ is the correlation between the selection criterion and the *TBV* of the trait of interest.) Below we present accuracies for mass selection, group selection and sib selection schemes. In all cases, the individuals that provide the information for selection are kept in groups of n members, but the selection criterion may differ.

3.3.1 Mass Selection

With mass selection, individuals are selected by truncation based on their own phenotypic record. For traits affected by social interactions, accuracy of mass selection equals (Ellen et al. 2007)

$$\rho_{\mathrm{mass}} = \frac{\sigma^2_{A_D} + (n-1)\sigma_{A_{DS}} + r(n-1)[\sigma_{A_{DS}} + (n-1)\sigma^2_{A_S}]}{\sigma_{\mathrm{TBV}}\sigma_P} \tag{7}$$

This result shows that, in contrast to the classical case, accuracy of mass selection depends on relatedness, r, between individuals in a group and on group size, n. When groups consist of unrelated individuals, $r = 0$, Equation (7) reduces to $\rho_{\mathrm{mass}} = \frac{\sigma^2_{A_D}+(n-1)\sigma_{A_{DS}}}{\sigma_{\mathrm{TBV}}\sigma_P}$. The numerator of this expression, $\sigma^2_{A_D} + (n-1)\sigma_{A_{DS}}$, is a covariance and can take negative values when direct and social effects have a negative genetic covariance, $\sigma_{A_{DS}} < 0$. (A negative $\sigma_{A_{DS}}$ indicates that individuals that have positive direct breeding values have on average negative social breeding values.) As a consequence, the accuracy of mass selection can be negative, meaning that the population will respond in the direction opposite to that of selection. What happens in this case is that mass selection yields a negative correlated response in social effects, which exceeds the direct response in absolute value, leading to a negative net result, $|(n-1)\Delta\bar{A}_S| > \Delta A_D$. In other words, due to correlated response, mass selection may lead to increased competition, annulling the increase in the direct effect.

Equation (7) shows that an increase in relatedness between group members moves the accuracy in a positive direction. (Except when $r_{A_{DS}} < -(n-1)\sigma_{A_S}/\sigma_{A_D}$, which does not seem likely.) For example, with full relatedness, $r = 1$, the accuracy equals $\rho_{\text{mass}} = \frac{\sigma^2_{A_D}+2(n-1)\sigma_{A_{DS}}+(n-1)^2\sigma^2_{A_S}}{\sigma_{\text{TBV}}\sigma_P} = \frac{\sigma^2_{\text{TBV}}}{\sigma_{\text{TBV}}\sigma_P} = \frac{\sigma_{\text{TBV}}}{\sigma_P}$. This result is an analogy of the square root of heritability, h, and is always positive. Thus, a breeding strategy to counteract an increase in competition is to keep individuals in groups consisting of relatives, e.g. full sibs, which was already suggested by (Griffing 1976).

3.3.2 Group Selection

With group selection, the entire group of individuals is either selected or rejected as a whole, based on mean performance of individuals in the group. Thus selection is by truncation among groups, and the selection criterion is the mean group performance, $\bar{P}_{\text{grp}}$. Accuracy of group selection equals (Ellen et al. 2007)

$$\rho_{\text{grp}} = \frac{[(n-1)r+1]\sigma_{\text{TBV}}}{n\sigma_{\bar{P}_{\text{grp}}}} \tag{8}$$

This result shows that accuracy of group selection is always positive, because both numerator and denominator of Equation (8) are positive. Moreover, the accuracy increases by increasing relatedness among group members, because this increases the numerator of ρ_{grp}. (Though it also affects $\sigma_{\bar{P}_{\text{grp}}}$, but to a lesser extent.) Thus selecting between groups and using groups composed of relatives are breeding solutions to prevent negative response due to increased competition.

3.3.3 Sib Selection

This section investigates response to selection based on sib information, using two different strategies. First, selection based on mean performance of sibs, $\bar{P}_{\text{sibs}}$, kept in groups of *unrelated* individuals. For example, selection based on mean performance of 20 full sibs kept in 20 distinct groups, in which the group members of a particular full sib are unrelated to the full sib. Second, selection based on mean performance of sibs kept in *family* groups. For example, selection based on mean performance of 20 full sibs kept in 5 distinct groups, each consisting of four full sibs. The accuracy of selection when using sib information coming from groups of unrelated individuals equals

$$\rho_{\text{sibs}} = \frac{r[\sigma^2_{A_D} + (n-1)\sigma_{A_{DS}}]}{\sigma_{\text{TBV}}\sigma_{\bar{P}_{\text{sibs}}}} \tag{9}$$

The numerator of this expression is proportional to that of mass selection with unrelated groups Equation (7). The accuracy takes negative values when $\sigma^2_{A_D} + (n-1)\sigma_{A_{DS}} < 0$, which may occur when direct and social effects have a negative genetic covariance, $\sigma_{A_{DS}} < 0$. This result shows that selection based on sib information, coming from groups of unrelated individuals, may lead to an increase in competition that exceeds the direct response, yielding a negative net response.

The accuracy of selection when using sib information coming from groups composed of related individuals is an analogy of the classical accuracy for traits not affected by interactions among individuals. Without interactions among individuals, accuracy of sib selection is given by (Falconer and Mackay 1996) $\rho = rh/\sqrt{t + (1-t)/N}$, in which r is relatedness between the candidate and its sibs, h the square root of heritability, t the intraclass correlations between the sibs, $t = r_{br}h^2$ with r_{br} denoting relatedness between the sibs, and N is the total number of observations. For example, when using a group of full sibs that are half sibs of the candidate, i.e. a group of offspring of the sire of the candidate but of a single other dam, then $r = \frac{1}{4}$ and $r_{br} = \frac{1}{2}$. With social interactions and groups composed of related individuals, accuracy is an analogy of the classical expression (Ellen et al. 2007),

$$\rho = \frac{r\eta}{\sqrt{\tau + (1-\tau)/N}} \tag{10}$$

In Equation (10), η is an analogy of the square root of heritability, and τ is an analogy of the intraclass correlations between sibs. The $\eta^2 = \sigma^2_{\text{TBV}}/\sigma^2_{\text{TPV}}$ and $\tau = r_{br}\eta^2$, in which σ^2_{TBV} is the total heritable variation in trait value (Equation 5), and σ^2_{TPV} is the variance in Total Phenotypic Values of individuals, $\text{TPV}_i = (A_{D,i} + E_{D,i}) + (n-1)(A_{S,i} + E_{S,i})$. The TPV_i represents the total phenotypic effect of an individual on the trait values of its group (including itself), of which an amount $(A_{D,i} + E_{D,i})$ is expressed in its own phenotype and an amount $(A_{S,i} + E_{S,i})$ in the phenotypes of each of its $n-1$ group members. (Note that the TPV differs from the observed phenotypic value, P, of an individual.) Analogous to Equation (5), the variance of Total Phenotypic Values is given by $\sigma^2_{\text{TPV}} = (\sigma^2_{A_D} + \sigma^2_{E_D}) + 2(n-1)(\sigma_{A_{DS}} + \sigma_{E_{DS}}) + (n-1)^2(\sigma^2_{A_S} + \sigma^2_{E_S})$. Investigation of Equation (10) shows that the limiting accuracy for a large number of sib groups is identical to that in the classical case. For large N, the accuracy approaches $r/\sqrt{r_{br}}$ which equals 0.5 for half sibs information, 0.71 for full sib information, and 1 for progeny information. These values are the same as for classical selection based on sibs or progeny in the absence of social interactions. Equation (10), therefore, shows that accuracies for traits affected by social interactions can be similar to accuracies for classical traits. The key issue is that the sibs need to be kept in family groups.

3.4 Parameter Estimation

The selection methods described above can be applied without knowledge of the genetic parameters. Without knowledge of the genetic parameters, however, it is unclear whether those selection strategies are needed at all, and what the expected response to selection is. Moreover, most applied breeding programs make use of BLUP for estimating breeding values, which requires knowledge of the genetic parameters. Thus there is a need to estimate the genetic parameters. This involves estimating the covariance between direct and social effects, $\sigma_{A_{DS}}$, and the additive genetic variance in social effects, $\sigma^2_{A_S}$. Those parameters can be estimated by extending the usual mixed model equations, in a manner similar to estimating maternal genetic effects. This results in the following mixed model (Muir 2005; Bijma et al. 2007b)

$$\mathbf{y} = \mathbf{Xb} + \mathbf{Z}_D\mathbf{a}_D + \mathbf{Z}_S\mathbf{a}_S + \mathbf{e}, \tag{11}$$

in which $\mathbf{y}$ is the usual vector of observations and $\mathbf{Xb}$ the usual fixed effects. The $\mathbf{Z}_\mathrm{D}\mathbf{a}_\mathrm{D}$ represent the usual (direct) additive genetic effects; $\mathbf{a}_\mathrm{D}$ is a vector of DBVs, with incidence matrix $\mathbf{Z}_\mathrm{D}$ linking phenotypic values of individuals to their DBV. The $\mathbf{Z}_\mathrm{S}\mathbf{a}_\mathrm{S}$ represent the social additive genetic effects; $\mathbf{a}_\mathrm{S}$ is a vector of SBVs with one element for each individual, with incidence matrix $\mathbf{Z}_\mathrm{S}$ linking phenotypic values of individuals to the SBVs of their associates. When considering four individuals per group, for example, the $\mathbf{Z}_\mathrm{D}$ has a one at the position of the individual producing the record, and the $\mathbf{Z}_\mathrm{S}$ has three ones, one for each cage member of the individual producing the record. The covariance structure of the random genetic terms is: $\mathrm{Var}\begin{bmatrix}\mathbf{a}_D\\ \mathbf{a}_S\end{bmatrix} = \mathbf{C} \otimes \mathbf{A}$, where $\mathbf{C} = \begin{bmatrix}\sigma^2_{A_D} & \sigma_{A_{DS}}\\ \sigma_{A_{DS}} & \sigma^2_{A_S}\end{bmatrix}$ and $\otimes$ indicates the Kronecker product of matrices. (Note that the element y_i of $\mathbf{y}$ in Equation (11) corresponds to P_i in Equation (2).) Thus, analogous to a maternal effects model, Equation (11) is a bivariate model involving the estimation of two variances, $\sigma^2_{A_D}$ and $\sigma^2_{A_S}$, and one covariance, $\sigma_{A_{DS}}$.

The residual term of Equation (11) requires special attention. Because social interactions may have a non-heritable component, denoted E_S in Equation (2), records of group members will be correlated for non-genetic reasons. Such non-genetic correlations between group members need to be accommodated in the model, to avoid biased estimates of the genetic parameters. It follows from Equation (2) that the non-genetic covariance between phenotypic values of group members equals $2\sigma_{E_{DS}} + (n-2)\sigma^2_{E_S}$, and that the residual variance equals $\sigma^2_e = \sigma^2_{E_D} + (n-1)\sigma^2_{E_S}$. Residuals of group members, therefore, have a correlation equal to $\rho_e = [2\sigma_{E_{DS}} + (n-2)\sigma^2_{E_S}]/[\sigma^2_{E_D} + (n-1)\sigma^2_{E_S}]$. Depending on $\sigma_{E_{DS}}$, $\sigma^2_{E_S}$ and n, this correlation may take either positive or negative values. The covariance structure of the residual is, therefore, block diagonal,

$$\mathrm{Var}(\mathbf{e}) = \mathbf{R}\sigma^2_e, \text{ with } R_{ii} = 1, \tag{12}$$

$R_{ij} = \rho_e$, when i and j are in the same group, $i \neq j$
$R_{ij} = 0$, when i and j are in different groups.

This structure can be fitted in ASREML by using the CORU statement in the description of the R-structure (Gilmour et al. 2002).

With large group size (n), the residual correlation is likely to be positive due to the term $(n-2)\sigma^2_{E_S}$ in the residual covariance. In that case, an equivalent but simpler solution is to fit a random group effect instead of a correlated residual. It follows from $\mathrm{Cov}_{\mathrm{within}} = \mathrm{Var}_{\mathrm{between}}$ that the between group variance, i.e. the variance of the random group effect, equals $\sigma^2_{\mathrm{group}} = 2\sigma_{E_{DS}} + (n-2)\sigma^2_{E_S}$. When the residual covariance is positive, both models are equivalent and their relationship is given by $\rho_e = \sigma^2_{\mathrm{group}}/\sigma^2_e$. When groups are of equal size, there is no information in the data to separate $\sigma_{E_{DS}}$ from $\sigma^2_{E_S}$; one can estimate either a random group effect or a residual correlation, but not identify their components. In most cases this will not be important because interest is in the genetic parameters. The reason to include a residual correlation or random group effect is to avoid bias in the estimated genetic variance components; not an interest in the underlying non-genetic parameters.

3.4.1 Varying Group Size

In our model, each individual has a social effect. Consequently, in groups of size n, each individual receives $n-1$ social effects and the total heritable variation has a term $(n-1)^2\sigma^2_{A_S}$ Equation (5). When the magnitude of social effects would be independent of group size, the total heritable variation would become infinite for large n, which is not realistic. It is likely that the social genetic variance, $\sigma^2_{A_S}$, will be a decreasing function of group size, so that σ^2_{TBV} increases less with increasing group size. Thus, Equation (5) applies to a specific group size. For other group sizes, the values of $\sigma_{A_{DS}}$ and $\sigma^2_{A_S}$ may be different. A more general formulation of Equation (5) would, therefore, be $\sigma^2_{\mathrm{TBV}} = \sigma^2_{A_D} + 2(n-1)\sigma_{A_{DS(n)}} + (n-1)^2\sigma^2_{A_{S(n)}}$, which makes explicit that $\sigma_{A_{DS}}$ and $\sigma^2_{A_S}$ are functions of group size. Hence, there will be a kind of GxE-interaction when group size varies. Extrapolation of estimates obtained from a specific group size to populations with different group sizes will require knowledge of the relationship between $\sigma_{A_{DS}}$, $\sigma^2_{A_S}$ and n. This relationship may be non-linear. For example, when the trait of interest is mortality due to cannibalism in four bird cages versus eight bird cages, there may be little dependency of $\sigma_{A_{DS}}$ and $\sigma^2_{A_S}$ on n. This is because a single individual has sufficient opportunity to kill all group members since their number is limited. The situation is different, however, when comparing four bird cages to floor systems with thousands of individuals. In that case, a single individual can never interact with all others. Consequently, $\sigma^2_{A_S}$ may be fairly independent of group size as long as groups are not too large, and then decrease with group size as groups become larger.

When data contain a single group size, there is no need to consider the relationship between $\sigma^2_{A_S}$ and group size when analysing data. The estimated $\sigma^2_{A_S}$ simply applies to the group size considered. When data contain groups of different size, the relationship between $\sigma^2_{A_S}$ and group size can be investigated using random regression models (Hadfield and Wilson 2007). In such models, the social effect of an individual can be modelled as polynomial function of group size, e.g. using an intercept and a term linear in group size ("slope"). In this case, however, the use of random regression models is a complex solution to a simple problem. A random regression model allows the social effect of each individual to have a different sensitivity to group size, whereas the only thing required is for $\sigma^2_{A_S}$ to depend on group size. A much simpler solution is to weigh the A_S in the model by a function of group size, and find the function of group size that best fits the data (Arango et al. 2005). This involves estimating a single additional fixed effect only. For example, when individuals compete for a fixed total amount of resources, one may evaluate a model in which the magnitude of social effects is inversely proportional to $n - 1$; $P_i = A_{D,\,i} + E_{D,\,i} + \frac{1}{n_i - 1}\left[\sum_{i \neq j}^{n_i} A_{S,j} + \sum_{i \neq j}^{n_i} E_{S,j}\right]$, in which n_i is the size of the group of individual i. In this case the total heritable variation is independent of group size.

The variance components for social effects, $\sigma_{A_{DS}}$ and $\sigma^2_{A_S}$, may not be estimable in all data structures occurring in practise (Bijma et al. 2007b). The $\sigma_{A_{DS}}$ and $\sigma^2_{A_S}$ are estimable when groups are composed at random with respect to family, but not when groups are entirely composed of full sibs or half sibs.

3.5 Discussion

The above has argued that social effects are potentially very relevant for livestock genetic improvement. This is most clear from Equation (5), which shows that social interactions may contribute substantially to the heritable variation in trait values. Whether or not social interactions are of interest to breeders will depend on the trait considered and on the environment in which individuals are kept. Results on mortality in non-beak trimmed layers kept in four-bird cages illustrate the relevance of social genetic effects in that situation (Bijma et al. 2007b). In that study, two-third of the total heritable variation in survival days originated from social effects. Our preliminary results in pigs indicate that social genetic effects are important for growth rate and feed intake. In contrast, results of Arango et al. (2005, 2006) in growing pigs indicate that social genetic effects may be difficult to estimate and not deviate significantly from zero. At present there is little knowledge of the importance of social genetic effects in livestock genetic improvement. In fish, the large size differences among individuals indicate that competition is important and that it affects uniformity, but it is unknown whether such competition has a heritable component. It would be interesting to see whether fish breeders can achieve something similar to plant breeders, who have effectively reduced competition among individuals in wheat, which has resulted in high yielding varieties with short stems. In cases where groups are large, such as in broilers, sheep or beef cattle, it will be difficult to estimate genetic parameters. One solution for such cases would be to invest in data collection on a large number of smaller groups. This would, however, require substantial investment, and social effects in smaller groups may not correspond accurately to those in larger groups. In species of small animals with short generation intervals, such as broilers, rabbits or certain fish species, a small scale selection experiment, in which small groups are used for selection and results are evaluated in larger groups, might be more promising. Though social genetic effects may not always be important and may be difficult to use in some cases, the promising results observed in layer chickens and pigs should be sufficient incentive for further research in this area.

References

Arango, J., Misztal, I., Tsuruta, S., Culbertson, M. and Herring, W.O. 2005. Estimation of variance components including competitive effects of Large White growing gilts. *J. Anim. Sci.* **83**: 1241–1246.

Arango, J., Misztal, I., Tsuruta, S., Culbertson, W., Herring, W.O. and Long, T. 2006. Competitive genetic effects in large white growing gilts. *Proc. 8th World Congr. Genet. Appl. Livest. Prod.* Comm. No. 06–24.

Bijma, P., Muir, W.M. and Van Arendonk, J.A.M. 2007a. Multilevel selection 1: quantitative genetics of inheritance and response to selection. *Genetics* **175**: 277–288.

Bijma, P., Muir, W.M., Ellen, E.D., Wolf, J.B. and Van Arendonk, J.A.M. 2007b. Multilevel selection 2: estimating the genetic parameters determining inheritance and response to selection. *Genetics* **175**: 289–299.

Denison, R.F., Kiers, E.T. and West, S.A. 2003. Darwinian agriculture: when can humans find solutions beyond the reach of natural selection? *Quart. Rev. Biol.* **78**: 145–168.

De Jong, G. 1993. Covariances between traits deriving from successive allocation of resources. *Functional Ecol.* **7**: 75–83.

Ellen, E.D., Muir, W.M., Teuscher, F. and Bijma, P. 2007. Genetic improvement of traits affected by interactions among individuals: sib selection schemes. *Genetics* **176**: 489–499.

Falconer, D.S. and Mackay, T.F.C. 1996. Introduction to quantitative genetics, Pearson Education Ltd., Harlow, Essex, UK.

Frank, S.A. 1998. *Foundations of social evolution*, Princeton University Press, Princeton, NJ, USA.

Gilmour, A.R., Gogel, B.J., Cullis, B.R., Welham, S.J. and Thompson, R. 2002. *ASReml User Guide Release 1.0*, VSN International, Hemel Hempstead, UK (http://www.vsn-intl.com/).

Griffing, B. 1967. Selection in reference to biological groups. I. Individual and group selection applied to populations of unordered groups. *Australian J. Biol. Sci.* **20**: 127.

Griffing, B. 1976. Selection in reference to biological groups. VI. Use of extreme forms of non-random groups to increase selection efficiency. *Genetics* **82**: 476–490.

Hadfield, J.D. and Wilson, A.J. 2007. Multilevel selection 3: modeling the effects of interacting individuals as a function of group size. *Genetics* **177**: 667–668.

Hamilton, W.D. 1964. The genetical evolution of social behaviour. *I. J. Theor. Biol.* **7**: 1–16.

Lynch, M. and Walsh, B. 1998. Genetics and analysis of quantitative traits, Sinauer Associates, Sunderland, MA, USA.

Meuwissen, T.H.E. 1997. Maximizing the response of selection with a predefined rate of inbreeding. *J. Anim. Sci.* **75**(4):934–940.

Meuwissen, T.H.E. and Goddard, M.E. 1996. The use of marker haplotypes in animal breeding schemes. *Genet. Sel. Evol.* **28**: 161–178.

Meuwissen, T.H.E. and Goddard, M.E. 1997. Selection for non-linear traits and profit. *Anim. Sci.* **65**: 1–8.

Meuwissen, T.H.E., Hayes, B.J. and Goddard, M.E. 2001. Prediction of total genetic value using genome-wide dense marker maps. *Genetics* **157**: 1819–1829.

Michod, R.E. 1999. *Darwinian dynamics: Evolutionary transitions in fitness and individuality*, Princeton University Press, Princeton, NJ, USA.

Muir, W.M. and Aggrey, S.E. 2003. *Poultry genetics, breeding and biotechnology*, CABI Publishing, Wallingford, Oxon, UK.

Muir, W.M. 1996. Group selection for adaptation to multiple-hen cages: selection program and direct responses. *Poult. Sci.* **75**: 447–458.

Muir, W.M. 2005. Incorporation of competitive effects in forest tree or animal breeding programs. *Genetics* **170**: 1247–1259.

Rauw, W.M., Knap, P.W., Verstegen, M.W.A. and Luiting, P. 2002. Food resource allocation patterns in lactating females in a long-term selection experiment for litter size in mice. *Genet. Sel. Evol.* **34**: 83–104.

Rice, S.H. 2004. *Evolutionary theory; mathematical and conceptual foundations*, Sinauer Associates, Sunderland, MA, USA.

Sonesson, A.K. and Meuwissen, T.H.E. 2000. Mating schemes for optimum contribution selection with constrained rates of inbreeding. *Genet. Sel. Evol.* **32**: 231–248.

Thrall, P.H. and Burdon, J.J. 1997. Host-pathogen dynamics in metapopulation context: the ecological and evolutionary consequences of being spatial. *J. Ecol.* **85**: 743–753.

Van der Waaij, E.H. 2004. A resource allocation model describing consequences of artificial selection under metabolic stress. *J. Anim. Sci.* **82**: 973–981.

Weber, K.E. and Diggins, L.T. 1990. Increased selection response in larger populations. II. Selection for ethanol vapor resistance in *Drosophila melanogaster* at two population sizes. *Genetics* **125**: 585–597.

West-Eberhard, M.J. 2003. *Developmental plasticity and evolution*, Oxford University Press, Oxford, UK.

Wilham, R.L. 1963. The covariance between relatives for characters composed of components contributed by related individuals. *Biometrics* **19**:18–27.

Wolf, J.B., Brodie, E.D. III, Cheverud, J.M., Mooreand, A.J. and Wade, M.J. 1998. Evolutionary consequences of indirect genetic effects. *Trends Ecol. Evol.* **13**: 64–69.

Discussion

Michael Goddard

There was a lively discussion of the three papers in this session by Hill, Walsh and Bijma.

Bruce Walsh argued that it is possible to define an almost unlimited number of traits but that, due to genetic correlations, genetic variation is not equally plentiful in all possible directions in this high dimensional space. In particular, natural selection decreases genetic variance in directions closely related to fitness and thus we expect to find low genetic variance in these directions. This conclusion is superficially in disagreement with Bill Hill's argument that mutation maintains genetic variance in all traits, including fitness, unless effective populations size (Ne) is very small. The two views may perhaps be reconciled by concluding that all dimensions will show variation but some more than others. For instance, in dimensions closely related to fitness, natural selection will eliminate variation and mutation will introduce variation, although much of this variation will be caused by deleterious alleles derived by recent mutation. However, for dimensions not closely related to fitness, mutations will still introduce variation but the strength of selection eliminating the variation may be less and so more standing variation is maintained. This reconciled view implies that some polymorphisms affect phenotype but are effectively neutral and so selection is not eliminating either allele. We need experimental evidence as to whether or not such neutral polymorphisms are common.

Another reason for reduced genetic variance in some dimensions might be that the genetic correlations that determine the pattern of variation may themselves evolve. For instance, it may be important to fitness that an animal maintains certain traits in balance. A simplistic example might be the need to maintain fore- and hind-legs of similar length. Then natural selection would lead to a genetic correlation between length of fore- and hind-legs and consequently a reduction of variance in the dimension (length of fore-legs minus length of hind legs). This process can be described as evolution of the genetic correlation but it can also be described as selection towards an optimum value of the trait (length of foreleg – length of hindleg); that is stabilising selection which works partly by eliminating mutations

---

M. Goddard (✉)
University of Melbourne and Department of Primary Industries, Victoria
e-mail: mike.goddard@dpi.vic.gov.au

J. van der Werf et al. (eds.), *Adaptation and Fitness in Animal Populations*,

that alter the trait away from existing current, optimum value. This is very similar to the process described in the previous paragraph for selection eliminating variation in fitness except that, in the case of stabilising selection, mutations which either increase or decrease the trait are selected against.

In domestic animal populations selection is not for natural fitness but for some combination of traits decided by the breeder. Bruce Walsh's argument implies that if selection occurs for long enough in the same direction, then genetic variance for the objective would decline. This is not what has been observed in livestock but selection has not been for a consistent objective over a long time period.

The practical implications for livestock breeders are as follows. Genetic variation is likely to exist for any breeding objective but there may well be unfavourable genetic correlations among some of the traits that need to be recognised. Selection for this objective (including fitness traits) should be combined with maximised Ne in an optimum manner. A low Ne will lead to inbreeding depression and compromise long term selection response. However, the costs of decreasing Ne are a smooth function of 1/Ne and so there is no value of Ne which is either adequate or inadequate but an optimum balance should be reached between the costs and benefits of increasing Ne.

Piter Bijma explained a model in which an animal's genes affect its own phenotype (direct breeding value) and the phenotype of other animals in its group (social breeding value). This model is similar to the maternal effects model where a female animal's genes affect the phenotype of her offspring. As in the maternal effects model, one needs to estimate the variance of direct BV and social BV and the correlation between them. To estimate these parameters the group size cannot be too large (e.g. < 12). A dataset roughly 4 times bigger than needed for estimating direct genetic variances is needed. Random allocation of animals to groups is a satisfactory but not optimal design for the estimation of these genetic variances and correlation. However, if family members are put in the same group the genetic variances are confounded and cannot be estimated. In contrast, once genetic parameters are known, selection for the combined effect of direct and social BV is most effective if family members are grouped together and selection is partly on family mean. Plant breeders often use this design when they select on the mean performance of a plot containing many plants that are from the same clone. However, animals can also be randomly allocated to groups (provided they are not too large) and BLUP used to estimate both direct and social BVs.

This model with distinct direct and social BVs is a generalisation of other models that have been used. For instance, in infectious diseases such as nematode parasitism, animals that are susceptible to the disease may also increase the number of pathogens to which their group mates are exposed. This model, in its simplest form, implies that the correlation between direct and social BV is 1.0 because animals with a high parasite burden are also the greatest source of infection to others. However, the full social BV model allows for the possibility that the correlation is not 1.0; that is, that an animal might excrete many infectious pathogens but not be highly affected itself.

Another subset of the social BV models involves competition for resources. In a pure competition model, an animal which has a high direct BV inevitably decreases the performance of its group mates. That is, the correlation between direct and social BV is -1.0. However, Piter said that in their experiments with pigs, the correlation between direct and social BV was close to zero. In other words some pigs have a high BV for growth rate without a detrimental effect on the growth rate of their group mates.

These competition effects have been investigated in the past by comparing performance when pigs are kept in single animal pens as opposed to when they are housed in groups. These environments may differ in other ways as well as in the competition from group mates, so a model that treats performance in the two different sorts of pen as different traits is not equivalent to a social BV model. When animals are selected on their own performance in either environment they are selected for their direct BV although this may be different in the two environments. However, if the response to some generations of selection is measured by grouping together in a pen animals of the selected line, then their mean performance will be affected by their social BV. If direct BV for growth rate in a pen is negatively correlated with social BV, then the selection for direct BV will have caused a negative correlated response in social BV, and the response to selection will be less than expected on the basis of direct BVs alone. It is an empirical question whether direct BV in a single pig pen has the same correlation with social BV in a group pen as direct BV in a group pen does. However, it is plausible that direct BV in a group situation is correlated with social dominance and that this is negatively correlated with social BV. Now that we understand the social BV model it will be possible to do experiments which estimate the parameters and answer these questions.

The genetic parameters of the social BV model depend on the environment just as the genetic parameters of a conventional model do. One environmental variable that is expected to influence the genetic parameters is the group size. The way Piter has formulated the model, if the genetic parameters did not change, the variance in phenotype due to the social BV of group mates would increase linearly with the number of group mates. Therefore it is likely that the variance of social BV will decrease as the group size increases but we need experimental evidence to determine this – it is not a theoretical issue but one of empirical estimation of parameters.

Part III
The Genetic Basis of Adaptation

Some Evolutionary Consequences of Niche Construction with Genotype-Environment Interaction

Kathleen Donohue

Abstract "Niche construction" is the ability of organisms to determine the environment they inhabit. Niche construction can occur through direct modification of the environment, habitat choice, or passive alterations to the environment, such as resource use or depletion. When organisms predictably alter the environment they inhabit, and when niche construction has a genetic basis, it can influence evolutionary responses to selection. Each component of evolutionary responses to selection—the strength of selection on a trait, the phenotype and phenotypic variance of a trait, and the genetic variance of a trait—can change with the environment. Examples of niche construction in plants are discussed, and a simple univariate model demonstrates that when niche construction alters phenotypic expression (via phenotypic plasticity) and the expression of genetic variation or heritability (via genotype-environment interaction), it can cause slower or faster responses to selection, less or more sustained responses to selection, or even negative responses to selection. In particular, genotype-environment interaction can counteract or augment phenotypic plasticity to the constructed environment in its effects on evolutionary responses to selection. Thus, genotype-environment interaction that results in environment-dependent genetic parameters influences evolutionary trajectories with niche construction.

Keywords Genotype-environment interaction · habitat selection · indirect genetic effects · response to selection · phenotypic plasticity

1 Introduction

One of the primary goals of empirical quantitative-genetics, especially as applied to breeding programs, is to distinguish the contributions of genetic versus environmental factors to trait variation. However, when organisms can alter the environment they are exposed to, and when this ability has a genetic basis, the environment too

K. Donohue (✉)
Department of Biology, Duke University, Box 30338, Durham, NC 27708, USA
e-mail: k.donohue@duke.edu

J. van der Werf et al. (eds.), *Adaptation and Fitness in Animal Populations*,

can have a genetic component, can evolve, and can influence the dynamics of trait evolution (Wolf 2003; Cheverud 2003). The ability of organisms to determine the environment they experience has been termed "niche construction" (Odling-Smee et al. 1996), which, broadly interpreted, includes what have come to be known as "indirect genetic effects" (Moore et al. 1997; Wolf et al. 1998), and more traditionally "habitat selection" (Levins 1968; Holt 1987; Rosenzweig 1987). Niche construction can occur by direct habitat modification of the environment, habitat choice, dispersal habits, or through passive processes such as resource use or depletion (Odling-Smee et al. 1996; Laland et al. 1999; Day et al. 2003; Odling-Smee et al. 2003). Animals can practice very sophisticated modes of niche construction, through the building of shelter, cultivating or hoarding of food supplies, complex behaviors of dispersal, or the creation of social (or anti-social) environments. Other organisms, such as plants, also practice niche construction through modifications of growth structure (e.g. clonal foraging and shade avoidance), environmentally regulated phenological switches such as germination or reproductive timing, or dispersal strategies (reviewed in Bazzaz 1991; Donohue 2003,2005). Many maternal effects (Falconer 1965; Kirkpatrick and Lande 1989) can also be considered to be niche construction, since the maternal parent creates the environment in which progeny develop (Wade 1998). Broadly speaking, all organisms in some sense determine the environment they inhabit and thereby are subject to the dynamics of niche construction.

Niche construction can influence evolutionary dynamics (Fig. 1). Responses to selection are determined by the strength of selection on a trait, and the proportion of the phenotypic variance of the trait that is genetically determined, or the heritability of the trait. The environment that organisms experience can influence all of these components of evolutionary responses to selection. In this manner, niche construction can influence evolutionary outcomes and trajectories.

First, the environment can influence the agents of natural selection to which an organism is exposed (Laland et al. 1999; Day et al. 2003; Donohue 2003; Schwilk 2003). Through niche construction practices that alter exposure to existing selective influences, or that create new agents of selection, organisms can alter the strength, direction, and mode of natural selection on a trait. Most investigations of the evolutionary consequences of niche construction have focused on this very important pathway, and these investigations comprise the literature on habitat selection or habitat choice. Perhaps the most interesting conclusions of these classic studies is that habitat selection can alter the dynamics of specialization. Specifically, when habitat selection increases exposure to one environment and decreases exposure to other environments, adaptation to the first environment can be faster, and adaptation can be closer to the optimum for that environment (Levins 1968; Holt 1987; Rosenzweig 1987; Brown 1990; Whitlock 1996); that is, habitat selection facilitates specialization (Via and Lande 1985; Schlichting 1986; Van Tienderen 1991; Scheiner 1993). In animal breeding programs, this aspect of niche construction can be important if traits that evolve in response to artificial selection result in new agents of natural selection, such as aggressive environment, intensity of competition, or probability of infection, for example.

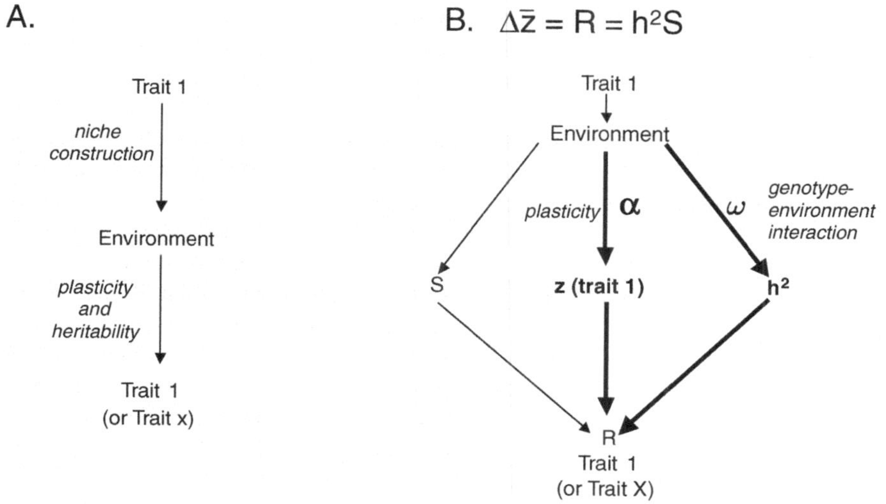

Fig. 1 (A) Niche construction. One trait influences the environment experienced by the organism, which in turn influences the evolution of that same trait or a different trait ("Trait X"). (B) The breeder's equation, showing pathways whereby effects of niche construction on phenotypic expression, α, and the heritability of the trait, ω, can influence total response to selection of the trait. $z =$ the phenotypic value of the trait; h^2 is the heritability of the trait; S is the strength of selection on the trait; α is the factor by which the environment that is determined by the trait influences the phenotypic value of the trait, via phenotypic plasticity; ω is the factor by which the environment that is determined by the trait influences the heritability of the trait. The trait is directly associated with the environment that the organism experiences and can therefore be considered to be a metric of the environment itself. In this case, the trait influences the environment, which influences its own expression and heritability or the expression and heritabilty of a different trait ("Trait X")

Second, the environment can influence phenotypic expression via phenotypic plasticity (Schlichting and Pigliucci 1998). Phenotypic plasticity is the ability of a given genotype to alter its phenotype in response to the environment that it experiences. Phenotypic plasticity thereby can alter adaptive dynamics because it can determine how close to the optimum the phenotype of that genotype is, and plasticity can act in the same or opposite direction as that favored by selection. For example, selection for less aggressive behaviour might have the unintended consequence of depriving an animal of food, which, through plasticity in response to reduced nutrition, would result in decreased body mass. In addition, the degree of phenotypic variance via plasticity to microenvironmental factors can influence the degree of environmental variance of traits, which is an important component of trait heritabilities.

Third, the environment can influence the expression of genetic variation (Via and Lande 1987). Just as a given genotype can alter its phenotype according to the environment, when different genotypes respond to the environment differently, the differences among them can be environment-dependent. In other words, with genotype-environment interaction, the magnitude of genetic variance can depend on the environment. The effect of the environment on phenotypic and genetic variances can cause changes in the heritability of traits and thereby alter the evolutionary potential of the trait. In the previous example, not only may body mass decline as a result of food deprivation, but heritability of body mass may also decline, making improvement of that trait more difficult.

This last path of influence—through genotype-environment interaction—has not received attention in the literature on niche construction or indirect genetic effects. Environment-dependent heritabilities and genetic variances are so widely documented that they are fundamental components of any study that attempts to predict responses to selection (Via and Lande 1987; Mazer and Wolfe 1992; Dorn et al. 2000; Schlichting and Pigliucci 1998). Usually, the environment is considered to be a fixed condition of the breeding program or ecology. In the literature on specialization and phenotypic plasticity, the environment varies between two or more states, and the organism experiences each of these states with a given frequency. When attributes of the organism itself determine the environment experienced by the organism, however, and when these attributes themselves can evolve, then the environment that the organism experiences can also evolve over time (Donohue 2003; Wolf 2003; Wolf et al. 2004). The evolution of this environment, in turn, is expected to result in predictable changes in the genetic architecture of traits—that is, changes in the genetic variances and covariances of traits—due to genotype-environment interaction.

In the simplest case, one trait can influence the environment that organisms experience, and this environment can influence the phenotypic expression and heritability of this same trait (Fig. 1). In more complex cases of interacting traits, one trait can influence the environment that organisms experience, and that environment can influence the phenotypic expression and heritability of other traits (reviewed in Donohue 2003, 2005). In multivariate cases, phenotypic and genetic covariances are also likely to be environment-dependent, contributing even more complexity to the dynamics.

In both univariate and multivariate cases, either positive or negative feedbacks can occur and are expected to influence evolutionary outcomes and trajectories. Here I discuss empirical examples of niche construction in plants and present a heuristic model of the simplest case of these dynamics: the univariate case in which one trait determines the environment that then influences its own phenotype and heritability. Even in this simplest case, niche construction via heritable (and evolvable) traits can cause faster, slower, or even reversed responses to selection, and it can constrain or sustain evolutionary responses over time. In particular, genotype-environment interaction can counteract negative or positive effects of phenotypic plasticity to the newly evolved environment, and it must therefore be considered when attempting to predict evolutionary outcomes with niche construction.

2 An Empirical Context of Niche Construction

In a volume on animal breeding, I will discuss examples of niche construction in plants, but suggest that these dynamics are frequently even more readily discernible in animals. Indeed, many historical and recent theoretical investigations of niche construction have been within the context of social evolution and habitat selection in animals (Moore et al. 1997; Wolf et al. 1998; Wolf 2003). Examples of behaviors that can be interpreted as "niche construction" are so abundant (Odling-Smee et al. 1996) that some have complained that almost anything at all could be interpreted as "niche construction", thereby undermining the utility of the concept. While it is perhaps true that one can be fully occupied by simply characterizing diverse phenomena in terms of the omnipresent "niche construction," I would like to counter that just because something is ubiquitous does not imply that it is unimportant. Yes, social aggression; habitat contamination by waste, toxins, or disease; construction of nests, lodges, or webs; thermoregulation; these can all be interpreted as niche construction, and I leave it to the reader to decide whether consideration of the consequences of niche construction is useful for their purposes. My purpose is simply to point out that traits of organisms frequently alter the environments they experience in ways that influence the expression of those or other traits, and in ways that influence the expression of genetic variation of those traits.

3 Phenotypic Plasticity to the Constructed Environment: Interactions Between Flowering and Germination Time

The following example of niche construction through phenological plasticity demonstrates the significance of niche construction and plasticity to the "constructed" environment to plant life histories. Variation in the phenology of germination and flowering determines overall life-history expression in many annual plants, including *Arabidopsis thaliana*. First, the germination season determines the seasonal conditions experienced by seedlings and rosettes (Weinig 2000; Donohue 2002). It is well known in *A. thaliana* and many other species that the seasonal cues of photoperiod and temperature, and especially cold vernalization of rosettes, interact to determine flowering time (Koornneef et al. 1991; Nordborg and Bergelson 1999; Mouradov et al. 2002; Simpson and Dean 2002). In temperate climates, long days and cold vernalization usually accelerate flowering. Germination phenology determines which of these seasonal factors a rosette will be exposed to.

Likewise, the flowering season determines the seasonal conditions experienced by seeds during maturation and soon after dispersal. Seed maturation conditions strongly influence seed dormancy, and post-dispersal seasonal environments determine exposure to various dormancy-breaking factors, such as cold (reviewed in Baskin and Baskin 1998). Therefore, flowering phenology strongly determines germination phenology.

In experimental studies of *A. thaliana*, we found that the season of seed dispersal strongly influenced natural selection on germination time, phenotypic expression of germination time, and the expression of genetic variation for germination time (Donohue et al. 2005a, b, c). Seeds dispersed after a summer flowering season experienced strong stabilizing selection favoring intermediate germination timing in mid October, and seeds dispersed after an autumn flowering season experienced weak directional selection favoring early germination. More dramatically, in the later cohort, non-dormant seeds had the highest fitness, but in the earlier cohort, non-dormant seeds had zero fitness. Phenotypic expression of germination timing also varied: seeds dispersed after a summer flowering season germinated slowly throughout the summer and autumn, while seeds dispersed after an autumn flowering season germinated very quickly that same autumn. The evolutionary potential of germination timing also strongly depended on the season of seed dispersal, with the heritability of germination timing being much higher for seeds dispersed after a summer flowering season than for seeds dispersed after an autumn flowering season.

Germination timing, in turn, influenced reproductive phenology by determining the seasonal environment experienced by young rosettes. In particular, seeds that germinated in the spring reproduced much earlier in development and at a smaller size than seeds that germinated in the autumn.

This mutual interaction between the two niche-constructing traits of flowering time and germination time leads to some interesting consequences for life-history expression. First, variation in germination timing accounts for the difference between the winter annual and spring annual life history, with winter annuals germinating in the autumn and spring annuals germinating in the spring. In addition, spring germinants responded plastically to the season of germination by accelerating their reproduction. Therefore, niche construction through germination timing, and plasticity to the constructed environment, can account for variation between spring and winter annual life histories.

Moreover, when seasonal variation in reproductive and dispersal phenology is present, a bivoltine life-history is possible. In particular, autumn flowering conditions in turn accelerated germination and enabled germination by spring, and spring germination conditions enabled spring flowering. These interactions can actually enable two generations to be completed within a single year instead of the typical one generation. This novel bivoltine life history is the outcome of one niche constructing character influencing the seasonal environment experienced by a second niche-constructing character, and vice versa. Importantly, phenotypic plasticity to the "constructed" seasonal environment plays a key role and in this example altered overall life-history expression in a manner that can influence generation time.

4 Plasticity and Environment-Dependent Heritability: Seed Dispersal as a Simple Case of One Niche-Constructing Character Influencing Itself

This example explores the consequences of niche construction on the expression of genetic variation. Seed dispersal provides an example of a simple case of one niche-constructing character influencing itself. Seeds are the most mobile life stage in most plants, excepting pollen, and their dispersal offers the opportunity to escape from adverse conditions such as pathogens, predators, or sibling competition at the maternal home site (reviewed in Howe and Smallwood 1982; Willson and Traveset 2000). Perhaps the most predictable environmental consequence of seed dispersal is reduced conspecific density experienced by efficiently dispersed seeds in open habitats, and highly competitive conditions experienced by poorly dispersed seeds (e.g. Janzen 1978; Baker and Dowd 1982; Augspurger 1983; Rees and Brown 1991; Augspurger and Kitajima 1992; Donohue 1999).

In many species, efficient seed dispersal leads to lower sibling densities after dispersal. It is therefore important to know how density in turn influences dispersal ability and the expression of genetic variation for dispersal ability. To investigate this, we conducted a quantitative genetic study using morphologically diverse inbred ecotypes of *Arabidopsis thaliana* (Wender et al. 2005; Donohue et al. 2005d). We grew plants at different densities, measured seed dispersion patterns under controlled conditions, and estimated the heritability of post-dispersal density. We found that, plants grown under high density dispersed seeds to lower sibling density than plants grown at low density. Significant genetic variation for post-dispersal density was detected, but only when the maternal plants were grown at high density, and this was due to altered environmental and genetic variances.

This simple result has some interesting potential evolutionary consequences. Consider a population of plants growing at high density, but in which natural selection favors lower post-dispersal density, as it does in many species (e.g. Burdon and Chilvers 1975; Augspurger and Kitajima 1992; Donohue 1999). Our results predict that an evolutionary response to such selection is possible, since genetic variation for dispersal ability is expressed at high density. However, once low post-dispersal density is achieved, genetic variation for dispersal will cease to be expressed. Such a negative feedback pathway would be a constraint on the further evolution of dispersal. Note that this genetic constraint is not caused by the elimination by natural selection of inappropriate genotypes, which would reduce genetic variation. Rather, this genetic constraint is caused by plasticity alone, causing less genetic variation to be expressed, even when the genotypes are identical. The evolution of the post-dispersal density environment and the density-dependent genetic expression for dispersal are adequate to impose this constraint.

In this particular example, the dynamics operated so as to reduce the expression of genetic variation, but for other niche-constructing traits, they may increase the expression of genetic variation. In such cases, the evolution of the niche-constructing character could facilitate its further evolutionary responses to selection by enabling genetic variance to be expressed in the newly evolved environment. Thus while selection may reduce genetic variation for a trait over time, the dynamics accompanying niche construction may act to provide additional genetic variation to enable more sustained responses to selection. Therefore the ability of organisms to determine the environment they experience, and the environment-dependent genetic variation for that ability, can cause unexpected evolutionary dynamics that can either constrain or facilitate the evolution of such characters. More generally, because environment-dependent genetic expression is so commonly observed (e.g. Mazer and Wolfe 1992; Donohue et al. 2000; Munir et al. 2001; Dorn et al. 2000), niche construction has the potential to alter the evolutionary potential of many traits whose genetic variation may depend on the "constructed" environment.

5 Consequences of Environment-Dependent Heritability to Evolutionary Dynamics with Niche Construction

A heuristic model is presented that demonstrates that niche construction, combined with genotype-environment interaction that alters trait heritability, can influence evolutionary trajectories. The model follows the example of seed dispersal discussed above, and examines the simplest case of one niche-constructing trait influencing itself. It focuses exclusively on the interactions between phenotypic plasticity and environment-dependent trait heritability. Selection on the trait is assumed to be constant, as is likely to be the case in breeding programs. I assume purely directional selection in which a fixed proportion of the population sample is retained for breeding purposes, and in which the intensity of selection is measured in terms of a standardized trait (that is, the change in the mean population before and after selection is measured in standard deviation units). This assumption simplifies the model, allowing the focus to remain on the interaction between phenotypic plasticity and genotype-environment interaction. Indeed, this assumption greatly oversimplifies the dynamics if the conclusions were to be applied to natural populations, in which natural selection itself can be expected to vary with changes in trait means and variances, as discussed below.

The breeder's equation of quantitative genetics describes the evolutionary response to selection, R, as the change in phenotype from one generation to the next, as a function of the strength of selection on the trait, S, and the heritability of the trait h^2 [which is the proportion of phenotypic variance of the trait that is (additive) genetically based variance, V_A/V_P].

$$R = h^2 S \tag{1}$$

In this simplest, univariate case, let the trait of interest, z, be the niche-constructing trait, which directly predicts the environment experienced by the organism. As such, the trait, z, can itself be a measurement of the environment that the organism experiences. For example, if a heritable attribute of the organism determines the conspecific density, the quality of light, or the temperature that the organism experiences, then the environment experienced by the organism can be said to have a heritable component, and the trait of interest can be a measurement of the density, light quality, or temperature that the organism experiences. All other components of that environmental factor are assumed to be random (unpredictable), and therefore do not contribute to the directional change in the environmental factor experienced by the organism.

From this point forward, R refers to the response to selection that would be predicted without any change to the environment due to niche construction. In contrast, $\Delta\bar{z}$ refers to the change in mean phenotype from one generation to the next, $z' - z$, when niche construction also occurs.

The environment determined by the trait can influence the heritability of that trait. The new environment created by niche construction can alter the environmental variance (V_E, where $V_A + V_E = V_P$) or the genetic variance (V_A). Changes in V_A occur because different genotypes can respond differently to the altered environment, leading to genotype-environment interactions and environment-dependent genetic variation. With changes in either V_P or V_A, the heritability itself can change. Let ω measure the effect of the new phenotype on the heritability of that phenotype, such that the heritability is altered by a factor proportional to the mean phenotype, z. When niche construction alters heritability, the response to selection is:

$$\Delta\bar{z} = (h^2 + \omega z h^2)S = R + \omega z R \tag{2}$$

The response to selection, $\Delta\bar{z}$, will be faster than R when $\omega > 0$, and it will be slower when $\omega < 0$, provided $z > 0$. The response will equal R when $\omega = 0$. There will be no evolutionary response when $\omega = -1/z$. The phenotype after selection and inheritance is:

$$z_1' = R + z(1 + \omega R) \tag{3}$$

It should be noted that the term $(h^2 + \omega z h^2)$ in equation (2) is not strictly the new heritability of the trait, but rather the predicted regression between parents and offspring. The anomaly of the revised "heritability" being able to exceed unity or even become negative can be understood in terms of genetic correlations across parent and offspring environments (Via and Lande 1985; Fry 1992; Windig 1997). Consider the extreme case when a trait is already perfectly heritable ($h^2 = 1$). If genetic variances are higher in the offspring environment than in the parental environment, the regression between parents and offspring can actually exceed one (Fig. 2), and if the genetic variances are lower in the offspring environment then the regression would become less than one. In an extreme case, if the genetic correlations across parent and offspring environments are negative, then the regression between parents and offspring can actually be negative. While this may seem unlikely, we actually have no empirical data on these dynamics. For the purposes of this model, I assume the simplest function of altered parent-offspring regressions: a linear change of heritability with a unit change in the environment that is given by the mean phenotype z. While other functions may be possible, we as yet have no empirical basis for preferring a more complex function. I will also restrict consideration to the more plausible parameter values of positive parent-offspring regressions within the range of 0 and 1.

When niche construction alters the environment that the organism experiences, the phenotype expressed by the organism may also change because of phenotypic plasticity. Let α measure the proportional change in the mean phenotype of the organism caused by phenotypic plasticity in response to the environment created by niche construction. Again, we have very little empirical data on how phenotypes change with unit changes in an evolving environment, so I assume the simplest function: a linear change of phenotype with a unit change of the environment that is given by the phenotype mean z. This is also in accordance with models of maternal effects (with "m" describing a similar proportional change in offspring

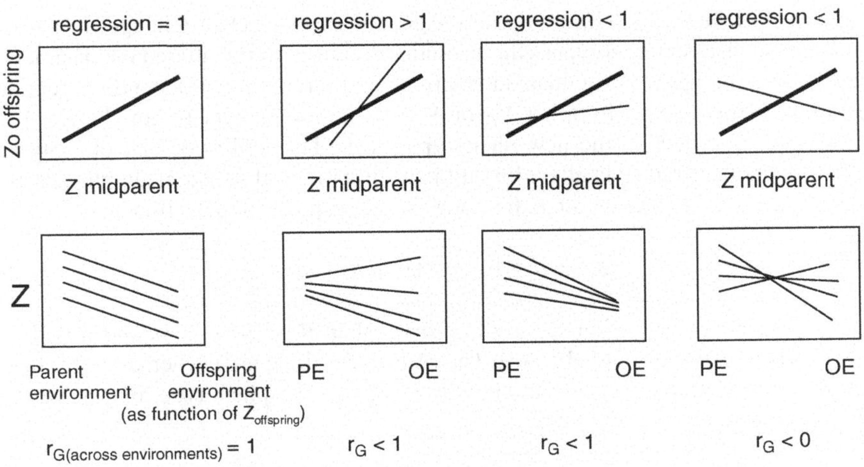

Fig. 2 Environment-dependent genetic expression. *Upper panels* show parent-offspring regressions. *Lower panels* show the reaction norms of a set of four genotypes, with the lines connecting the mean phenotype of each genotype as it is expressed in the parent (*left*: PE) and offspring (*right*: OE) environments. r_g is the genetic correlation across parent and offspring environments. The left-most panel shows an extreme case of a perfect heritability, with the parent-offspring regression equal to one. The second panel shows a case when genotype-environment interaction causes increased genetic variance in the offspring environment. The third panel shows a case in which genotype-environment interaction causes decreased genetic variance in the offspring environment. The fourth panel shows the case when genotype-environment interaction does not change genetic variance in the offspring environment, but the genetic correlation across parent and offspring environment is negative

phenotype as a function of the maternal phenotype, or more specifically the non-genetic phenotypic correlation between mothers and offspring; Falconer 1965; Kirkpatrick and Lande 1989) and indirect genetic effects (with Ψ describing the proportional plasticity of the focal trait in response to the evolving social environment; Moore et al. 1997).

Considering phenotypic plasticity to the newly evolved environment, first, selection causes an evolutionary change in the phenotype to $[R + z_1(1 + \omega R)]$, as described above. Second, because the phenotype, z'_1, is also a measure of the environment, an additional change in the phenotype can result because of plasticity to the new environment. This additional change is measured as $\alpha z'_1$, which is $\alpha[R + z(1 + \omega R)]$. Thus the phenotype after selection, inheritance, and plasticity is revised to be:

$$\begin{aligned} z' &= R + z(1 + \omega R) + \alpha[R + z(1 + \omega R)] \\ &= (R + z) + \omega z R + \alpha(R + z + \omega z R) \end{aligned} \quad (4)$$

The first term, $(R + z)$, is the phenotype expected with no effect of the newly constructed environment. The second term, $\omega z R$ is the deviation from that expectation

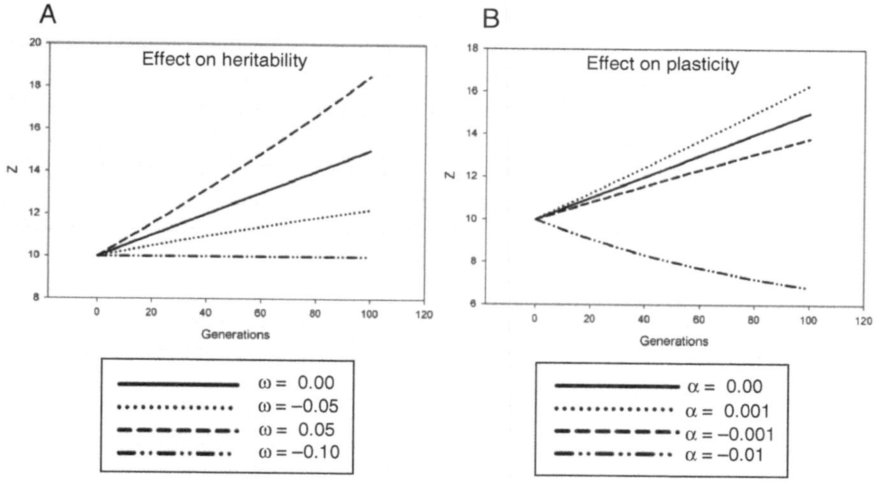

Fig. 3 Response to selection when the constructed environment increases or decreases only the trait heritability (**A**) or only the plasticity of a trait (**B**). The y-axis is the mean phenotype, z, of a trait that is associated with niche construction, and the x-axis is the number of generations of selection. Selection favors an increase in the phenotypic value, z. $h^2 = 0.5$, $s = 0.1$ in both panels. In A, $\alpha = 0$. In B, $\omega = 0$

caused by a change in the heritability of the trait. The third term, $\alpha(R + z + \omega z R)$, is the deviation caused by plasticity of the trait to the newly constructed environment. The total evolutionary response to selection is:

$$\Delta \bar{z} = z' - z = R(1+\alpha) + \omega z R(1+\alpha) + \alpha z \tag{5}$$

Environmental effects on heritability alone: First consider only the effects of the newly constructed environment on the heritability of the trait (Fig. 3a). When the newly constructed environment decreases the expression of genetic variation for the trait, evolutionary responses are slower than when the environment exerts no effect on heritability. In the extreme case of $\omega = -1/z$, no response to selection occurs. Conversely, when the newly constructed environment increases the expression of genetic variation, evolutionary responses to selection are faster. Over long periods of time, the former case causes faster depletion of genetic variation, and eventually inhibits responses to selection altogether (note the non-linear, diminishing increase in z over time). The latter case, in contrast, can cause more sustained responses to selection. Thus niche construction, and the accompanying effects of the environment on trait heritabilities, can either impose constraints on long-term responses to selection, or they can enable more sustained responses to selection.

Effects of phenotypic plasticity with environment-dependent heritability: The newly constructed environment can influence not only the heritability of the trait, but it can also influence the phenotypic expression of that trait through phenotypic plasticity (Fig. 3b). Consider the case when selection favors an increase in a trait.

If an increase in the trait causes a change in the environment that causes a further increase in that trait because of phenotypic plasticity ($\alpha > 0$), then the total response to selection will be greater than the case without niche construction and plasticity to the constructed environment. Conversely, if an increase in the trait causes a change in the environment that causes a decrease in the trait due to phenotypic plasticity ($\alpha < 0$), then the total response to selection will be less than the case without niche construction and plasticity to the constructed environment. If the opposing plasticity is very strong compared to the strength of selection (see equation (11) below) the total change in the phenotype can be in the opposite direction to that favored by selection.

Now consider the effects of the newly constructed environment on both the phenotypic expression and the heritability of the trait (Fig. 4). First, no response to selection will occur when:

$$\alpha = -R(1+\omega z)/[R(1+\omega z)+z] \text{ or when } \omega = -[R(1+\alpha)+\alpha z]/zR(1+\alpha) \quad (6)$$

More significantly, no response to selection will occur when:

$$z = -Sh^2(1+\alpha)/[\omega Sh^2(1+\alpha)+\alpha] \quad (7)$$

which, when $\alpha = 0$, reduces to $z = -1/\omega$, as mentioned above. What this implies is that the phenotype, z, can reach an equilibrium, z^*, with no further change, provided that the phenotype evolves towards z^*, given in equation (7). If it evolves in the opposite direction, then the equilibrium is unstable. The direction and magnitude of evolution depends on the distance from this equilibrium value. In the simple case of $\alpha = 0$ and $z^* = -1/\omega$, it follows that whether the phenotype evolves towards or away from the equilibrium z^* depends on the signs of S and ω, since, from equation (2), $\Delta\bar{z} = h^2 S\omega(z - z^*)$. If $z > z^*$, then z evolves towards the equilibrium only when $S\omega$ is negative, and if $z < z^*$, then z evolves towards the equilibrium only when $S\omega$ is positive. Thus whether evolution proceeds in the direction of the equilibrium also depends on the starting value of z.

Next, if the newly constructed environment simultaneously increases heritability and causes a change in phenotypic expression in the same direction that selection favors (Fig. 4, dotted line), then the total response to selection is faster and more sustained. In contrast, if the newly constructed environment decreases heritability and causes a change in phenotypic expression opposite to the direction favoured by selection (Fig. 4, short-dashed line), then the total response to selection is slower.

It is also possible that the newly constructed environment influences heritability and phenotypic expression in opposite directions. For example, it may increase heritability but cause a change in the phenotype in the direction opposite to that favored by selection (Fig. 4, dash-double dot line); or it may decrease heritability but cause a change in the phenotype in the same direction favored by selection (Fig. 4, long-dashed line). As before, for a given magnitude of α ($\alpha = -0.001$ in Fig. 4), a slower response results when $\omega < 0$, and a faster response results when $\omega > 0$. Likewise for a given value of ω, the response is faster when $\alpha > 0$, and the response is slower

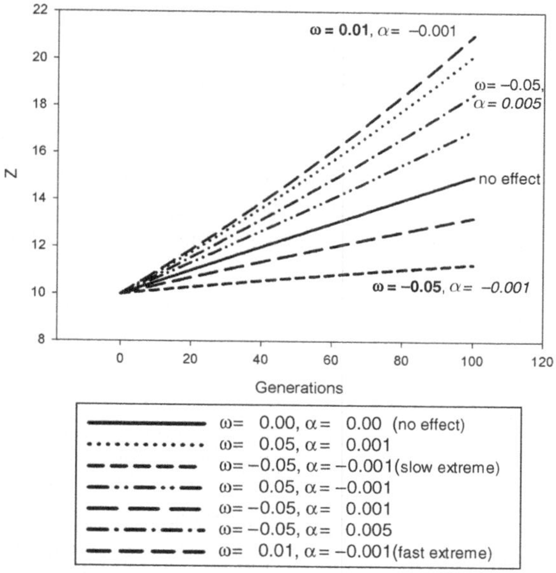

Fig. 4 Response to selection when the constructed environment influences both the heritability and the phenotypic expression of the trait. The y-axis is the mean phenotype, z, of a trait that is associated with niche construction, and the x-axis is the number of generations of selection. Selection favors an increase in the phenotypic value, z. $h^2 = 0.5$, $s = 0.1$ "no effect" refers to the case in which the newly constructed environment created by the trait has no effect on the phenotypic expression or the heritability of the trait (when $\alpha = 0$ and $\omega = 0$). Direct comparisons of the effects of different magnitudes of ω are shown in bold. Direct comparisons of effects of different magnitudes of α are shown in italics.

when $\alpha < 0$. Moreover, the effect of a change in magnitude of ω (from -0.05 to $+0.05$) depends on the value of α. For example in Fig. 4, the phenotype after 100 generations changed by 6.9 units when $\alpha = 0.001$, but with the same change in ω, the phenotype changed by 5.7 units when $\alpha = -0.001$. Likewise, when α changed from -0.001 to $+0.001$, the change in phenotype after 100 generations was 3.2 when $\omega = 0.05$, but it was 2.0 when $\omega = -0.05$. Thus, one value of ω enhanced the effect of changed α, while another value masked that effect, and vice versa. Whether niche construction causes a faster or slower response to selection therefore depends on the relative magnitudes of α and ω. When the two effects oppose one another, the effect on heritability can exactly balance the effect of phenotypic plasticity when, from equation (6):

$$R(1+\alpha) + \omega z R(1+\alpha) + \alpha z = R \qquad (8)$$

which occurs when:

$$\omega = [-\alpha(R+z)]/[zR(1+\alpha)] \qquad (9)$$

or when:

$$\alpha = -\omega z R/(\omega z R + R + z) \quad (10)$$

A negative response to selection can occur when the effect of the constructed environment causes a large plastic change in the direction opposite to that favored by selection, even when the environment also causes an increase in the heritability of the trait (Fig. 5). Specifically, a negative response to selection can result when, from equation (5):

$$R(1+\alpha) + \omega z R(1+\alpha) + \alpha z < 0 \quad (11)$$

which occurs when:

$$\alpha < -R(1+\omega z)/(R + z + \omega z R) \quad (12)$$

The magnitude of the negative response to selection also depends on whether the environment alters heritability (Fig. 5a). If it increases heritability, then the negative response is slower, whereas if it decreases heritability, the negative response is faster than when the environment exerts no influence on heritability.

The magnitude of the negative response also depends on the strength of selection on the trait (Fig. 5b). Specifically, selection on the trait will override plasticity in the opposite direction when:

$$S > -\alpha z/\left\{h^2[(1+\alpha) + \omega z(1+\alpha)]\right\} \quad (13)$$

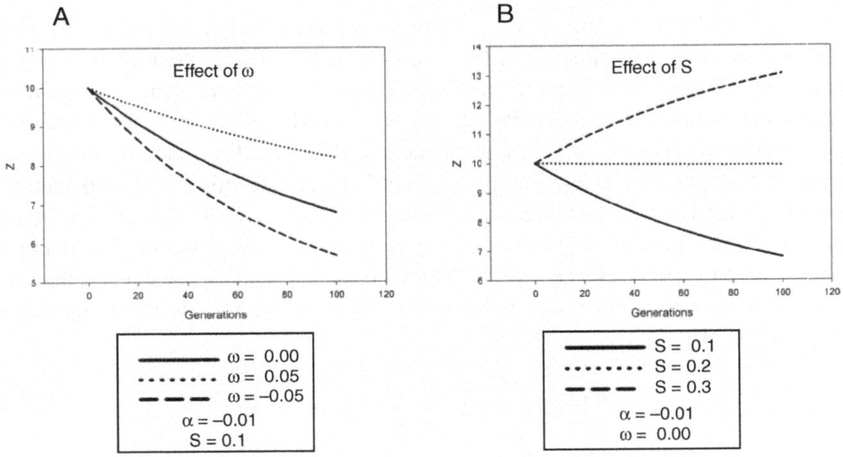

Fig. 5 Cases in which niche construction causes a negative response to selection. *Panel A* shows effects of different magnitudes of ω. $\alpha = -0.01$, $S = 0.1$, $h^2 = 0.5$. *Panel B* shows effects of different strengths of selection (*S*) favoring an increase in z. $\alpha = -0.01$, $\omega = 0$, $h^2 = 0.5$

In summary, when the newly evolved trait is associated with a change in the environment experienced by the organism—that is, when niche construction occurs—that new environment can influence both phenotypic expression, via phenotypic plasticity, and the heritability of the trait under selection. Whether these dynamics accompanying niche construction accelerate, retard, or reverse evolutionary responses to selection depends on the relative magnitudes and directions of the effect of the newly constructed environment on plasticity and the expression of genetic variation. Importantly, environment-dependent heritability can actually override the effects of "negative plasticity", or plasticity in the opposite direction from that favored by selection. Thus genotype-environment interaction, which results in environment-dependence of heritability and other genetic parameters, is important to consider when trait evolution causes a change in the environment.

6 Discussion

This univariate model shows that the dynamics of niche construction can influence evolutionary responses to selection. They can accelerate, retard, or reverse responses to selection. They can also impose genetic constraints on responses to selection, or they can enable more sustained responses to selection.

Accelerated, retarded, or reversed responses to selection can occur through the effects of the environment on phenotypic expression alone. These dynamics have been demonstrated in previous literature on maternal effects, in which the magnitude of the maternal effect measures the degree of correlation between maternal and offspring phenotype that is not due to genetic inheritance but rather due to phenotypic plasticity (Kirkpatrick and Lande 1989; Wade 1998). Similar dynamics have also been reported more generally in models of "indirect genetic effects", in which the social environment (genetic composition of groups of interacting individuals) evolves over time and also influences plastic social behaviour (Moore et al. 1997; Wolf 2003; Wolf et al. 1998, 1999, 2004). The basic dynamic occurs when selection causes a response in a trait mean, and when the trait causes a plastic response in that same trait or a different trait. The results presented here recapitulate those dynamics.

Also included in the model presented here are effects of the environment on the expression of trait heritabilities. In the case of environmental effects on trait heritabilities alone, the environment alters rates of evolutionary change. What is particularly interesting is that environmental effects on trait heritability can impose a severe constraint on evolutionary responses if an evolutionary change in the mean trait causes an opposite effect on the heritability of the trait, as for example, when an adaptive increase in the trait causes lower heritability of the trait (Donohue 2005; Donohue et al. 2005d). It should be emphasized that this reduction in heritability is due not to the selective elimination of maladaptive genotypes causing a reduction in genetic variance. Instead, it is caused by genotype-environment interaction whereby even the exact same sample of genotypes can be less phenotypically distinct (exhibit less genetic variance) in the new environment.

Conversely, an evolutionary increase in the population mean trait may increase heritability of that trait. In this case, more sustained responses to selection may be possible, even when maladaptive genotypes are being eliminated from the population and the number of distinct genotypes actually decreases. If the phenotypic differences among the remaining genotypes become more pronounced in the new environment, then the depletion of genetic variation due to the selective elimination of genotypes can be counteracted.

The model presented above excludes some components that would be important especially for predicting evolutionary responses under natural, as opposed to artificial, selection. First, with changes in the phenotype and phenotypic variances, natural selection would also be expected to change as a direct consequence. When the intensity of selection is measured in units of the raw trait as opposed the standardized trait (as may be more relevant for cases of hard selection), a scalar change in phenotype (by magnitude α) would result in a scalar change in directional selection (literally, the difference in the mean of the unstandardized trait before and after selection). The departures are even more apparent when natural selection is stabilizing as opposed to purely directional, since increases or decreases in trait variance due to plasticity (with phenotypic variance scaling as α^2) would directly affect the intensity of stabilizing selection. These effects on selection would be further complicated by additional changes to genetic variances caused by genotype-environment interaction (ω). Greater resolution of the dynamics might also be possible by considering environmental effects on genetic versus environmental variances of traits separately, and this would be especially important for a multivariate analysis, in which phenotypic correlations influence patterns of multivariate selection. Finally, the model here considers only the univariate case. Most examples of niche construction probably occur with multiple interacting traits, as in the example of germination and flowering phenology discussed above. In such cases, environment-dependent covariances among traits are likely to alter the dynamics appreciably. In the case of maternal effects, interactions among correlated traits can cause "cycles" of feedbacks that lead to oscillating responses to selection (Kirkpatrick and Lande 1989), which illustrates the importance of considering trait evolution in multivariate terms. Future investigations of genotype-environment interactions with niche construction will need to incorporate their consequences to natural selection and will need to consider these dynamics within a multivariate context.

When the constructed environment influences both trait expression and trait heritability, the evolutionary trajectories and outcomes depend on the relative magnitudes of the effect of the environment on these two factors. In particular, environmental effects on heritability can counteract or augment effects of phenotypic plasticity. To predict these dynamics requires knowledge of how environments influence both trait expression and the heritability traits, the traits with which organisms influence their exposure to particular environmental factors, and the evolutionary potential of these niche-constructing traits.

Experimental demonstrations of these dynamics are lacking, but departures from expectation in empirical evolutionary studies, including breeding programs, may be due in part to unmeasured changes in the environment accompanying evolutionary change. More explicit studies of the dynamics of niche construction would be useful for predicting responses to selection in both artificial and natural populations. Acknowledging the manner in which all organisms alter their exposure to environmental factors, moreover, can increase the precision of evolutionary predictions and contribute to our understanding of evolutionary dynamics more generally.

Acknowledgments I would like to thank Julius van der Werf and Cedric Gondro for organizing the Symposium on "Fitness and Adaptation", as well as the participants of that symposium for the many interesting conversations and very useful commentary that helped improve this manuscript. I also thank two anonymous reviewers for comments that significantly improved this paper. The research was supported in part by NSF grant # IBN-0235662 to KD.

References

Augspurger, C.K. 1983. Offspring recruitment around tropical trees: changes in cohort distance with time. *Oikos* **40**: 189–196.

Augspurger, C.K. and Kitajima, K. 1992. Experimental studies of seedling recruitment from contrasting seed distributions. *Ecology* **73**: 1270–1284.

Baker, G.A. and Dowd, D.J.O. 1982. Effects of parent plant density on the production of achene types in the annual *Hypochoeris glabra. Journal of Ecology* **70**: 201–215.

Baskin, C.C. and Baskin, J.M. 1998. *Seeds: Ecology, biogeography and evolution of dormancy and germination*, San Diego: Academic Press.

Bazzaz, F.A. 1991. Habitat selection in plants. *The American Naturalist* **137**: S116–S130.

Brown, J.S. 1990. Habitat selection as an evolutionary game. *Evolution* **44**: 732–746.

Burdon, J.J. and Chilvers, G.A. 1975. Epidemiology of damping-off disease (*Pythium irregulare*) in relation to density of *Lepidium sativum* seedlings. *Annals of Applied Biology* **81**: 135–143.

Cheverud, J.M. 2003. Evolution in a genetically heritable social environment. *Proceedings of the National Academy of Sciences* **100**: 4357–4359.

Day, R.L., Laland, K.N. and Odling-Smee, J. 2003. Rethinking adaptation: The niche-construction perspective. *Perspectives in Biology & Medicine* **46**: 80–95.

Donohue, K. 1999. Seed dispersal as a maternally-influenced character: mechanistic basis of maternal effects and selection on maternal characters in an annual plant. *The American Naturalist* **154**: 674–689.

Donohue, K. 2002. Germination timing influences natural selection on life-history characters in *Arabidopsis thaliana. Ecology* **83**: 1006–1016.

Donohue, K. 2003. Setting the stage: Plasticity as habitat selection. *International Journal of Plant Sciences* **164**: S79–S92.

Donohue, K. 2005. Niche construction through phenological plasticity: Life history dynamics and ecological consequences. *New Phytologist* **166**: 83–92.

Donohue, K., Dorn, L.A., Griffith, C., Schmitt, J., Kim, E.-S. and Aguilera, A. 2005a. Environmental and genetic influences on the germination of *Arabidopsis thaliana* in the field. *Evolution* **59**: 740–757.

Donohue, K., Dorn, L.A., Griffith, C., Schmitt, J., Kim, E.-S. and Aguilera, A. 2005b. The evolutionary ecology of seed germination of *Arabidopsis thaliana*: Variable natural selection on germination timing. *Evolution* **59**: 758–770.

Donohue, K., Dorn, L.A., Griffith, C., Schmitt, J., Kim, E.-S. and Aguilera, A. 2005c. Niche construction through germination cueing: Life history responses to timing of germination in *Arabidopsis thaliana*. *Evolution* **59**: 771–785.
Donohue, K., Hammond-Pyle, E., Heschel, S.M., Messiqua, A. and Schmitt, J. 2000. Density-dependence and population differentiation of genetic architecture in *Impatiens capensis* in natural environments. *Evolution* **54**: 1969–1981.
Donohue, K., Polisetty, C.R. and Wender, N.J. 2005d. Genetic consequences of niche construction: Plasticity-mediated genetic constraints on the evolution of seed dispersal in *Arabidopsis thaliana*. *The American Naturalist* **165**: 537–550.
Dorn, L., Hammond-Pyle, E. and Schmitt, J. 2000. Plasticity to light cues and resources in *Arabidopsis thaliana*: Testing for adaptive value and costs. *Evolution* **54**: 1982–1994.
Falconer, D.S. 1965 Maternal effects and selection response. In *Genetics Today, Proceedings of the VI International Congress on Genetics* (S.J. Geerts, ed.), pp 763–774.
Fry, J.D. 1992. The mixed-model analysis of variance applied to quantitative genetics: biological meaning of the parameters. *Evolution* **46**: 540–550.
Holt, R.D. 1987. Population dynamics and evolutionary processes: the manifold roles of habitat selection. *Evolutionary Ecology* **1**: 331–347.
Howe, H.F. and Smallwood, J. 1982. Ecology of seed dispersal. *Annual Review of Ecology and Systematics* **13**: 201–228.
Janzen, D.H. 1978. The size of a local peak in a seed shadow. *Biotropica* **10**: 78.
Kirkpatrick, M. and Lande, R. 1989. The evolution of maternal characters. *Evolution* **43**: 485–503.
Koornneef, M., Hanhart, C.J. and van der Veen, J.H. 1991. A genetic and physiological analysis of late flowering mutants in *Arabidopsis thaliana. Molecular and General Genetics* **229**: 57–66.
Laland, K.N., Odling-Smee, F.J. and Feldman, M.W. 1999. Evolutionary consequences of niche construction and their implications for ecology. *Proceedings of the National Academy of Sciences of the United States of America* **96**: 10242–10247.
Levins, R. 1968. *Evolution in changing environments*, Princeton: Princeton University Press.
Mazer, S.J. and Wolfe, L.M. 1992. Planting density influences the expression of genetic variation in seed mass in wild radish (*Raphanus sativus* L.: Brassicaceae). *American Journal of Botany* **79**: 1185–1193.
Moore, A.J., Brodie, E.D.I. and Wolf, J.B. 1997. Interacting phenotypes and the evolutionary process: I. Direct and indirect genetic effects of social interactions. *Evolution* **51**: 1352–1362.
Mouradov, A., Cremer, F. and Coupland, G. 2002. Control of flowering time: Interacting pathways as a basis for diversity. *Plant Cell* **14**: S111–S130.
Munir, J., Dorn, L., Donohue, K. and Schmitt, J. 2001. The influence of maternal photoperiod on germination requirements in *Arabidopsis thaliana. American Journal of Botany* **88**: 1240–1249.
Nordborg, M. and Bergelson, J. 1999. The effect of seed and rosette cold treatment on germination and flowering time in some *Arabidopsis thaliana* (Brassicaceae) ecotypes. *American Journal of Botany* **86**: 470–475.
Odling-Smee, F.J., Laland, K.N. and Feldman, M.W. 1996. Niche construction. *American Naturalist* **147**: 641–648.
Odling-Smee, F.J., Laland, K.N. and Feldman, M.W. 2003. *Niche construction: The neglected process in evolution*, Princeton: Princeton University Press.
Rees, M. and Brown, V.K. 1991. The effect of established plants on recruitment in the annual forb *Sinapsis arvensis. Oecologia* **87**: 58–62.
Rosenzweig, M.L. 1987. Habitat selection as a source of biological diversity. *Evolutionary Ecology* **1**: 315–330.
Scheiner, S.M. 1993. Genetics and evolution of phenotypic plasticity. *Annual Review of Ecology and Systematics* **24**: 35–68.
Schlichting, C.D. 1986. The evolution of phenotypic plasticity in plants. *Annual Review of Ecology and Systematics* **17**: 667–693.
Schlichting, C.D. and Pigliucci, M. 1998. *Phenotypic evolution: A reaction norm perspective*, Sunderland, MA: Sinauer Associates, Inc.

Schwilk, D.W. 2003. Flammability is a niche construction trait: Canopy architecture affects fire intensity. *American Naturalist* **162**: 725–733.

Simpson, G.G. and Dean, C. 2002. *Arabidopsis*, the Rosetta stone of flowering time? *Science* **296**: 285–289.

van Tienderen, P.H. 1991. Evolution of generalists and specialists in spatially heterogenous environments. *Evolution* **45**: 1317–1331.

Via, S. and Lande, R. 1985. Genotype-environment interaction and the evolution of phenotypic plasticity. *Evolution* **39**: 505–522.

Via, S. and Lande, R. 1987. Evolution of genetic variability in a spatially heterogeneous environment: Effects of genotype-environment interaction. *Genetical Research* **49**: 147–156.

Wade, M.J. 1998. The evolutionary genetics of maternal effects. In *Maternal Effects as Adaptations* (T. Mousseau and C. Fox, eds.), Oxford University Press, Oxford, pp 5–21.

Weinig, C. 2000. Differing selection in alternative competitive environments: Shade-avoidance responses and germination timing. *Evolution* **54**: 124–136.

Wender, N.J., Polisetty, C.R. and Donohue, K. 2005. Density-dependent processes influencing the evolutionary dynamics of dispersal: A functional analysis of seed dispersal in *Arabidopsis thaliana* (Brassicaceae). *American Journal of Botany* **92**: 88–98.

Whitlock, M.C. 1996. The red queen beats the jack -of-all-trades: The limitations on the evolution of phenotypic plasticity and niche breadth. *The American Naturalist* **148**: S65–S77.

Willson, M.F. and Traveset, A. 2000 The ecology of seed dispersal. In *Seeds: The Ecology of Regeneration in Plant Communities* (M. Fenner ed.), CABI, New York, pp 85–110.

Windig, J.J. 1997. The calculation and significance testing of genetic correlations across environments. *Journal of Evolutionary Biology* **10**: 853–874.

Wolf, J.B. 2003. Genetic architecture and evolutionary constraint when the environment contains genes. *Proceedings of the National Academy of Sciences of the United States of America* **100**: 4655–4660.

Wolf, J.B., Brodie, E.D.I., Cheverud, J.M., Moore, A.J. and Wade, M.J. 1998. Evolutionary consequences of indirect genetic effects. *Trends in Ecology and Evolution* **13**: 64–69.

Wolf, J.B., Brodie, E. III, and Moore, A.J. 1999. Interacting phenotypes and the evolutionary process. II. Selection resulting from social interactions. *The American Naturalist* **153**: 254–266.

Wolf, J.B., Wade, M.J. and Brodie, E.D.I. 2004. The genotype-environment interaction and evolution when the environment contains genes. In *Phenotypic Plasticity. Functional and Conceptual Approaches* (T.J. DeWitt and S.M. Scheiner, eds.), Oxford University Press, Oxford.

Genotype by Environment Interaction in Farm Animals

John W. James

Abstract Interactions between genotype and environment are discussed in relation to genetic improvement of farm animals. Different types of classification are considered, with emphasis on their implications. A variety of statistical methods which can be used for the analysis of interactions between genotype and environment are discussed. The use of correlated response theory to predict responses in the selection environment and in other environments is described. Some examples are given in illustration, but no attempt has been made at an exhaustive compilation of experimental data. The comparison of different strategies such as development of a separate strain for each environment and improvement of a single strain for average performance over all environments is summarised. The effect of selection on sensitivity to environmental change is discussed. Adaptation to, or fitness in, specific environments is often an important part of such interactions.

Keywords Genotype-environment interaction · scale · joint-regression · reaction norm · correlated response · selection strategy

1 Introduction

In the broadest sense, interaction between genotype and environment is ubiquitous, being at the base of adaptation of different species to their environments. However, this is generally taken for granted, and genotype by environment interaction is regarded as referring to a restricted set of genotypes and environments. Genotype differences are limited to those within a species, and environments do not extend beyond those which are "normal" for that species. Nevertheless, the possible (probable?) existence of such interactions is of vital importance to both plant and animal breeders, since they are seeking to adapt their populations to the relevant production and marketing environments. If genotype by environment interactions are important

J.W. James (✉)
Centre for Advanced Technologies in Animal Genetics and Reproduction (Reprogen), Faculty of Veterinary Science, University of Sydney, NSW 2006, Australia
e-mail: jwj@vetsci.usyd.edu.au

J. van der Werf et al. (eds.), *Adaptation and Fitness in Animal Populations*,

causes of the observed variation, it is possible that wrong selection decisions will be made when the interactions are ignored. Here, I shall not attempt to provide a comprehensive survey of experimental observations on the prevalence and nature of genotype by environment interactions in farm animals, but will concentrate on some principles, illustrated by a few examples.

2 Types of Interaction

When genotype by environment interactions (G $\times$ E) occur, they can be classified in various ways. For example, McBride (1958) classified G $\times$ E according to whether the environments were macro-environments or micro-environments, and the genotypes were inter-population or intra-population in nature. Dunlop (1962) made a similar classification, depending on whether differences between genotypes and environments were large or small. These classifications are essentially descriptive, being convenient for splitting observed interactions into classes in order to survey the field. It is not always obvious where to draw the line in classifying genetic or environmental differences as large or small. For example, small genetic differences occurring within a population might be increased by selecting separate lines in two environments, and in time the difference between lines could become large. At some point the small difference would become large, but there is no clear way of saying when this occurs. However, though there is no unequivocal way to make such decisions, this does not make the classification valueless.

A different type of classification, which has practical consequences, is the division of G $\times$ E into scale-type and rank-type interactions. Scale-type interactions are those in which the differences between genotypes change in magnitude, but not in sign, with changes in the environment, or equivalently differences between environments change in magnitude but not in sign with changes in genotype. In a stricter sense, a scale-type interaction is one which can be removed by transformation of the scale of measurement, but many cases in which such a scale transformation has not been identified can usefully be regarded as being of scale type. On the other hand, rank-type interactions are those in which genotype A may be superior to genotype B in environment 1, but the reverse is true in environment 2. The importance of this distinction is obvious, since with scale-type interactions a genotype selected as best in one environment will be best in the other environments considered, only the magnitude of the superiority being affected, while with rank-type interactions the best genotype selected in one environment may perform poorly in another. Thus the distinction between these types of G $\times$ E has important implications for the design of breeding programs.

A third classification concerns the question of whether the genotypes and environments should be regarded as fixed or random effects. Genetic effects such as breeds, strains or specific quantitative trait loci, that is, genetic classes which are identifiable and thus easily controllable and repeatable, can be regarded as fixed effects. Random genetic effects are those such as genetic differences between individuals or sires within a breed or strain. Fixed genetic effects in essence afford the opportunity for a one-time selection, such as the choice of a particular breed to use in a specific production-marketing system, while random genetic effects are those which can form the basis of a continuing breeding program with cumulative genetic improvement. Fixed environments include such factors as locations, feeding levels, presence or absence of specific diseases, while random environments include such factors as years and farms practising nominally similar husbandry systems. These "random" factors may not be strictly random, as successive years in a pastoral environment may not validly be regarded as a random sample from a population of all possible seasonal conditions, but are nevertheless best regarded as random. The reason for this is that the animal population needs to be adapted to the population of environments to which it will be exposed, and it does not make sense to suppose that a Merino sheep breeder, for example, could use a separate flock depending on weather forecasts for the following year, whereas it can reasonably be supposed that a cereal grower could choose to sow a particular variety to suit the expected rainfall pattern for the following season, provided that the rainfall predictions were sufficiently accurate. However, a beef cattle breeder might well decide to choose a particular strain as suited to feedlot production, or choose to breed for improved performance in that specific environment. Essentially, the distinction is between the need to choose genotypes for performance in a specific environment or for average performance over a population of environments. In actuality both types of decision must be made, but with respect to a particular G × E one or the other is relevant.

For a fixed environment, the breeder can direct his attention to the specific controllable factors in a straightforward way, while for random environments he must make sure that performance is evaluated over a sufficiently large sample of the population of environments to give a reliable estimate of average performance over all environments. Therefore these two types of interaction raise different problems for the breeder.

A somewhat different classification refers to whether G × E occurs for specific traits in a particular set of production environments, or whether it refers to overall economic efficiency in a set of production-marketing systems. While there are real differences between these cases, it is sufficient here to note that the consequences for breeding programs will be essentially the same.

3 Falconer's Contribution

An important contribution to our understanding of G×E was made in the early 1950s by Douglas Falconer. This was partly experimental (Falconer and Latyszewski 1952), but mainly conceptual (Falconer 1952). The experimental work was done following some suggestions by Hammond (1947). Hammond claimed that response to selection would be most rapid if conducted in the environment which is most favourable for the expression of the character to be improved, and that improvement obtained under good conditions will be maintained if the improved strain is transferred to a poorer environment. Falconer argued that each case would need to be considered on its merits, since it could not be certain that these predictions would hold in every case.

The experiment was done by selecting mice for weight at six weeks on either *ad libitum* feed or on a diet of identical composition, but restricted to about 75 per cent of normal consumption. As predicted by Hammond, the rate of response in the mice selected on full feeding was greater than in those selected on a restricted diet, because the phenotypic standard deviation was about 1.6 times as high on the full as on the restricted diet. On the other hand the heritability was about 1.5 times as high on the restricted diet, so that the combined effect was a small difference in favour of response on the full diet. Because of sampling variation, these values are not very accurate, but there is no striking advantage for selection on the full diet. However, when mice bred by selection on full feed were given a restricted diet, they grew little if any faster than unselected mice, while mice bred on a restricted diet were almost as heavy on full feed as were mice bred on a full diet. This definitely contradicted Hammond's claim.

Fatness of mice from both lines was compared for mice given the full diet, and it was found that a greater proportion of body weight consisted of fat in the mice selected on the full diet, this difference in fat weight being larger than the difference in total body weight, so that weight of non-fat tissues was greater in the restricted than the full diet lines. Later work showed that selection on *ad libitum* feeding had led to increased appetite, which could not happen on restricted feeding since the amount fed was less than what would be eaten voluntarily. Thus improved growth on a restricted diet was based on more efficient conversion of feed intake to body tissue, while growth on a full diet resulted from a larger food intake, more of which was converted to fat. These results suggested that in livestock improvement selection would be better carried out in "the worst conditions that the improved breed is likely to encounter", but laboratory work could give only general guidance: applications would need to be based on work with farm animals.

In seeking to explain the results of the experiment, Falconer (1952) realised that it made sense to consider performance in two environments as two different traits with appropriate variances and correlations. Unlike traits such as fleece weight and body weight in sheep, these traits could not both be measured on the same individual, but by raising relatives in both environments it would be possible to estimate the genetic correlation between the traits by making use of standard procedures. Since these relatives would have no environmental factors in common, there would be no environmental correlations between the traits, but in other respects the standard quantitative genetic apparatus could be applied. Thus the theory of correlated response could be used to predict the response in one environment from selection in another. After this point had been made it seemed entirely obvious, and Falconer's insight has been the basis of a great deal of quantitative genetic analysis in the past fifty years, though it did not fully account for the mouse experiment results. It may be noted that scale-type interactions would give genetic correlations close to unity, whereas lower genetic correlations would be associated with rank-type interactions.

4 Statistical Approaches to Analysis of G × E Interactions

The standard approach to analysis of G × E is by analysis of variance of a table of mean phenotypes cross-classified by genotype and environment, with the simplest case having only one factor for environments. The model for analysis of the data is of the form

$$E[\bar{Y}_{ij}] = \mu + G_i + E_j + GE_{ij} \quad (1)$$

where μ is the overall mean, G_i is the mean effect over all environments of the *i*th genotype, E_j is the mean effect over all genotypes of the *j*th environment, and GE_{ij} measures the extent to which the mean of genotype *i* in environment *j* departs from additivity of these effects.

The interaction term, as mentioned before, may be caused by a scale effect, which can be removed by transformation, or by other effects such as changes in rank of genotypes from one environment to another. It is clear from this model that the presence or absence of G × E depends on the ranges of genotypes and environments in the data set. Adding or removing a genotype or environment may change the G × E status.

As a simple example, let us consider an example from Havenstein et al. (1994) which gives a comparison between broiler chicken strains from 1957 (a control strain) and 1991 (an industry strain) raised on diets typical of 1957 and 1991 broiler raising. Table 1 shows the mean six-week weights for the four groups.

On the 1957 diet the strains differ by 1266 g, while on the 1991 diet they differ by 1506 g, indicating an appreciable interaction. However, the ratios of 1991 strain to 1957 strain weights are 3.405 on the 1991 diet and 3.403 on the 1957 diet, so that an analysis on the logarithmic scale would give almost exact additivity. So in this particular case the interaction is essentially entirely of the scale type. In the same experiment, twelve-week weight did not show such a clear scale effect, the corresponding ratios being 2.676 and 2.985. It is possible that a different transformation from the logarithm would remove the interaction in this case. Indeed, with a 2 × 2 table of genotypes and environments it should usually be possible to find some transformation which will render the effects additive, though this may be implausible, especially if the genotypes change rank from one environment to the other.

In this example, a more practically relevant check on G × E could be argued to be a comparison of feed conversion ratios at 12 weeks on 1957 rations and 6 weeks

Table 1 Six-week weights (g) of two strains of broiler chickens raised on two diets (Havenstein et al. 1994)

	Strain	
Diet	1957	1991
1957	508	1774
1991	626	2132

Table 2 Feed conversion ratios of two broiler strains in two commercial environments (see text for details). Data from Havenstein et al. (1994)

Strain	Commercial environment	
	1957	1991
1957	4.06	2.52
1991	3.14	2.04

on 1991 rations, these corresponding approximately to commercial environments in 1957 and 1991. Then we would obtain the results shown in Table 2. Again we see that on the original scale there is an interaction, but $4.06/3.14 = 1.29$ and $2.52/2.04 = 1.24$, so that on a logarithmic scale there is little if any interaction. This example has been given to emphasise that it may be relevant to differentiate environments by two or more factors (age and diet here) rather than just a single one.

More generally, with several genotypes and environments involved, it is possible to partition the interaction into removable (by transformation) and non-removable non-additivity. The relevant procedure is to use the method of Tukey (1949). This is illustrated by Snedecor and Cochran (1967, page 331) and enables a check on whether there is residual non-additivity after scale effects have been removed, and also gives a little guidance on the nature of the transformation required to remove scale effects. It would usually be considered good practice to remove scale effects before further analyses of G × E. Such further analysis might involve computation of correlations between performance in different pairs of environments if there are enough genotypes present, following in the footsteps of Falconer.

An alternative approach is based on the work of Yates and Cochran (1938) which was brought into prominence by Finlay and Wilkinson (1963). This is a "joint-regression" analysis, which fits the model

$$E[\bar{Y}_{ij}] = \mu + G_i + (1 + b_i)E_j. \tag{2}$$

where E_j is the quality of environment j, taken as the mean performance of all genotypes.

As can be seen from (2), the interaction term GE_{ij} in (1) has been replaced by $b_i E_j$ suggesting that the interaction is due to linear regressions of performance on environmental quality differing between genotypes. The actual regressions are $(1+b_i)$, this form being used because the average regression coefficient is of necessity unity, and so the absence of G × E implies that all values of b_i are zero. This can be estimated and tested, and the residual interaction variation can also be checked. To do the analysis, the values of E_j are unknown, but are estimated from the mean of all genotypes in the jth environment. When the slope $(1 + b_i)$ for the ith genotype is estimated, the observed genotypic values for genotype i are the dependent variable, but are also included in the calculation of the independent variable, but if there are a large number of groups in the experiment this is of little consequence, and in any case allowance can be made for this dependence (e.g. Freeman 1973). The value of this approach, if an appreciable amount of the G × E variation can be accounted for by the joint-regression component, is that the regression slopes of genotypic performance on environmental mean give estimates of the sensitivity of the different genotypes to environmental quality, those with large slopes being very sensitive, while a zero slope implies that a genotype is unaffected by the range of environments encountered. Of course, there will usually be a considerable quantity of variation of genotypic performances about the regression lines, which is to be expected because the environments will usually differ in a large number of factors, and it would be surprising if a genotype reacted to changes in all of these factors in the same manner. For instance, two genotypes with the same value for the slope could have very different GE_{ij} values, with one genotype's means clustered closely about the regression line while the other genotype's means were scattered widely about the regression.

As an example of this joint-regression approach in livestock, we may consider the work of McGuirk et al. (1978) on crossbreeding of Merino and Border Leicester sheep. Animals of both pure breeds and reciprocal crosses were compared for many traits over five years which are the relevant environments. For three traits, weaning weight, hogget weight and clean fleece weight, McGuirk et al. (1978) calculated regressions of annual means of the purebreds on annual mean performances of the crossbreds. This was done to avoid inclusion of the purebred means in estimates of yearly environmental quality, since only four genotypes were involved, the crossbred means being then an independent measure. The Tukey (1951) method of using variance components allowed for sampling error in the crossbred means. Border Leicester lambs increased their weaning weight by 1.04 kg per kg increase in crossbreds, compared with 0.69 kg/kg in Merinos. The corresponding figures for hogget weight were 1.04 and 0.72, while for fleece weight they were 0.85 and 0.82. These results suggest that for body weight Border Leicesters responded more to improvement in the environment than did Merinos, though only the weaning weight regressions differed significantly, but for fleece weight both responded in the same manner. There seems a possibility that these regressions could be accounted for by scale effects, since if we calculate ratios of overall means of Border Leicester and Merino to crossbred we find that for weaning weight they are 1.09 and 0.79, for hogget weight 1.04 and 0.76 and for clean fleece weight 0.87 and 0.91. It seems possible that on a logarithmic scale the regressions might not have differed between purebreds. This emphasises that the joint-regression approach works differently from the removable non-additivity approach. Although the regressions in this case might have turned out differently on a logarithmic scale, it is still true that on the observed scale, which is after all the economically relevant scale, the Border Leicester and Merino appear to differ in their body weight response to improved conditions.

With Falconer's approach, it is necessary to estimate genetic and phenotypic parameters in each environment and also the genetic correlations between phenotypic values in the two or more environments. This involves using data on relatives in the different environments. In perhaps the simplest case we can consider a number of sires each progeny tested in two environments. We can then estimate the between-sire and within-sire variances in both environments and the covariances between sires in both environments by standard methods. It has sometimes happened that the genetic correlation between performance in two environments has been estimated from the correlation between family means, but to avoid bias between-sire variance and covariance components must be used. With the development of modern computing procedures it is now straightforward to estimate parameters using REML programs, and it is necessary only to have related animals recorded in the environments of interest in order to obtain estimates, though unless the data is of good quality it may be that the estimates are quite inaccurate. Cameron (1993) has discussed some of the issues involved in parameter estimation with more than one environment.

Carrick and van der Werf (2007) have recently analysed Australian Merino field data, defining environments as quintiles on contemporary group means for three traits, and found that extreme quintiles tended to show G × E for body and wool weights, but not for wool quality traits. It could be interesting to try such analyses using contemporary group means in a joint regression analysis.

5 Reaction Norms

In joint-regression analysis, a linear regression of genotype mean performance in each environment on the average performance of all genotypes in that environment is fitted. It is of course possible that a non-linear relation would fit better than a straight line, and this could be accommodated if thought advisable. But a limiting factor in this approach is the point made above, that in general environments differ in several factors, such as temperature, rainfall and disease incidence, and there is no reason to assume that a genotype's reaction to all of these would be the same. In some situations, particularly in experimental work, the environments may be differentiated by controlled variation in only one factor, such as temperature. It is then possible to plot the phenotypic mean of each genotype against the corresponding value of the environmental variable, the resulting function being generally known as a reaction norm. These reaction norms may be of arbitrary form, and G × E is indicated by lack of parallelism of the reaction norms.

Such a reaction norm can be represented in the form

$$E[\bar{Y}_{ij}] = \mu + G_i + f_i(E_j) \tag{3}$$

where the expected mean is partitioned into an overall mean, an average genotype effect and a term dependent on the level of the environmental factor, this term being possibly different for each genotype. Strictly, only the last term is required, since as a general function it can include the first two terms which are used only to emphasise that the G × E depends on different reactions to the environmental factor. The joint regression model (2) can be seen as a special form of (3). The absence of G × E would mean that the last term was the same for each genotype, so that

$$E[\bar{Y}_{ij}] = \mu + G_i + f(E_j) \tag{4}$$

and the lines for the different genotypes would be parallel with heights above the axis determined by the values of G_i.

More generally, we may regard the phenotypic value of a genotype in a particular environment as a general function of the variables (possibly including indicator variables) which characterise the environment. If we denote the values of the variables in the *j*th environment as $\theta_{1j}, \theta_{2j}, \ldots \theta_{nj}$, the performance of the *i*th genotype in the *j*th environment can be represented as a general function $f_i(\theta_{1j}, \theta_{2j}, \ldots \theta_{nj})$. This form is so general as to be of little use, since it could refer to a single individual at a single moment of its existence, with further environmental variables arising from moment to moment, and the performance could be a large set of trait values arranged in a vector.

Nevertheless, this suggests an alternative to the joint-regression analysis. Instead of using the mean of all genotypes as a measure of the quality of the environment and the linear regression of performance on environmental quality, it may be possible to identify a number of factors which differentiate the environments and whose values are known. The other θ_{kj} values in the general function would then be assumed to be the same for all environments in the data set or else randomised across genotypes. The constant factors would then correspond to mean values for the genotypes averaged over the population of randomised factors, and the identified variables would then be used for reaction norm analysis. That is, the model would become

$$E[\bar{Y}_{ij}] = \mu + G_i + f_i(\theta_{1j}, \theta_{2j}, \ldots, \theta_{mj}) \tag{5}$$

where there are m identified environmental variables. Of course the data set would need to be large enough to allow estimation of the various parameters. The forms of the functions f_i would need to be chosen so as to be estimable from the data but also to give a reasonable description of the relationships observed. In practice this would probably be done by trying linear regressions in the first place, with quadratic terms added if the linear terms appeared inadequate. However, except in data sets much larger than likely to be available in farm animals, a detailed exploration of such functions will not be feasible. This can be regarded as an application of random regression methods.

6 Strain Evaluation over Several Environments

When making a decision on which breed and/or strain to use in commercial production, the farmer would like to know how the various types of animals will perform in the system of husbandry, including location, management methods and markets for output, for which they are intended. In fact, this information will very rarely, if ever, be available. The sheer scale of operations required for a thorough evaluation is enormous. For a discussion of some of the questions which arose in the cattle breed study of the US Meat Animal Research Center, see Dickerson (1969). Even with a large amount of commercial experience to draw on in support of experimental work, the question of which breed or strain is best is unlikely to be easily answerable in many cases.

Comparison between genetic groups under one set of conditions will give reliable guidance to relative performance in another environment only if G × E is unimportant for the traits which are of major significance. Many tropical countries have been misled by the high production of improved breeds in temperate conditions and assumed that such high-producing stock will improve productivity in the tropics, only to find that heat, lower feed availability and tropical diseases have rendered the "high-producing" animals virtually useless. Experience has now led to a general awareness of this problem, and such mistakes are now rarer than they once were. But the problem of breed and strain comparisons remains, because this is vital information for industry productivity. If the decision is to be made by a very large company it may be possible to test the most likely breeds or strains on the company's own farms, making sure that the tests are carried out over a sufficiently large sample of random environmental factors, and find the most appropriate genotype by direct evaluation. For a smaller producer, however, the decision will have to be made on publicly available information, which may well be inadequate. It could well be necessary to make use of genotype comparisons made in a single environment, supplemented by information from other sources on the possible occurrence of G × E in traits of importance. A degree of guesswork is unavoidable, though the guessing should be restricted to the minimum necessary.

In making strain comparisons there can be unexpected difficulties. James (1975) gave an example from laying hens involving competition effects. The data came from a paper by Tindell and Craig (1959), who measured social dominance ability and egg production in six strains of hens. Egg production was measured in pure flocks of each strain and in a mixed flock containing birds from all six strains. Table 3 shows some of the results of this study. It is notable that the first three strains, which were more socially dominant, did better in mixed than in pure flocks, while the reverse was true for the less socially dominant last three strains.

In a mixed flock strains 1 and 3 did best, while in pure flocks strains 4 and 1 were best. The effect of competition led to the suppression of production in hens at the bottom of the social ladder, and in mixed flocks these low-status hens were predominantly of strains 4 to 6, while in pure flocks all strains had hens at the bottom of the social order. Thus, although it might seem good experimental practice to make comparisons with all hens intermingled, in fact strain comparisons should be made with replicated pure flocks. In this example competition is an important environmental factor leading to G × E because the strains differed markedly in their

Table 3 Social dominance scores and per cent egg production in pure and mixed flocks of six strains for hens surviving to 47 weeks. Data from Tindell and Craig (1959)

Strain	Dominance	Pure	Mixed
1	63	63	67
2	56	48	56
3	56	60	66
4	46	64	54
5	43	58	55
6	29	58	51

social dominance abilities, but if strains of equal social status had been used the competition effect would have been equally distributed over the strains in mixed flocks, and there would have been no G × E involving flock structure. This illustrates again that the occurrence of G × E depends on the particular genotypes and environments under consideration. Bijma et al. (2007) have modelled social effects in groups of animals.

7 Strain Improvement Across Environments

Much of the importance of G × E in farm animals stems from the problem of making genetic improvement in productivity of commercial livestock. For various reasons, it is common for the environment in studs to be rather different from the conditions in which commercial animals are kept. Another consideration is that much animal breeding is now globalised, with animals selected in one country being used for breeding in other countries. This occurs in the dairy industry, where semen from northern hemisphere bulls is sold in Australia and New Zealand, where production is largely based on pasture rather than concentrates. It would then be expected that G × E is likely, and in fact it is found that the genetic correlation between production in the north and south is significantly less than unity, though still relatively high at about 0.8. For this reason it is necessary to adjust estimated breeding values (EBVs) for overseas bulls to make them relevant to local conditions. Once daughters have been tested locally, these data can be used in conjunction with overseas data to improve EBVs. In the dairy industry the large data sources available make this process relatively uncomplicated, but in some other industries estimation of the relevant parameters is still a problem.

Experience in the dairy industry suggests that a worldwide breeding system can be successful, but a key question remains, namely, what is the best way to operate a breeding program when there is an appreciable amount of interaction between genotype and environment? James (1961), taking the case of two environments and limiting the discussion to a single trait, suggested three possible procedures:

(i) selection in one environment
(ii) separate selection of two strains in two environments
(iii) selection on an index combining performance in both environments.

He developed the appropriate equations for these options and compared their efficiencies under somewhat restrictive assumptions. For example, he assumed that selection intensities and facilities available were the same for all three approaches, whereas it is likely that if two separate strains were developed, each would have a smaller set of resources than if only one population were being improved. Also, the methods necessary for index selection would involve ensuring that cross-environment links were maintained, and this requirement would probably have an effect on the selection program.

If selection is to be conducted in only one environment (A or B) and the resulting animals are to be used in both, the overall gains, defined as $\Delta_A g$ or $\Delta_B g$ according to which environment is used for selection, have the relative value, if the breeding program is otherwise identical,

$$\frac{\Delta_B g}{\Delta_A g} = \frac{h_B}{h_A}\left[r_G + \frac{1 - r_G^2}{wH + r_G}\right] \tag{6}$$

where w is the economic importance of environment A relative to that of B, and H is the ratio of genetic standard deviations in the two environments (σ_{GA}/σ_{GB}) (James, 1961). If $r_G = 1$, this ratio is simply equal to h_B/h_A and selection should be done in the environment with the higher heritability, and the value of wH is irrelevant. On the other hand, if w is extremely large, environment B is of little or no economic importance, and the ratio tends to the value $r_G h_B/h_A$ which is the standard result for indirect selection. Under the assumptions made by James, it was highly probable that index selection would be better than selection in one environment only, and would be better than selection of two separate strains provided $r_G > 0.7$, though these conclusions were based on a limited range of numerical comparisons.

Recently, these problems have been addressed quite thoroughly in a dairy cattle breeding context by Mulder and Bijma (2006) and Mulder et al. (2006). Mulder and Bijma (2006) considered the effects of G × E on the advantages of cooperation between two breeding programs. If there is no G × E, the opportunity to select bulls from among a larger number of candidates will be expected to favour cooperation. This is especially likely to be the case for a small program cooperating with a much larger one. If there are large genetic differences between animals in the two programs there is a bigger advantage to the genetically inferior stock, while there may be no advantage to the better one. When there is G × E, the magnitude of the genetic correlation between the breeding goals of the two programs is important. Even if there is no G × E in individual traits different economic values may lead to imperfectly correlated objectives. In the long term, cooperation is favoured only if this correlation is at least of the order of 0.8 to 0.9 and this criterion was very little affected by heritability and family size unless these were extremely different in the two environments. Even when r_G is quite low (0.4 to 0.5) cooperation could be worthwhile in the short term, but as selection proceeded the populations would diverge to such an extent that one program could not find desirable bulls bred in the other. At this point, cooperation would cease and the two programs would go their separate ways.

Mulder et al. (2006) studied optimisation of dairy breeding programs for two environments in the presence of G × E using strategies which included one or two environments in the breeding objective, used either one or two breeding programs and progeny tested bulls in either a single or both environments. This was an extension of the study by James (1961) with a more detailed analysis and a more careful definition of how the breeding programs were to be run, together with use of more up-to-date genetic evaluation methods (pseudo-BLUP).

Broadly, the conclusions were in line with those of James; for example a single breeding program using progeny testing in both environments gave most overall gain when r_G was greater than 0.61 (cf. $r_G > 0.7$ in James (1961)), while separate breeding programs for improvement in each environment were superior when r_G was lower. The ranking of the breeding strategies was sensitive to selection intensity, with separate breeding programs being preferable when r_G was as high as 0.7 to 0.8 for very intense selection. If one of the environments was much less important than the other, say having only 10 per cent relative importance, the strategy of having two breeding programs was, perhaps unsurprisingly, more difficult to justify. Variation in heritability or progeny numbers had relatively little influence on which strategy was best, nor did minor changes in relative importance of the two environments. These conclusions are based on the criterion of best overall performance, which may conflict with the interests of commercial producers in one environment, since genetic progress in that environment may be less than could be achieved if attention were to be concentrated on that environment. Nevertheless, there may be other advantages to a single breeding program, such as a reduced rate of inbreeding due to larger breeding population size or the ability of the breeding organisation to invest more in a superior testing and selection regime because of the greater returns obtained by selling semen in two markets rather than one. These are matters which need to be dealt with on a case by case basis, but Mulder et al. (2006) have given valuable guidance.

When the environments are best regarded as random, there is no alternative to aiming at breeding for the best average performance over all environments. For the case of a breeder who finds that there is interaction between sire families and farms, the question is one of progeny testing potential sires on a number of farms, with the total number of progeny recorded being divided among a number of test locations. If there is no G×E then all testing can be done on one farm, but as r_G between farms decreases the number of farms should increase. However, testing costs for a given number of progeny would also increase with number of farms and this would also need to be taken into account in a practical selection program. Dickerson (1962) has discussed such problems.

A similar problem arises in the design of sire reference schemes, where progeny tests are done on several test stations or farms. If there is no G × E, then there is a comparatively simple question of design in order to minimise the prediction error variances, as discussed by Miraei Ashtiani and James (1991). When there is interaction, the expression for prediction error variance is rather more complicated (Miraei Ashtiani 1994). If there are t tests, each with q reference sires having m progeny each and p test sires with n progeny each, and letting $k = (1 - \frac{1}{4}h^2)/h^2$, the average prediction error variance for differences between estimated breeding values for overall performance of test sires can be written as

$$\frac{2V_{\bar{G}}}{n+k}\left[k + n(1-r_G) + n^2 r_G \left(\frac{t(p-1)}{tp-1}\right)\left(\frac{k+m(1-r_G)}{qm(n+k)+pn[k+m(1-r_G)]}\right)\right] \tag{7}$$

which reduces to the value in Miraei Ashtiani and James (1991) when $r_G = 1$. Provided that r_G is not too far below unity, this expression should give results close to those for no interaction, while if the correlation is much lower it is questionable whether such a testing program is desirable.

8 Discussion

It has often been stressed since the pioneering discussion of Falconer (1952) that in considering the implications of G × E for livestock improvement each case needs to be considered on its merits, since there is no universal rule. Despite this, there have been some patterns which have emerged. For example, Dunlop and Young (1966) reviewed sire by environment interactions in clean wool weight of Australian Merino sheep run in five locations. Previous work such as that of Hull and Gowe (1962) had suggested that G × E was more likely when the environmental differences were large, and on this basis Dunlop and Young predicted that interactions of sire with station would be more important than those with year and drop. This expectation was borne out by the data, with the average ratio of the ram by environmental factor variance to ram variance being 0.527 for station, 0.065 for year and 0.014 for drop. This would give an average genetic correlation between performance at different stations as $1/1.527 = 0.65$ which is close to the lower limit at which single strain selection seems to be desirable. Of course, this is only for a single trait, and the total economic value would need to be considered before recommending any specific strategy.

Woolaston (1987) reviewed information on G × E in Australian sheep, and found that while there was clear evidence on interaction in some cases, there was also a lack of data on which to base decisions in relation to many traits, genotypes and environments. However, he did identify environmental factors of great importance in choosing between strains. Firstly, there are strains which are unsuited to environments favourable to the development of fleece rot and fly strike, since G × E has been found to be very important for these traits (Dunlop and Hayman 1958). This is, of course, even more important with recent increased focus on animal welfare. Secondly, the relative importance of fleece weight, fibre diameter and sale stock in profitability can be vital in determining which strain should be used in a particular environment. Thirdly, strains with high fecundity are suited only for environments which can support high lamb survival and good milk production.

The case of fleece rot has some similarities to mouse growth rate on *ad libitum* and restricted feeding, In each case there is genetic variation in a factor of importance in one environment but this factor is entirely missing in the other environment, so that this genetic variation is unexpressed, and cannot provide response to selection. In one case, the factor is susceptibility to wetting of the fleece, which does not arise if the requisite degree of wetting does not occur, while in the other it is appetite, which has no variation if all mice are fed at a level at which all mice consume all food offerred. A similar situation arises in disease resistance, since in environments where the disease is absent no variation in susceptibility will be found. It is perhaps to be expected that environmental differences which suppress the expression of variation will very often be sources of G × E, since if there is a genetic component to such variation we would expect interaction to be present.

When discussing reaction norms we noted that some genotypes changed their performance more than others as the environment was altered. It is of interest to ask what is expected to happen to reaction norms when a population is subjected to selection for performance in a particular environment. Jinks and Connolly (1973) stated that sensitivity to change in the environment is reduced by selection for high performance in a bad environment, and by selection for poor performance in a good environment. This can be analysed by treating performance in the two environments as correlated traits and using results for direct and correlated response. Thus suppose selection is for high performance in a bad environment. The direct response to selection will be

$$R_{+B} = ih_B^2\sigma_{PB} \tag{8}$$

while the correlated response in a good environment will be

$$CR_{+G} = ir_G h_B h_G \sigma_{PG} \tag{9}$$

in an obvious notation. The difference between performance in good and bad environments will therefore change by $CR_{+G} - R_{+B}$, and this equals

$$ih_B(r_G\sigma_{GG} - \sigma_{GB}) \tag{10}$$

which may be positive or negative depending on the relative values of the genetic standard deviations in the two environments. There is a similar result for selection for low performance. If the data are scaled so that genetic standard deviations are equal in the two environments, then any genetic correlation below unity will lead to a reduction in sensitivity. However, on the original scale, the rule may or may not be valid. Falconer (1990) reviewed a number of experiments involving *synergistic* and *antagonistic* selection, where these terms were defined so that synergistic selection was selection for high performance in a good environment or low performance in a bad environment, while antagonistic selection was for high performance in a bad environment or low performance in a good environment. These experiments involved fungi, plants, weevils, mice and pigs. Out of 42 cases, 30 agreed with the Jinks-Connolly rule and 12 did not. This is certainly evidence that the rule is correct more often than not, but it also agrees with the theory that it will not always be true. Falconer also considered a modified version of the rule, namely that sensitivity would be less after antagonistic than after synergistic selection, and found that this was true in all 21 cases. Analysis along the lines of equations (8) to (10) shows that this rule will be correct if

$$(h_B - r_G h_G)\sigma_{GB} + (h_G - r_G h_B)\sigma_{GG} > 0 \tag{11}$$

and while this condition is not necessarily true, it is likely to be true more often than the condition in equation (10). This does therefore appear to be a useful generalisation, even if not universal.

It has often been suggested, following the work of Falconer and Latyszewski (1952) and Fowler and Ensminger (1960) that average response in two environments might be greater under antagonistic selection than under synergistic selection, and Falconer (1990) used the above data to check this prediction. Of the 21 comparisons, 14 agreed with the prediction, 5 disagreed and in 2 cases the mean responses were equal. Although the rate of agreement is significantly greater than 0.5, the experimental support for this idea is not very strong. Thus there seems to be little generality in the nature of G × E, and indeed, every case must be treated on its merits.

References

Bijma, P., Muir, W.M. and van Arendonk, J.A.M. 2007. Multilevel selection 1: quantitative genetics of inheritance and response to selection. *Genetics* **175**: 277–288.

Cameron, N.D. 1993. Methodologies for estimation of genotype with environment interaction. *Livestock Production Science* **35**: 237–249.

Carrick, M.J. and van der Werf, J.H.J. 2007. Sire by environment interaction in sheep may re-rank sires for some traits. *Proceedings of the Association for the Advancement of Animal Breeding and Genetics* **17**: 248–251.

Dickerson, G.E. 1962. Implications of genetic environmental interaction in animal breeding. *Animal Production* **4**: 47–64.

Dickerson, G.E. 1969. Experimental approaches in utilising breed resources. *Animal Breeding Abstracts*, **37**: 191–202.

Dunlop, A.A. 1962. Interactions between heredity and environment in the Australian Merino. 1. Strain X location interactions in wool traits. *Australian Journal of Agricultural Research* **13**: 503–531.

Dunlop, A.A and Hayman, R.H. 1958. Differences among Merino strains in resistance to fleece rot *Australian Journal of Agricultural Research* **9**: 260–266.

Dunlop, A.A. and Young, S.S.Y. 1966. Interactions between heredity and environment in the Australian Merino. III. Sire × environment interaction. *Australian Journal of Agricultural Research* **17**: 227–235.

Falconer, D.S. 1952. The problem of environment and selection. *American Naturalist* **86**: 293–298.

Falconer, D.S. 1990. Selection in different environments: effects on sensitivity (reaction norm) and on mean performance. *Genetical Research* **56**: 57–70.

Falconer, D.S. and Latyszewski, M. 1952. The environment in relation to selection for size in mice. *Journal of Genetics* **51**: 57–80.

Finlay, K.W. and Wilkinson, G.N. 1963. The analysis of adaptation in a plant breeding programme. *Australian Journal of Agricultural Research* **14**: 742–754.

Fowler, S.H. and Ensminger, M.E. 1960. Interaction between genotype and plane of nutrition in selection for rate of gain in swine. *Journal of Animal Science* **19**: 434–449.

Freeman, G.H. 1973. Statistical methods for the analysis of genotype-environment interaction. *Heredity* **31**: 339–354.

Hammond, J. 1947. Animal breeding in relation to nutrition and environmental conditions. *Biological Reviews* **22**: 195–213.

Havenstein, G.B., Ferket, P.R., Scheideler, S.E. and Larson, B.T. 1994. Growth, livability and feed conversion of 1991 versus 1957 type broilers when fed "typical" 1957 and 1991 broiler diets. *Poultry Science* **73**: 1785–1794.

Hull, P. and Gowe, R.S. 1962. The importance of interactions detected between genotype and environmental factors for characters of economic significance in poultry. *Genetics* **47**: 143–159.

James, J.W. 1961. Selection in two environments. *Heredity* **16**: 145–152.

James, J.W. 1975. Genetical considerations in large field experiments. In *Developments in field experiment design and analysis* (V.J. Bofinger and J.L. Wheeler, eds.), Commonwealth Agricultural Bureaux, Farnham Royal, Slough UK, pp 155–167.

Jinks, J.L. and Connolly, V. 1973. Selection for specific and general response to environmental differences. *Heredity* **30**: 33–40.

McBride, G. 1958. The environment and animal breeding problems. *Animal Breeding Abstracts* **26**: 349–358.

McGuirk, B.J., Bourke, M.E. and Manwaring, J.M. 1978. Hybrid vigour and lamb production. 2. Effects on survival and growth of first-cross lambs, and on wool and body measurements of hogget ewes. *Australian Journal of Experimental Agriculture and Animal Husbandry* **18**: 753–763.

Miraei Ashtiani, S.R. 1994. PhD Thesis. University of New South Wales, Kensington, Australia.

Miraei Ashtiani, S.R. and James, J.W. 1991. Efficient use of link rams in Merino sire reference schemes.*Proceedings of the Australian Association for Animal Breeding and Genetics* **9**: 388–391.

Mulder, H.A. and Bijma, P. 2006. Benefits of cooperation between breeding programs in the presence of genotype by environment interactions. *Journal of Dairy Science* **89**: 1727–1739.

Mulder, H.A., Veerkamp, R.F., Ducro, B.J., van Arendonk, J.A.M. and Bijma, P. 2006. Optimization of dairy cattle breeding programs for different environments with genotype by environment interaction. *Journal of Dairy Science* **89**: 1740–1752.

Snedecor, G.W. and Cochran, W.G. 1967. *Statistical Methods* (sixth edition), The Iowa State University Press, Ames, Iowa, USA.

Tindell, D. and Craig, J.V. 1959. Effects of social competition on laying house performance in the chicken. *Poultry Science* **38**: 95–105.

Tukey, J.W. 1949. One degree of freedom for non-additivity. *Biometrics* **5**: 232–242.

Tukey, J.W. 1951. Components in regression. *Biometrics* **7**: 33–69.

Woolaston, R.R. 1987. Genotype × environment interactions and their possible impact on breeding programs. In *Merino Improvement Programs in Australia* (B.J. McGuirk, ed.), Australian Wool Corporation, Melbourne, Australia, pp 421–435.

Yates, F. and Cochran, W.G. 1938. The analysis of groups of experiments. *Journal of Agricultural Science* **28**: 556–580.

Drosophila and Selection in Nature: From Laboratory Fitness Components to Field Assessments

Ary A. Hoffmann

Abstract Studies on *Drosophila* model organisms are producing a wealth of information on the genetic basis of complex traits and on patterns of genetic divergence in specific genes. This emerging information is interpreted within an evolutionary framework but direct connections with fitness in nature are usually tenuous. *Drosophila* fitness under natural conditions can be difficult to define partly because of the small size of flies and immature stages, the tendency of flies to use ephemeral resources, and the interaction between *Drosophila*, yeasts and plant tissue. Here I briefly examine new ways in which some aspects of fitness have been measured in *Drosophila* under semi-natural and natural conditions. I highlight some of the insights that can be gained from combining genomic and other – omic assessments when studies are interpreted within a fitness framework. These types of studies can ultimately provide insights into the way selection produces evolutionary diversity within *Drosophila* and more generally.

Keywords Drosophila · diversity · fitness measurements

1 Introduction

Drosophila species continue to be widely used as model systems for understanding the connections between genetic and phenotypic variation. Variation in many complex traits, such as aggressive behaviour, olfaction, stress resistance, mating behaviours (Dierick and Greenspan 2006; Drapeau et al. 2006; Gerber and Stocker 2007; Orengo and Aguade 2007; Pool and Aquadro 2007; Van Naters and Carlson 2007) is being dissected genetically at increasingly detailed levels. Comparative genomic information is becoming available at both the within and between species levels. Parts of the genome under selection are being defined from patterns of genetic disequilibrium across single nucleotide polymorphisms and other genetic

A.A. Hoffmann (✉)
CESAR – Centre for Environmental Stress and Adaptation Research,
Department of Genetics, University of Melbourne Vic 3010, Australia
e-mail: ary@unimelb.edu.au

J. van der Werf et al. (eds.), *Adaptation and Fitness in Animal Populations*,

markers. Variation in traits is being increasingly linked to genes through characterising effects on transcription and metabolism (Sinclair et al. 2007; Sorensen et al. 2007; Yang et al. 2007).

Although genetic and phenotypic variation is invariably interpreted within an evolutionary framework, there is very limited information available on natural selection acting on the traits being used in laboratory assays and on their underlying genes. A key assumption in many studies is that the phenotypes being investigated have direct links with fitness. For instance, thermal resistance in *Drosophila* is studied because there is an interest in how organisms adapt to different climates; yet the many types of traits used to define thermal resistance are rarely related to conditions experienced in the field (Hoffmann et al. 2003). There is almost no information on the selection gradients for thermal resistance traits under field conditions (Kristensen et al. 2007a). This criticism also applies to selection on starvation resistance which has been widely linked to longevity and other life history traits (Rion and Kawecki 2007). While *Drosophila* is a good model system for studying mating and sexual interactions in the laboratory, there is very little information on selection for the relevant traits in the field (Hine et al. 2004). Variation in evo-devo genes is studied to understand selection on morphological traits leading to evolutionary divergence among species (Prud'homme et al. 2007), but the way morphological traits influence field fitness has been considered in only a few cases (Hine et al. 2004).

When I am faced with yet another clever study on the genetic dissection of a complex trait in *Drosophila*, I am left pondering the significance of the findings. Can a line with a very high level of heat resistance in the laboratory really perform better under hot temperatures in nature? Are the alleles that increase remating frequency in laboratory lines or populations likely to be carried by rapidly remating flies in the field? Do the large phenotypic effects of alleles in the laboratory really translate into large selection coefficients in the field? Do pleiotropic effects of alleles on different traits really influence evolutionary trajectories in natural populations? How easily can we move from making statements about areas of the genome that have undergone selective sweeps to understanding the natural selection on that region responsible for that sweep?

These issues are relevant to animal breeding as well as evolutionary questions. Frankham (this volume) has already considered connections between current and ancestral environments in animal breeding, which is a problem faced by *Drosophila* researchers attempting to link results from lab stocks to natural field populations. In addition, animal breeders often face a situation where the performance of lines selected in one set of environments is linked to different environments, which links to the problem of whether the fitness of *Drosophila* lines established in one environment can be used to make inferences about fitness effects across other environments.

To provide answers to such questions, a link needs to be established between genetic and phenotypic variation with fitness under natural conditions. Yet translating genetic and phenotypic variation into fitness variation in natural *Drosophila* populations is hard work. Flies move. Adults select favourable microclimates for resting, and they locate food, mates and breeding sites by actively responding to olfactory, tactile and visual cues. Even the relatively immobile larval stage moves and selects the environments experienced by pupae; in some cases they can move spectacularly by jumping along a surface. Because of their movement and small size, individual *Drosophila* adults are nearly impossible to follow in nature except for short periods by filming. The movement of adults can therefore only be followed indirectly, such as through the use of micronized dust (Moth and Barker 1975) or genetic markers (Barker et al. 1989). Moreover *Drosophila* larval and adult feeding diets are difficult to define because they tend to feed on yeasts and bacteria that grow on plant tissue, although sugars, toxins, and various metabolites in the plant tissues also influence *Drosophila*. The environment occupied by larvae and adults is further complicated by wasp parasitoids of different life cycle stages, ecto- and endoparasites including nematodes, predators including ants, spiders and vertebrates, and a range of drosophilid and non-drosophilid competitors using the same feeding and breeding sites. Many of these complications were recognised by Stuart Barker and others several years ago and led to research on the interaction between *Drosophila* and their natural yeasts (e.g. Barker 1992; Barker et al. 1994).

Here I briefly consider some types of studies that can help researchers understand selection on traits in *Drosophila* and extend lab findings to selection in the field. An outline of the different assays used to link traits to fitness is provided in Table 1. I suggest ways in which the burgeoning molecular information from lab studies could be incorporated into these assays. This overview is not meant to be comprehensive, but to provide pointers for the future.

Table 1 Some approaches for assessing fitness in *Drosophila*

Laboratory or field?	Type of measure	Examples
Laboratory based	fitness components	strain comparisons for fitness components like larval viability (competitive, non competitive), mating speed, fecundity, fertility or longevity
	cage experiments	allele frequency changes in cages over one or multiple generations
Semi-field	small field enclosures	clinal variation in overwintering ability in small field cages
	large field enclosures	allele frequency changes across seasons in large enclosures
	open environments	strain comparisons for life history traits in *Drosophila* breeding on fruit
Field	adult releases	Strain comparisons with marked fly releases
	larval comparisons	characterising genotypes from different microenvironments from the field
	repeated sampling	allele frequency changes over time or space in field populations
	perturbations	allele frequency changes in isolated field populations like isolated cactus rots or winery cellars

2 Constant to Variable Laboratory Environments

In a lab environment fitness can often be accurately defined by measuring a series of fitness components. Following the classic work of Prout (1971) and others, the fitness of specific genotypes can be defined by its effects on larval viability, female reproductive success, and male reproductive success. These components can be further expanded into others like development time, feeding performance, fecundity, fertility and longevity. By measuring these components, expected changes in gene and genotype frequencies can be defined. Moreover, the relative importance of fertility and viability selection can be estimated and compared as originally shown by Prout (1971).

There are several other elegant and indirect ways of computing the laboratory fitness of *Drosophila* genotypes. For instance, an approach developed by Sved (1975) considers the competitive fitness of an entire *Drosophila melanogaster* chromosome when competed against other genotypes. This type of assay can be modified so that the total competitive fitness of *Drosophila* within a laboratory environment can be estimated (Gardner et al. 2005; Jungen and Hartl 1979).

These approaches are typically applied using constant laboratory environments. However, there is ample evidence that the fitness of genotypes can change rather dramatically with environmental conditions. This environmental dependency of fitness in *Drosophila* is perhaps most clearly illustrated in experiments that consider inbreeding effects under different environmental conditions. Sets of lines with a particular level of inbreeding when compared under a range of environmental conditions (Bijlsma et al. 1999) show that poor performance of an inbred line under one set of conditions often does not translate into poor performance under a different set of conditions. Also the time taken for inbred lines to become extinct varies markedly with environmental conditions (Bijlsma et al. 2000). Much of the fitness variation in *Drosophila* inbreeding effects – and heterosis (Ferreira and Amos 2006) varies with environmental conditions.

The environmental dependency of fitness effects is also apparent from single marker polymorphisms; fitness of the classic allozyme and inversion polymorphisms in *Drosophila* depends on the environment wherein they are measured (for example Betran et al. 1998; Kamping and van Delden 1999; Yadav and Singh 2007). QTL mapping indicates that genetic regions contributing to the same lab fitness components often change with environmental conditions (Mackay 2004). As the environmental range increases, genetic correlations between strains for fitness related traits often decline. Genetic correlations among life history traits can change in magnitude and even in sign when different environments are considered (Sgrò and Hoffmann 2004). Thus patterns of interaction between traits and fitness components change very markedly as the environment is altered – there is unlikely to be a constant difference between a high and a low fitness genotype across environments, even for components like viability. Fitness of particular alleles is simply not meaningful without an environmental context, total fitness is fairly much meaningless as it can only apply to one environmental state, and any attempts to measure the fitness consequences of particular trait values or genes must consider the range of natural environments to which an organism is exposed. Except for some generally deleterious genes, "universal" fitness measures are meaningless. Any fitness effects of alleles and trait values established in one environment may only extrapolate to a narrow range of other environments.

Because of these problems, it is time that we use terms like viability, longevity, survival, fertility and mating success much more carefully, by adding a fuller description to the measurements we are making. The term "lab male mating success with virgins" is much more appropriate than the term "mating success" for describing a common assay where males are confined with several highly receptive females in a small empty container. Success in this assay is completely unrelated to other assays where males might be successful with females because they defend patches of food in larger containers (Dierick and Greenspan 2006). The term "viability" might be more accurately defined by the food source, larval density and other conditions used for measurement, perhaps along with information on the microorganism community found on the food surface.

To counter the problem of environmental variation on genetic effects, experiments need to be undertaken in environments relevant to field conditions. There have been numerous attempts to make *Drosophila* laboratory experiments more relevant to the field. For instance, selection experiments can be undertaken that investigate specific responses to *Drosophila* parasitoids as in the case of *Drosophila nigrospiracula* recently selected for resistance to an ectoparasitic mite (Luong and Polak 2007), or to use levels of stress that reflect those found in nature, as in the case of lines selected for exposure to levels of ethanol found in nature. These approaches are more likely to capture important aspects of field environments if there are variables that have particularly large effects on fitness. Plant toxins represent one such class of variables, because one *Drosophila* species might be excluded from a habitat occupied by a related species when a particular toxin is present (McBride 2007). Another option is to move towards environments that more realistically reflect conditions in nature.

3 Selection in Outdoor Cages

To investigate field fitness, it is sometimes feasible to confine flies in small outdoor population cages when measuring performance. Two examples of this approach are Junge-Berberović (1996) and Mitrovski and Hoffmann (2001). The former used small enclosures (50 mm diameter, height 110 mm) or vials to investigate life history variation in *D. melanogaster* and *D. subobscura* to link fecundity, development and lifespan with temperature variation over 5 months. Cages and vials were placed in a shaded position under a tree. Differences in temperature responses were detected between the species and linked to their ecology. The second study used population cages ($400 \times 300 \times 400$ mm) with around 7000 adults to investigate overwintering in *Drosophila* by monitoring survival of adults and development of immature stages. Cages were started with populations collected from along the cline and then placed outside in temperate Melbourne over winter. While survival was similar for adults from the different populations, there were differences in fecundity and the timing of egg production. Flies from southern temperate populations produced their eggs relatively later in the season, whereas flies from northern populations produced eggs relatively earlier. Thus field cages were used to establish clinal patterns and isolate a trait contributing to overwintering fitness. This pattern is probably related to diapause induction characterised in laboratory studies from flies collected in eastern North America (Schmidt and Conde 2006).

Small cages have been used to investigate fitness of Wolbachia infected *D. melanogaster* and *D. simulans* (Olsen et al. 2001; Turelli and Hoffmann 1995). For the *D. simulans* work, cages were placed under a shade house and used to rear flies that were then tested for their ability to cause cytoplasmic incompatibility. Wolbachia bacteria are maternally transmitted and cause incompatibility (failure of egg hatch) when infected males from the cages were mated to uninfected females, a phenomenon that changes from almost complete incompatibility (total egg mortality) to low incompatibility (near normal hatch rates) as males age. At the same time, incompatibility was tested in free living females. The free living females showed incompatibility levels which matched those produced by males from cages when these males were around 2–3 weeks old, suggesting that males obtaining most of the field matings were relatively old. In *D. melanogaster*, small field cages have been used to show that Wolbachia effects on fecundity depend on the nuclear background and location (Olsen et al. 2001).

Large cages have the advantages that microhabitats can be mimicked and that populations can be left for multiple generations. Oakeshott et al. (1988) considered changes in allozyme genotypes over 12 generations in cubic cages with 1 m sides containing rotting fruit, maintained as multiple discrete generations. Genotypic patterns suggested that there was heterozygous advantage at two of the loci measured. Drift appeared to contribute to changes in allele frequencies over winter, but at other times population sizes were large enough to preclude drift as an explanation. An even larger cage for monitoring changes in genetic markers consisted of a confined 50 m^3 area within a greenhouse, followed between 1972 when a population of *D. melanogaster* was introduced until 1997 (Kamping and van Delden 1999). As temperatures fluctuated seasonally, there were repeatable changes in the frequency of the genetic markers that could be linked to patterns of variation in natural populations of *D. melanogaster*.

The above studies illustrate that field cages can be applied in a variety of contexts to investigate selection effects in *Drosophila*. Most of the cages used are small, allowing replication but limiting their use in simulating natural environments where there is the potential for microhabitat variation and long distance food/mate location. These fitness components can only be realistically addressed in assays carried out under field conditions.

4 Selection in Open Field Environments: Adults

A diversity of approaches has been used to investigate the performance of flies under field conditions (Tables 1 and 2). Some of these approaches have not considered the relative performance of different variants in the field, and others have compared

Table 2 Examples of treatment comparisons undertaken with *Drosophila* in natural populations

Approach	Reference	Treatments compared
Releases of adults	Hoffmann and Turelli 1985	effect of physiological condition on movement
	Jaenike 1986	association between habitat response and use
	Hoffmann and O'Donnell 1992	effect of capture host on F1 behaviour
	Hoffmann and Loeschcke 2006	effect of larval density on adult capture success
	Loeschcke and Hoffmann 2007	effect of heat hardening on adult capture success
	Kristensen et al. 2007	capture success of selected lines
	Feder et al. 2000	adult heat exposure of manipulated lines
Local measures on field individuals and offspring	Partridge et al. 1987	association between size and mating success
	Stalker 1980	inversion effects on size traits
Field – F1/Fx lab comparisons	Sokolowski et al. 1986	pupal site selection by genotypes on pear
	McKenzie and Parsons 1974	ethanol tolerance of flies inside and outside cellar
	Hoffmann and McKechnie 1991	adult and larval responses to wine-related resources
	Nevo et al. 1998	multiple traits on stocks from opposite slopes of a canyon
Field perturbations	Barker and East 1980	changes in allele frequency in cactus rots
	McKenzie et al. 1994	changes in allele frequency in cellar

environmental rather than genetic effects (Table 2), but they are all included here because they indicate ways of evaluating field fitness components. Most approaches are focused on adult fitness – larval fitness has rarely been evaluated even though larval traits associated with resource utilisation are likely to be particularly important and laboratory studies have shown that larval performance can carry over to adult fitness (Chippindale et al. 1998).

Adult releases have been used widely to investigate movement rates and survival in *Drosophila* (Moth and Barker 1975; Rosewell and Shorrocks 1987) but they have also been used to compare environmental treatments such as releases with flies starved for different times, reared at a range of densities, held under conditions triggering acclimation, originating from different locations or reared on different resources (Hoffmann and Loeschcke 2006; Hoffmann and Turelli 1985; Kristensen et al. 2007a; Loeschcke and Hoffmann 2007). Typically in these experiments flies are marked with fluorescent dust and then released at one location. Capture rates at baits are then used to assess the success of flies in locating resources or in going to a particular resource when there are different types of resources available such as different types of fruit. Flies can respond to micro-environmental cues and use flight to locate resources. The difference in environmental conditions experienced by free flying insects as opposed to caged individuals is illustrated by releases carried out by Feder et al. (2000) with a strain of *Drosophila* carrying a *β-galactosidase* gene that was expressed in response to heat shock. When flies were released on days when temperatures (°C) reached the mid-30s, some heat shock was experienced by the flies as reflected in an increase in gene expression, although the majority of adults appeared to avoid heat stress successfully. In contrast a large proportion of adults confined in a cage did experience heat stress.

In field releases, capture success has been used as an estimate of fitness, particularly when capture points are at least several meters away from release points. Baits may be placed in a line extending away from the release point (Loeschcke and Hoffmann 2007), placed in a regular grid around a release point (Hoffmann and Turelli 1985) or placed in a concentric circle around a release point (Hoffmann and Loeschcke 2006). Capture rates can be high depending on conditions: under cool or mild conditions, more than 20% of the flies released can be captured within a day or a couple of days, and in some cases up to 75% of the flies have been captured (Hoffmann and Turelli 1985). When conditions become hot in the high 30s, capture rates may decrease substantially to less than 1% (Loeschcke and Hoffmann 2007).

A simple measure of fitness is probability of capture of a fly at bait. Where there are no other food or breeding resources in an environment, this measure provides an indication of the ability of adults to survive and locate resources. If traps are placed in a line or in a grid, the probability of flies being caught near the release point as opposed to more distantly might also be examined. Flies from distant traps might simply have been more disturbed following release, although these flies tend to have higher wing/thorax ratios, as expected if they have an increased flight ability (Hoffmann et al. 2007). Disturbance of flies is always a potential confounding factor in release experiments. Flies are disturbed when they are dusted and also when they are held at a high density in vials just prior to release. Steps can be taken to minimise these problems, such as by measuring out dust to reduce the amount placed on the flies, and also by allowing flies a preening period after dusting but before release. During a release, flies can be left in containers in the shade and allowed to leave containers on their own rather than being actively forced from containers. They can also be protected from predation, and released under conditions when they are likely to be less active, such as in darkness or in the morning when conditions are cool.

When trapping grids or lines are set up and *D. melanogaster* are tested, the vast majority of flies are typically caught near the release point, and almost no flies are caught more than 60–80 m from this point, particularly under cool conditions (Kristensen et al. 2007a). Because capture probability falls off sharply from near the release point, these designs appear to measure fitness components associated with finding food rather than simply capturing a non-dispersing fraction of the release population. Comparisons of different treatments suggest that flies expected to have a relatively high fitness have an increased capture success. Males and females starved prior to release tend to be captured relatively more often on poor quality fruit than on high quality fruit, when compared to flies that have not been starved (Hoffmann and Turelli 1985). On hot days, flies of both sexes exposed to heat hardening prior to release have a much greater probability of capture than flies that have not been hardened, consistent with the fitness benefits of heat hardening expected in hot conditions (Loeschcke and Hoffmann 2007). This trend is reversed on cold days when heat hardening might have a fitness cost. Finally, releases with inbred lines suggest that inbred flies have a reduced capture success compared to those from outbred crosses, and that effects of inbreeding detected in field assays are much larger than those detected under lab conditions (Kristensen et al. 2007b).

Apart from these inbred line comparisons, capture success has perhaps surprisingly rarely been used to evaluate the field fitness of different genotypes. Kristensen et al. (2007a) evaluated the capture success of flies from lines artificially selected for increased cold or heat resistance under a range of temperature conditions. They found that heat selected lines had a higher capture success under hot conditions than controls, whereas controls performed relatively better under cold conditions. The heat resistant lines performed more poorly under cold conditions, but the reverse was not the case for the cold resistant lines, which performed as well as heat resistant lines under hot conditions. Laboratory performance in stress assays could therefore be only partly related to field conditions. There were also differences in the relative performance of replicate control lines that was not evident from laboratory assays.

Genetic differences among lines have been linked to adult responses to breeding resources. In an experiment with *D. melanogaster*, flies were released in the field and collected on baits with either of two types of fruit, then reared in the laboratory and offspring released to assess whether they tended to return to the same resources as their parents (Hoffmann and O'Donnell 1992). There was strong resource fidelity in this experiment, reflecting genetic variance in the tendency of flies to be attracted to particular fruit types in the field, and simulations also suggested that the genetic basis of this fidelity had to be relatively simple (Hoffmann and O'Donnell 1992). Releases with strains of *D. tripunctata* have indicated heritable differences in response to resource type (tomato, mushroom) in the field (Jaenike 1986). In this case, there was no association between attraction of strains to the resources and their tendency to oviposit on those resources, although another study on responses to yeast in *Drosophila buzzatii* has found such an association (Barker et al. 1994).

In some cases, it is possible to measure mating success of *Drosophila* directly and body size has been related to mating success in field experiments. Partridge et al. (1987) collected mating *D. melanogaster* and *D. pseudoobscura* from a rotting fruit heap, as well as randomly selected males and compared these groups for wing size. Mating males were consistently larger than the randomly selected males. This type of experiment has since been repeated on other insects and *Drosophila* species, although a mating advantage for large males has not always been found. Adult body size variation has been widely studied at the clinal level in several *Drosophila* species. Typically clines occur across broad geographic areas and are maintained when flies are reared under constant conditions, suggesting that fitness associations with size depend on factors that vary geographically (James et al. 1997). However, the selective factors that produce size clines in the field are unclear although field releases suggest antagonistic selection between wing and thorax size (Hoffmann et al. 2007).

It is possible to demonstrate selection on molecular polymorphisms in the field. In a few instances, fitness effects of genetic polymorphisms under field conditions have been explored in perturbation experiments. Barker and East (1980) perturbed allozyme frequencies at three loci in *D. buzzatii* at an isolated site with cactus rots by allowing adults with defined genotypes to emerge from cages placed near rots. They then followed allozyme frequencies after the perturbation and observed frequencies returning to pre-perturbation levels, reflecting the likely effects of selection on chromosome regions marked by the allozymes, although migration may also have contributed to the results. Perturbation experiments have also been used to demonstrate selection on allozyme polymorphisms in *D. melanogaster* (McKechnie and McKenzie 1983, McKenzie and Parsons 1974). These results and others illustrating that large genotypic differences can emerge under field conditions highlight the likely success of attempts to demonstrate field selection on polymorphisms.

5 Selection in Open Field Environments: Other Life Stages

There is very little information on environmental conditions experienced by larvae in the field and even less about how these conditions influence genotypic fitness. A couple of elegant experiments have investigated temperatures experienced by larvae indirectly by monitoring effects on gene and phenotypic expression. Feder et al. (1997) found that larvae in necrotic fruit could experience temperatures exceeding 40°C, and that levels of hsp70 in larvae from the field reached the maximum levels that were detected in the laboratory. These results highlighted that *D. melanogaster* larvae experience severe heat stress in nature and could be used to devise meaningful laboratory assays of larval heat resistance (Krebs and Feder 1997). In a different approach, Jones et al. (1987) monitored the temperatures experienced by pupae using a temperature sensitive mutation that changed the eye colour of adult flies depending on conditions experienced by pupae. When compared to ambient conditions, pupae experienced less extreme temperatures, suggesting habitat selection by larvae.

In an investigation of temperature effects on body size and other morphological traits, Warren et al. (2006) used an open set up where nectarines were placed on a table and held either in the shade or in sunshine. Nectarines were placed in a wire cage to exclude birds, but otherwise accessible to *Drosophila*. These were colonised by different *Drosophila* species and size. As well, wing abnormalities were found in emerging flies that had bred on the nectarines, although these were unrelated to the position of the fruit.

I am unaware of experiments that have directly assessed the fitness of genotypes at immature stages under field conditions. An indirect approach used by Sokolowski et al. (1986) involved taking pupae from different areas on or near necrotic pears and assessing larval behaviour of emerging offspring in the lab. Larvae from pupae collected on the soil surface tended to move more (follow a "rover" strategy) than those that had pupated on the fruit (these followed a "sitter" strategy). Because the two types of larval behaviour are genetically determined, these results suggest that larval genotype influences field behaviour. Links between field collection sites and laboratory genetic variation have also been made in a few other studies. Larval tolerance of ethanol in the lab has been shown to differ between flies collected inside and outside a cellar (McKenzie and Parsons 1974), although the exact selective factors involved are unclear and adults as well as larvae can show differences in use of winery resources (Hoffmann and McKechnie 1991).

As for adult traits, there is a substantial literature on clinal variation in larval traits characterised under lab conditions on lab reared stocks. While clinal patterns suggest selection acting on a trait, the challenge in these instances remains the linking of specific values of a trait with field fitness under a variety of conditions. A modification of this design is to consider variation along a much smaller scale and to test for parallel patterns in different species. For instance, Van der Linde and Sevenster (2006) characterised life history parameters of different species of *Drosophila* collected along a 15 km gradient in the Philippines. Development time showed parallel differences between the sites, suggesting that there was an adaptive basis to the genetic differences among species, although the selective factors producing these differences were not ascertained.

6 Concluding Remarks

Although it is difficult to study selection in field *Drosophila*, there are many ways in which molecular genetic variation can be linked to field selection using adult releases, field cages, larval manipulations of fruit and other resources, collections of mating pairs, or field development time comparisons. In identifying selection on traits and polymorphisms, it is important to assess the same field fitness components under a wide range of conditions. This can be done by multiple releases across different seasons, or through repeated evaluations of larval fitness perhaps with different types of plant material or with different yeast inoculations. Once high quality sets of repeat experiments have been completed across different conditions, DNA extractions can be used to test interesting candidate genes that emerge in the future. Sets of replicated samples from different labs and with different types of genetic background could become available to researchers for this purpose.

References

Barker, J.S.F. 1992. Genetic variation in cactophilic *Drosophila* for oviposition on natural yeast substrates. *Evolution* **46**: 1070–1083.

Barker, J.S.F. and East, P.D. 1980. Evidence for selection following perturbation of allozyme frequencies in a natural population of *Drosophila*. *Nature* **284**: 166–168.

Barker, J.S.F., East, P.D. and Christiansen, F.B. 1989. Estimation of migration from a perturbation experiment in natural populations of *Drosophila buzzatii* Batterson and Wheeler. *Biological Journal of the Linnean Society* **37**: 311–334.

Barker, J.S.F., Starmer, W.T. and Fogleman, J.C. 1994. Genotype specific habitat selection for oviposition sites in the cactophilic species *Drosophila buzzatii*. *Heredity* **72**: 384–395.

Betran, E., Santos, M. and Ruiz, A. 1998. Antagonistic pleiotropic effect of second-chromosome inversions on body size and early life-history traits in *Drosophila buzzatii*. *Evolution* **52**: 144–154.

Bijlsma, R., Bundgaard, J. and Boerema, A.C. 2000. Does inbreeding affect the extinction risk of small populations? Predictions from *Drosophila*. *Journal of Evolutionary Biology* **13**: 502–514.

Bijlsma, R., Bundgaard, J. and Van Putten, W.F. 1999. Environmental dependence of inbreeding depression and purging in *Drosophila melanogaster*. *Journal of Evolutionary Biology* **12**: 1125–1137.

Chippindale, A.K., Gibbs, A.G., Sheik, M., Yee, K.J., Djawdan, M., Bradley, T.J. and Rose, M.R. 1998. Resource acquisition and the evolution of stress resistance in *Drosophila melanogaster*. *Evolution* **52**: 1342–1352.

Dierick, H.A. and Greenspan, R.J. 2006. Molecular analysis of flies selected for aggressive behaviour. *Nature Genetics* **38**: 1023–1031.

Drapeau, M.D., Cyran, S.A., Viering, M.M., Geyer, P.K. and Long, A.D. 2006. A cis-regulatory sequence within the yellow locus of *Drosophila melanogaster* required for normal male mating success. *Genetics* **172**: 1009–1030.

Feder, M.E., Blair, N. and Figueras, H. 1997. Natural thermal stress and heat-shock protein expression in *Drosophila* larvae and pupae. *Functional Ecology* **11**: 90–100.

Feder, M.E., Roberts, S.P. and Bordelon, A.C. 2000. Molecular thermal telemetry of free-ranging adult *Drosophila melanogaster*. *Oecologia* **123**: 460–465.

Ferreira, A.G.A. and Amos, W. 2006. Inbreeding depression and multiple regions showing heterozygote advantage in *Drosophila melanogaster* exposed to stress. *Molecular Ecology* **15**: 3885–3893.

Gardner, M.P., Fowler, K., Barton, N.H. and Partridge, L. 2005. Genetic variation for total fitness in *Drosophila melanogaster*: Complex yet replicable patterns. *Genetics* **169**: 1553–1571.

Gerber, B. and Stocker, R.F. 2007. The *Drosophila* larva as a model for studying chemosensation and chemosensory learning: A review. *Chemical Senses* **32**: 65–89.

Hine, E., Chenoweth, S.F. and Blows, M.W. 2004. Multivariate quantitative genetics and the lek paradox: Genetic variance in male sexually selected traits of *Drosophila serrata* under field conditions. *Evolution* **58**: 2754–2762.

Hoffmann, A.A. and Loeschcke, V. 2006. Are fitness effects of density mediated by body size? Evidence from *Drosophila* field releases. *Evolutionary Ecology Research* **8**: 813–828.

Hoffmann, A.A. and McKechnie, S.W. 1991. Heritable variation in resource utilization and response in a winery population of *Drosophila melanogaster*. *Evolution* **45**: 1000–1015.

Hoffmann, A.A. and O'Donnell, S. 1992. Heritable variation in the attraction of *Drosophila melanogaster* to fruit in the field. *Biological Journal of the Linnean Society* **47**: 147–159.

Hoffmann, A.A., Ratna, E., Sgrò, C.M., Barton, M., Blacket, M., Hallas, R., De Garis, S. and Weeks, A.R. 2007. Antagonistic selection between adult thorax and wing size in field released *Drosophila melanogaster* independent of thermal conditions. *Journal of Evolutionary Biology* **20**: 2219–2227.

Hoffmann, A.A., Sorensen, J.G. and Loeschcke, V. 2003. Adaptation of *Drosophila* to temperature extremes: bringing together quantitative and molecular approaches. *Journal of Thermal Biology* **28**: 175–216.

Hoffmann, A.A. and Turelli, M. 1985. Distribution of *Drosophila melanogaster* on alternative resources – Effects of experience and starvation. *American Naturalist* **126**: 662–679.

Jaenike, J. 1986. Genetic complexity of host-selection behaviour in *Drosophila*. *Proceedings of the National Academy of Sciences of the United States of America* **83**: 2148–2151.

James, A.C., Azevedo, R.B.R. and Partridge, L. 1997. Genetic and environmental responses to temperature of *Drosophila melanogaster* from a latitudinal cline. *Genetics* **146**: 881–890.

Jones, J.S., Coyne, J.A. and Partridge, L. 1987. Estimation of the thermal niche of *Drosophila melanogaster* using a temperature sensitive mutation. *American Naturalist* **130**: 83–90.

Junge-Berberović, R. 1996. Effect of thermal environment on life histories of free living *Drosophila melanogaster* and *Drosophila subobscura*. *Oecologia* **108**: 262–272.
Jungen, H. and Hartl, D.L. 1979. Average fitness of populations of *Drosophila melanogaster* as estimated using compound-autosome strains. *Evolution* **33**: 359–370.
Kamping, A. and van Delden, W. 1999. A long-term study on interactions between the Adh and alpha Gpdh allozyme polymorphisms and the chromosomal inversion In(2L)t in a semi natural population of *Drosophila melanogaster*. *Journal of Evolutionary Biology* **12**: 809–821.
Krebs, R.A. and Feder, M.E. 1997. Natural variation in the expression of the heat-shock protein Hsp70 in a population of *Drosophila melanogaster* and its correlation with tolerance of ecologically relevant thermal stress. *Evolution* **51**: 173–179.
Kristensen, T.N., Loeschcke, V. and Hoffmann, A.A. 2007a. Can artificially selected phenotypes influence a component of field fitness? Thermal selection and fly performance under thermal extremes. *Proceedings of the Royal Society B-Biological Sciences* **274**: 771–778.
Kristensen, T.N., Loeschcke, V. and Hoffmann, A.A. 2007b. Linking inbreeding effects in captive populations to field performance – field releases with replicated *Drosophila melanogaster* lines under different temperatures. *Conservation Biology* **22**: 189–199.
Loeschcke, V. and Hoffmann, A.A. 2007. Consequences of heat hardening on a field fitness component in *Drosophila* depend on environmental temperature. *American Naturalist* **169**: 175–183.
Luong, L.T. and Polak, M. 2007. Costs of resistance in the *Drosophila macrocheles* system: A negative genetic correlation between ectoparasite resistance and reproduction. *Evolution* **61**: 1391–1402.
Mackay, T.F.C. 2004. The genetic architecture of quantitative traits: lessons from *Drosophila*. *Current Opinion in Genetics & Development* **14**: 253–257.
McBride, C.S. 2007. Rapid evolution of smell and taste receptor genes during host specialization in *Drosophila sechellia*. *Proceedings of the National Academy of Sciences of the United States of America* **104**: 4996–5001.
McKechnie, S.W. and McKenzie, J.A. 1983. Polymorphism of alcohol-dehydrogenase (ADH) in a winery cellar population of *Drosophila melanogaster* – gene frequency association with temperature and genotypic difference in progeny production. *Evolution* **37**: 850–853.
McKenzie, J.A., McKechnie, S.W. and Batterham, P. 1994. Perturbation of gene frequencies in a natural population of *Drosophila melanogaster* evidence for selection at the ADH locus. *Genetica* **92**: 187–196.
McKenzie, J.A. and Parsons, P.A. 1974. Microdifferentiation in a natural population of *Drosophila melanogaster* to alcohol in environment. *Genetics* **77**: 385–394.
Mitrovski, P. and Hoffmann, A.A. 2001. Postponed reproduction as an adaptation to winter conditions in *Drosophila melanogaster*: evidence for clinal variation under semi-natural conditions. *Proceedings of the Royal Society of London Series B-Biological Sciences* **268**: 2163–2168.
Moth, J.J. and Barker, J.S.F. 1975. Micronized fluorescent dusts for marking *Drosophila* adults. *Journal of Natural History* **9**: 393–396.
Nevo, E., Rashkovetsky, E., Pavlicek, T. and Korol, A. 1998. A complex adaptive syndrome in *Drosophila* caused by microclimatic contrasts. *Heredity* **80**: 9–16.
Oakeshott, J.G., Wilson, S.R. and Knibb, W.R. 1988. Selection affecting enzyme polymorphisms in enclosed *Drosophila* populations maintained in a natural environment. *Proceedings of the National Academy of Sciences of the United States of America* **85**: 293–297.
Olsen, K., Reynolds, K.T. and Hoffmann, A.A. 2001. A field cage test of the effects of the endosymbiont *Wolbachia* on *Drosophila melanogaster*. *Heredity* **86**: 731–737.
Orengo, D.J. and Aguade, M. 2007. Genome scans of variation and adaptive change: Extended analysis of a candidate locus close to the phantom gene region in *Drosophila melanogaster*. *Molecular Biology and Evolution* **24**: 1122–1129.
Partridge, L., Hoffmann, A. and Jones, J.S. 1987. Male size and mating success in *Drosophila melanogaster* and *Drosophila pseudoobscura* under field conditions. *Animal Behaviour* **35**: 468–476.
Pool, J.E. and Aquadro, C.F. 2007. The genetic basis of adaptive pigmentation variation in *Drosophila melanogaster*. *Molecular Ecology* **16**: 2844–2851.

Prout, T. 1971. Relation between fitness components and population prediction in *Drosophila*. 1. Estimation of fitness components. *Genetics* **68**: 127–149.

Prud'homme, B., Gompel, N. and Carroll, S.B. 2007. Emerging principles of regulatory evolution. *Proceedings of the National Academy of Sciences of the United States of America* **104**: 8605–8612.

Rion, S. and Kawecki, T.J. 2007. Evolutionary biology of starvation resistance: what we have learned from *Drosophila*. *Journal of Evolutionary Biology* **20**: 1655–1664.

Rosewell, J. and Shorrocks, B. 1987. The implication of survival rates in natural populations of *Drosophila* – capture recapture experiments on domestic species. *Biological Journal of the Linnean Society* **32**: 373–384.

Schmidt, P.S. and Conde, D.R. 2006. Environmental heterogeneity and the maintenance of genetic variation for reproductive diapause in *Drosophila melanogaster*. *Evolution* **60**: 1602–1611.

Sgro, C.M. and Hoffmann, A.A. 2004. Genetic correlations, tradeoffs and environmental variation. *Heredity* **93**: 241–248.

Sinclair, B.J., Gibbs, A.G. and Roberts, S.P. 2007. Gene transcription during exposure to, and recovery from, cold and desiccation stress in *Drosophila melanogaster*. *Insect Molecular Biology* **16**: 435–443.

Sokolowski, M.B., Bauer, S.J., Waiping, V., Rodriguez, L., Wong, J.L. and Kent, C. 1986. Ecological genetics and behaviour of *Drosophila melanogaster* larvae in nature. *Animal Behaviour* **34**: 403–408.

Sorensen, J.G., Nielsen, M.M. and Loeschcke, V. 2007. Gene expression profile analysis of *Drosophila melanogaster* selected for resistance to environmental stressors. *Journal of Evolutionary Biology* **20**: 1624–1636.

Stalker, H.D. 1980. Chromosome studies in wild populations of *Drosophila melanogaster* 2. Relationship of inversion frequencies to latitude, season, wing loading and flight activity. *Genetics* **95**: 211–223.

Sved, J.A. 1975. Fitness of 3rd chromosome homozygotes in *Drosophila melanogaster*. *Genetical Research* **25**: 197–200.

Turelli, M. and Hoffmann, A.A. 1995. Cytoplasmic incompatibility in *Drosophila simulans* – Dynamics and parameter estimates from natural populations. *Genetics* **140**: 1319–1338.

Van Der Linde, K. and Sevenster, J.G. 2006. Local adaptation of developmental time and starvation resistance in eight *Drosophila* species of the Philippines. *Biological Journal of the Linnean Society* **87**: 115–125.

Van Naters, W.V.G. and Carlson, J.R. 2007. Receptors and neurons for fly odors in *Drosophila*. *Current Biology* **17**: 606–612.

Warren, M., McGeoch, M.A., Nicolson, S.W. and Chown, S.L. 2006. Body size patterns in *Drosophila* inhabiting a mesocosm: interactive effects of spatial variation in temperature and abundance. *Oecologia* **149**: 245–255.

Yadav, J.P. and Singh, B.N. 2007. Evolutionary genetics of *Drosophila* ananassae: evidence for trade-offs among several fitness traits. *Biological Journal of the Linnean Society* **90**: 669–685.

Yang, J., McCart, C., Woods, D.J., Terhzaz, S., Greenwood, K.G., Ffrench-Constant, R.H. and Dow, J.A.T. 2007. A *Drosophila* systems approach to xenobiotic metabolism. *Physiological Genomics* **30**: 223–231.

Discussion

Keith Hammond

The development of this session on future directions in adaptation work was set by its first contribution, where Dr Kathleen Donohue used a univariate niche construction model and research in *Arabidopsis thaliana* and *Cakile edentula* to clearly highlight the dynamism of the interaction between plants and their environment. It is possible for both the genetic and environmental variances of adaptive traits to vary with the environment, with changes in one or both of these variances resulting in environment-dependent heritabilities. The environment may also influence phenotypic expression through plasticity of traits of habitat selection, habitat modification or of more passive processes of resource use or depletion. The model was introduced in an evolutionary context but interpreted via a neat extension to quantitative genetic selection theory. The phenomenon draws our attention to the possibility of each component of selection—the intensity, the phenotype and phenotypic variance, and the genetic (co)variances of traits—changing with the environment. When niche construction alters phenotypic expression (via phenotypic plasticity) and the expression of genetic variation or heritability (not by selective elimination of maladapted genotypes but by genotype-environment interaction (GxE)), it can cause slower or faster responses to selection, less or more sustained responses to selection, or even negative responses to selection. In practice the norm will involve a complex of adaptive traits and also interactions between traits with environmental elements (in his chapter Dr Bruce Walsh raises multiple-trait plasticity). To predict these dynamics Donohue suggests requires knowledge of: the traits with which organisms influence their exposure to particular environmental factors, how environments influence trait expression and heritability, and the evolutionary potential of these niche-constructing traits. Experimental demonstrations of these dynamics are lacking.

Within animal production environments a range of opportunities exist for the animals, rather than the humans managing them, to determine the environment they experience and, when determination has a genetic basis, to alter the genetic variance expressed. General examples of these opportunities include the impact on the

K. Hammond (✉)
5B Coral Place, Canberra ACT 2612, Australia
e-mail: pk@pamandkeith.net

J. van der Werf et al. (eds.), *Adaptation and Fitness in Animal Populations*,

phenotype and genotype of preferential grazing, social impacts of grouping (also see the chapter by Dr Piter Bijma), multi-tiered nucleus breeding with environmental differences between the nucleus and multiplier breeding herds, and maternal effects can also be considered as niche construction. Of course, human intervention in managing production systems could be operating for or against trait niche construction.

Niche construction may have important implications, positive and negative, for the design and conduct of genetic improvement programs when substantial differences exist within elements of the environment. For particular programs it becomes important then to study the possible niche construction elements by clearly defining the environmental measures and (adaptive) animal traits involved in order to understand how the environment and genotypes impact their expression and their phenotypic and genetic (co)variances. The work required to do this well is not trivial.

Initial discussion of this contribution briefly considered reasons for the variation in life history being maintained in the experimental work by Dr Donohue, and more generally. With negligible experimental evidence interpretations vary: exposure to new environments, plasticity itself being adaptive, density-dependent selection? The cost to the organism of retaining this genetic flexibility will be genetic architecture dependent as will the trajectory of changes in the heritabilities from, for example, offspring on parent regressions.

Incomplete genetic correlations for adaptation traits between environments, the need to better define and measure these traits, and to understand and characterise the environmental and genetic variation associated with the traits led to the next contribution to this session, by Professor John James. This contribution is rich with examples, and James pays tribute to the important early and more recent work to understand GxE's by Professor Douglas Falconer.

James overviewed the various descriptions of GxE, and the analysis and interpretation of experiments concerned with GxE, mentioning the strengths, weaknesses and peculiarities of each method. Of particular relevance to improved definition and measurement of these traits is the notion that environmental differences which suppress the expression of variation will very often be sources of GxE, since if there is a genetic component to such variation we would expect interaction to be present. GxE can involve changes in scale, rank, or be described as fixed, to which genotypes and planning can adapt, and random where adaptation will be to the average. James stressed the care required during analysis, in separating scale effects from non-additivity, and mentioned the added value of the joint regression approach for examining adaptive trait data as the regression slopes of genotypic performance on environmental mean provide estimates of the sensitivity of the different genotypes to environmental quality. When adaptive trait environments are 'internal', such as social grouping impacts, strain-environment experimental designs need to utilise replicated pure-strain rather than mixed grouping.

James also addressed the design of genetic improvement programs when GxE's are a consideration. Past experimental attempts to validate various rules which have been proposed for use in the design of livestock improvement programs when GxE is expected, and the theoretical expectations for these rules were discussed. These rules are additional to those developed recently by Mulder and Bijma (2006) where breeding program design depends on the magnitude of the genetic correlation(s) between environments, the relative importance to overall production of each environment and, when planned selection intensity is high this too should be taken into account when deciding between one breeding program or one for each environment. With GxE and random environments it can be difficult to sample a sufficient number of environments, a potential difficulty with sire reference schemes, perhaps pointing to the need for such schemes to utilise extensively artificial insemination, comprehensive performance recording, and to take particular care when developing and executing the design.

During initial discussion of this contribution it was suggested that some pig and chicken breeding activity supplying particularly temperate production environments would have made substantially greater progress during the past half-century if GxE had not been considered. But of course, the attention required depends on the nature and magnitude of GxE's between and within breeds and production systems. As cost-price pressures increase and community demands tighten more of the many GxE's will become important to genetic improvement activity. Where the environments involved are spatially separate Dr Mike Goddard suggested that it may be desirable under at least some conditions to provide for GxE's by selecting across the environments whilst utilising different breeding objectives for each environment. This possibility enables identification of the best animals for use in each environment regardless of which environment generated them. Genomic selection will further enable this approach and this will serve to extend globalisation of breeding programs. Discussion also drew attention to the significant recent improvements in mixed model methods by plant breeders concerned with GxEs. Whilst plant data are commonly better structured it was suggested that these methods would be applicable to some animal data structures.

Accepting James' conclusion that there exists little generality in the nature of GxE with every case needing to be considered on its merits, and now given the potential impact of climate change on many animal production systems, there is a growing need to advance research on some elements fundamental to the successful genetic improvement of livestock. These are: the clear definition and measurement of adaptation traits as well as of the environments involved with both the generation of breeding stock and production, more seriously addressing characterisation of the genetic architecture for these traits within and between these environments and, associated with this research effort, developing the work to better understand the detailed system biology of these traits (assisted by the recent rapid progress in this area in the human field).

It is clear that irrespective of species and production environment, most animal adaptation and fitness traits will prove to be complex, commonly interplaying together as well as with the environments. Small, independent research projects aimed at understanding one or all elements of these complexes will generally be of low utility. The contribution to this meeting by Dr Ary Hoffmann and his team of national and international colleagues offers a fine example of the comprehensive research approach required and the careful planning and design needed to define and develop meaningful measurements of adaptation and fitness traits, of the environments involved (laboratory, experimental and field), and to understand the nature of the phenotypic expression and the genetic architecture for these complexes within and amongst the organisms concerned.

Unless one's direct interest is adaptation and fitness of *Drosophila*, the primary and substantial utility of Hoffmann's contribution is its clear demonstration of the nature and size of the approach required to progressively develop a clear understanding of animal adaptation and fitness, and how our armoury of tools can be effectively and efficiently combined in design, analysis and interpretation of this research directed at precise and otherwise cost-effective outcomes. For livestock development these outcomes will be utilised to target and manipulate adaptive variation between and within the available range of strains, breeds, species, and environments in the sector. Hoffmann also highlighted the importance of longitudinal studies to this type of strategic research. To properly develop and utilise its knowledge-base, design and planning needs to accommodate as best it can future advances in techniques. So, for example, the program of work by Hoffmann et al. is now beginning to incorporate powerful elements of biophysical modelling to: predict the conditions being experienced by genotypes along the 3000 kilometre cline as the environment changes, provide a spatial framework for predicting the outcome of changes in these conditions, develop individual-based physiological models (also providing for further extension to study of, for example, cellular systems biology), and provide the potential to interact with ecologists to better incorporate genetic variation into their activity.

Aspects of the protocols being employed in this work of particular interest to livestock genetic research include: further partitioning the adaptive challenges of (both high and lower) production environments being experienced by individuals, sieving candidate genes, and relating experimental and field definition and measurement of adaptation and fitness life-history, physiological and behavioural traits. The work also clearly demonstrates the level of care and input required to validate laboratory assays, including of course genomic measurements, within and between field environments. Rather than attempt from the outset a comprehensive understanding of complex adaptation traits, the design emphasis of the work is focused on locating and understanding those genes controlling *adaptive shifts* throughout the cline. This approach should also be useful for adaptation work in domestic livestock populations involving a number of specific breeds and different environments rather than substantial clines.

A short-term perspective of this comprehensive approach being used for future animal production research may be that it seems to be over-costly. That perspective as it is proposed is false. The need to better understand the biology of production, adaptation and fitness is now rapidly increasing in importance, to provide livestock sectors with the capability of successfully responding to the increasing global demand for quantity and quality of food and other goods from animals, and to community demands for change in animal welfare and reductions in the environmental impact of livestock. These responses must be achieved under further pressures; whilst the livestock sector continues to improve productivity to respond to the long-term deteriorating terms of trade in food and agriculture production, during a period when the sector begins to consider what spatial and genetic reorientation will be required to successfully meet the impact of climate change, and increasing its efforts to reduce its carbon footprint. The above-referred comprehensive approach to future livestock research directed at meeting this portfolio of demands has no useful alternative. Planned well, these research programs will offer over time a continuing supply of substantial outcomes which will be widely applicable nationally and, with suitable planning, internationally. Without such advances in our understanding the livestock breeding sectors are likely for example, to be increasingly paying for apparently sophisticated genomic assays with doubtful real benefit, and the gains mentioned above as necessary at the production level will fall far short of those required for long-term economic viability.

Given the need to understand livestock adaptation and fitness to genome (and eventually cellar system) level, the standard arguments for utilising divergent selection lines for adaptation work are weak. Better the work take the form of multidisciplinary research programs which utilise the broad range of actual production environments in the livestock sector being studied and the actual breeds/strains in use in these environments.

Discussion again highlighted the value of including amongst the essential tools in this work high-density SNP genotyping, for reliably and rapidly identifying the regions of the genome which are undergoing selection in the different environments (Hoffmann's team has not yet attempted to demonstrate that variation under selection can be reliably identified within a geographic region). Dr Mike Goddard commented on their efforts to identify genomic regions in "medium-sized" high and low lines selected for feed conversion efficiency, noting that whilst they were successful in identifying many significant changes between the lines, they were unsuccessful in their attempts to validate these in other populations. He suggested that unless the population size of selection lines is large only rarely will a SNP be sufficiently close to the causative gene to be detected, because of linkage drag.

It was agreed that the change in position detected along the latitudinal cline of the otherwise apparently stable (over 20 years) chromosomal inversion 3RP was likely an artifact, because of insufficient power in the particular experiment.

A discussion of the session as a whole followed, around the topic "evaluating models of phenotypic performance in different environments". The range of queries and comment clearly suggested there remain many opportunities for advancing this issue, and the meeting noted the somewhat different analytical approaches to the topic which are currently being taken in the plant, animal and human work.

There was particular interest to further explore animal traits for which the environment is partially genotype driven. For example, as the level of gastrointestinal parasite infection is reduced under selection, animals possessing higher resistance shed fewer eggs to pasture thereby reducing the opportunity for infection and likely also altering (co)variance components for the trait(s). A central issue here is the interaction between the parasite environment and the resistance trait(s), where particular combinations of the environment and genotypes can quite easily influence the phenotypes and genotypic expression. Is it feasible then to better characterise the parasite environment? For situations where the environment is being genotypically driven, Dr John Gibson queried the preferred prediction approach, parameterising what is observed in the population, such as the elegant modelling described in Session 2 by Dr Piter Bijma, or utilising a biological or epidemiological approach which would be aimed more at longer-term prediction? Bijma suggested it could be interesting to determine whether such modelling of interactions could be used to characterise additive genetic variation for social effects for multi-generational predictions.

In relation to the challenge of involving small-holder farms in developing country genetic improvement activity, Dr Sandy McClintock sought suggestions for grouping animal data in prediction based upon classing individuals into common environments and analysing for breed-type within environment. Of course, the definition of these classes would be simplified following detailed studies of the field production environments. Dr Brian Kinghorn queried the potential to address the GxE issue by dissecting phenotypes into their "driving components". These components may not exhibit so much interaction across environments, would likely possess higher heritabilities than the observed traits and possibly would be more robust across environments. To do this would require high data quality and quantity. Dr Donohue suggested that minimising the negative correlations for traits across environments by identifying and carefully utilising the particular gene loci involved, should enable genetic improvement without some of the trade-offs. This approach, concentrating on the driving components, would likely need to cover all important adaptation traits, which returns us to the earlier suggestion by Dr Mike Goddard concerning the desirability of providing for GxE's in breeding programs by selecting across the environments whilst utilising different breeding objectives for each environment.

The session ended with a brief discussion concerning the lack of clarity of terminology associated with adaptation. By way of example, the chair cited the variation in usage and overlap in the genetics and ecological literature of the terms sensitivity, phenotypic plasticity, and phenotypic resilience, suggesting also that climate change action is likely to very substantially raise the profile of adaptation work within and beyond the sciences, thereby raising the importance for clarity of terms. The ensuing discussion served to emphasise the issue.

Part IV
Strategies for Managing Diversity

Strategies to Exploit Genetic Variation While Maintaining Diversity

Brian P. Kinghorn, Robert Banks, Cedric Gondro, Valentin D. Kremer, Susan A. Meszaros, Scott Newman, Ross K. Shepherd, Rod D. Vagg and Julius H. J. van der Werf

Abstract How we should manage genetic diversity depends on why we want to manage it. The most generally useful strategy is to maintain variation across the genome, using methods that consider one or more of: population size, population structure, animal selections, mate allocations and information from genetic markers. A key reason to maintain genetic diversity is to facilitate longer-term genetic gains, and this means that most breeding programs need to consider genetic diversity as well as shorter-term genetic gains. This paper discusses these issues, and presents developments in methods to integrate genetic gains, genetic diversity and other issues within breeding programs.

Keywords Diversity · genetic variation · inbreeding · mate allocation

1 What is Genetic Variation?

This question seems quite easy to answer – genetic variation is the variation among individuals and/or populations in genetic constitution. In this context we often think of variation summed over single polymorphisms for mutations such as single nucleotide substitutions or small deletions. But we can move up the functional chain to consider co-variation among many polymorphisms, variation in functionality of biochemical pathways or other such biological networks, and on to variation in the impact of the whole (epi)genotype on phenotype in defined environments.

Taking another angle, we can use genome-wide markers to make inference about genetic variation and maintenance of genetic variation, and their association with selection history as driven by fitness – without using explicit phenotypic information (e.g. Borevitz et al. 2007).

How we should define genetic variation depends on why we want to manage it, and the future opportunities we might have to exploit it. The less we know about

B.P. Kinghorn (✉)
The Institute for Genetics and Bioinformatics, University of New England, Armidale NSW 2351, Australia
e-mail: bkinghor@une.edu.au

J. van der Werf et al. (eds.), *Adaptation and Fitness in Animal Populations*,

these things, the more we resort to defining genetic variation in simple terms based, for example, on data from neutral genetic markers, and/or on population structure. Alternatively, if we want to manage genetic variation for a given trait in a given environment, we may elect to focus purely on the prevailing genetic component of variance for that trait, as derived from phenotypes and relationships.

Consider the somewhat optimistic scenario where we have conducted a genome scan for the population of interest, with sufficient genotypes and phenotypes to develop a reliable handle on QTL or marker associations that explain almost all of the genetic variation in a trait of interest. It is reasonably clear how to use this information to make early genetic gains in a breeding program. But how would we use this information to help maintain genetic variation for this trait? Driving towards intermediate or optimal allele frequencies across multiple loci would take many generations, especially if we are focusing on more loci because of attention to multiple traits. Assortative mating on genotypes to generate extra divergence would help reveal variation, making it more accessible later on, but again that takes many generations, and in fact extra variation is being unlocked, rather than created.

Moreover, such direct strategies will be very challenged to handle properly the shifting epistatic landscape, which over time will alter the relative importance of each locus being tracked, and will cause new unmarked loci to become important. Therefore it will probably be difficult to beat the simple approach of maintaining genetic variation by using population size and structure targets to drive selections and mate allocations.

Such a conclusion can be supported by consideration of the potential outcomes of highly progressive breeding programs that use genetic markers to help give increased genetic gains (rather than to maintain genetic diversity): Consider what our domestic animals will be like in 25 to 50 years time, using an appropriate blend of progressive practice and improving technologies. Key characteristics of animals will probably change, with not even extreme outlier animals in today's populations being representative of the new types. Biology is complex, and such changes will involve complex interactions and shifting emphasis across redundant pathways. Many loci of importance today will no longer be the "weak links", and new loci will assume importance in the face of new opportunities in the biological network. This is the shifting window of genetic background, and how it impacts allele effects and epistasis (Barton and Keightley 2002). Indeed, 2-locus epistasis simulated under a classical statistical model leads to longer half-lives of selection response than a purely additive model (Jannink 2003). This likely underestimates the opportunities afforded by the true underlying situation, given the complexity of biological networks.

Furthermore, it is not only the biology that shifts. Consumer preferences, production systems and product niches are also subject to change, such that target traits and the emphasis among them will also change.

Accordingly, where the objective is to maintain genetic variation to help ensure ongoing opportunity to make genetic gains in domesticated populations, it is contended that specific attention to genetic markers that impact on phenotype is of limited value. But some attention is appropriate; most importantly in the use of strategies to help ensure that favourable mutations at low frequencies are not lost.

With these thoughts in mind attention will be focused on using population structure and animal relationships to help manage genetic variation. This approach is more generic and more genome-wide, but simpler and generally more reliable in ensuring the availability of appropriate forms of diversity that might come to be useful in the more distant future.

The nature of genetic variation *across* breeds/populations is somewhat different, as breed differences in gene frequencies are more stable, and therefore more useful for conservation decisions, than the differences in gene frequencies between lineages within an essentially panmictic population. Evidence supports the expectation that the genetic basis of population differentiation for fitness traits is non-additive, with different adaptive gene complexes evolved in each breed, making it important to conserve breeds as adapted resources for breeding programs (Barker 2001).

2 Why Maintain Genetic Variation?

Recombinant DNA technologies provide some power to create genetic variation directly. We can borrow sequences from other populations or species, and we can even aim to predict and ultimately generate ideal sequences. However, life is complex, and we do not have full understanding of the impact of such exercises. Moreover, in populations that are not fully inbred and are subject to ongoing genetic change, we need very good data and models to predict ideal genotypes to target. Such prediction of ideal genotypes is very complex and will require a detailed understanding of biology – it is a long way off. In this context, genetic variation is one of the key ingredients for understanding the biological systems involved. It is not an essential ingredient, but it will help a great deal.

However, the primary reason agriculturists maintain genetic variation is to exploit it directly in breeding programs. There are two scenarios in which maintaining genetic variation deserves special attention:

1. Closed lines of stock at the top of the breeding pyramid. Important loss of genetic variation in such lines is most simply remedied by importing from outside, and this may well bring significant compromises in genetic merit and health risks. Moreover, loss of genetic variation across populations may further increase genetic compromises.
2. Open breeding systems where there are globally accepted breeding objectives, with little variation in these, relatively uniform production environments across regions, progressive genetic evaluation with use of information from relatives, extreme fecundity in at least one sex, and little economic incentive at the enterprise level to maintain genetic variation.

The second scenario relates to the world Holstein dairy cattle population, and this constitutes a problem that is more social/business in nature, rather than technical.

Some measure of altruism may be required, as regulation across countries is challenging. However, changes in breeding objectives from strictly performance-based towards goals incorporating health, adaptation and longevity might well alleviate the problem.

It is difficult to be fully objective about how much predicted loss of genetic diversity is acceptable. Selection is a dynamic process, and many factors can impact on predicting the best strategies that maximise long-term selection response. Technical factors include loss of genetic variance, inbreeding depression in target traits and reproduction, and reduced selection intensity due to lowered reproduction. A practical factor is unforeseen opportunities to widen the genetic base of a selection line by importing competitive stock externally. Most practitioners handle these issues by setting a limit to the inbreeding rate, perhaps 0.5% or 1% per generation. A criterion that is less influenced by the pattern of mate allocation is the rate of increase in mean coancestry among parents, as noted later.

3 Simple Measures of Genetic Diversity

We need some measure of genetic diversity, and/or predicted changes in genetic diversity, to provide an objective value to maximise or minimise as a function of breeding strategy. However such measures are of much wider value because in agriculture we rarely seek to maximise genetic diversity alone. Our motivations to manage diversity compete with other issues, including operating costs and making genetic gains. So we need measures to include in a multi-objective implementation system of some kind.

With few breeding animals the pool of genetic material diminishes, and each allele has an increased chance of "meeting itself" in descendants, giving inbreeding depression and loss of diversity. The classic approach has been to maintain the effective breeding population size (N_e, Falconer and Mackay 1996) at a suitable level, usually under the assumption of no selection. The rate of loss of diversity is then linked to the rate of inbreeding: $\Delta F = 1/(2N_e)$. For implementation purposes, we can allow for variable contributions from parents by noting that $\boldsymbol{x}'\boldsymbol{x}/2 = 1/(2N_e)$, as illustrated in Table 1, where $\boldsymbol{x}$ is a vector of proportional genetic contributions from candidates for breeding, summing to $^1/_2$ for each sex and thus summing overall to 1.

Table 1 An illustration of equivalence of $x'x/2$ and $1/(2N_e)$

Example Scenario	x′	x′x/2	N_e	$1/(2N_e)$
2 males and 2 females	(0.25 0.25 ∣ 0.25 0.25)	1/8	4	1/8
1 male and 2 females	(0.5 ∣ 0.25 0.25)	3/16	$N_e = \frac{4N_m N_f}{N_m+N_f} = 8/3$	3/16
An optimal contributions scenario	(0.1 0.4 ∣ 0.1 0.1 0.1 0.2)	3/25	$N_{cm} = 1/(0.2^2 + 0.8^2) = 25/17$ $N_{cf} = 1/(3{*}0.2^2 + 0.4^2) = 25/7$ $N_e = \frac{4N_m N_f}{N_m+N_f} = 25/6$	3/25

$x'x/2$ is maximised by equalising contributions within each sex. However, we can see that this is suboptimal if, for example, we have three male candidates, one unrelated and two being full sibs. The full sibs should each be used less as they each contribute less to diversity than does the unrelated animal. They should in fact be used in a ratio of 3:2:2.

This optimisation can be achieved by using $x'Ax/2$, where A is the numerator relationship matrix, rather than $x'x/2$ (or $x'Ix/2$). Here $x'Ax/2$ is the weighted mean coancestry of selected parents, noting that $A/2$ is equal to the coancestry between the individuals concerned. This approach reduces the contributions from individuals that are more related to other selected candidates. It also takes account of the effects of selection on inbreeding and drift, in a manner that is missed by the classical approaches to calculating N_e.

Inbreeding coefficient in progeny is predicted by $x'Ax/2$ ***if*** full equilibrium is reached immediately, involving selfing, same-sex mating and random allocation. However, with the exclusion of selfing, etc., the relationship between $x'Ax/2$ and inbreeding level or rate is not as simple. Moreover, variation in $x'Ax/2$ among possible selection sets reduces over generations, as the elements of A asymptote towards 2, giving a need to increase the weighting on $x'Ax/2$ in a competitive index to give a constant ΔF. However, the augmented A of Grundy et al. (1998) removes this trend over generations by effectively redefining the base population at each generation, and so removes the need to increase weightings.

A practical measure of change in diversity is the rate of change in coancestry from generation *t-1* to generation *t*:

$$\Delta f = \frac{f_t - f_{t-1}}{1 - f_{t-1}} \ldots \text{where} f = x'Ax/2 \qquad (1)$$

This can be used to evaluate the impact of population structure, or indeed animal selection sets, on rate of change in genetic diversity. However, in a practical setting, the time frame over which to estimate Δf depends on balance between sampling error (longer is better) and relevance to current decisions (shorter is better). In contrast, mean parental coancestry (expressed as $x'Ax/2$ or something similar) is much simpler to calculate, and is becoming quite widely used as the actual evaluation criterion, as described later. Investigation of Δf values can be used to help set target values and or weightings related to $x'Ax/2$.

Measures of mean parental coancestry are complicated by overlapping generations. This is addressed by Grundy et al. (1998) and Meuwissen and Sonesson (1998), and an alternative approach is outlined later in this paper.

4 How to Maintain Genetic Variation

A key task for breeding programs within populations is to balance the relative emphasis on genetic gain and maintenance of diversity. This has now been done quite widely in the context of an objective function that includes measures of genetic merit and coancestry among selected parents. Wray and Goddard (1994) and Brisbane and Gibson (1994) included negative weighting on animal relationships in selection indices in order to help control the rate of inbreeding, and hence genetic diversity. Meuwissen (1997) formalised the optimisation of animal contributions, differentiating a function that includes genetic merit and animal coancestries, while handling some constraints, and accommodating individual contribution constraints in an iterative manner. Hinrichs et al. (2006) have improved speed in this method considerably, and Pong-Wong and Woolliams (2007) have illustrated a method that handles individual contribution constraints directly, and thus gives a guaranteed optimal solution.

The results from such methods generally give *optimal contributions* as proportions on a continuous scale. The practitioner uses these to help make selections and allocate numbers of matings to each breeding animal. This represents a powerful way to allocate more matings to higher-EBV animals in a manner that properly manages genetic diversity.

Progressive breeding programs are motivated by many other issues in addition to genetic gain and diversity, including progeny inbreeding, connection between herds, management of genetic marker frequencies and genotypes, multi-stage selection, various types of costs, and various logistical constraints. For such cases, contributions can be specified in numbers of matings to be allocated, and indeed the actual mate allocation set (e.g. Kinghorn 2000), but the optimal solution must be discovered, rather than derived, due to the complexity of the resulting objective function (Kinghorn and Shepherd 1999). Such problems can be solved using Evolutionary Computation (EC), and adaptations of Differential Evolution (DE, Storn and Price 1997) have been used for this. For the reader who is unfamiliar with this type of algorithm, Evolutionary Computation is a broad field of research in optimisation methods loosely based on biological evolutionary processes such as mutation, crossing over and selection. The central idea behind EC is to create populations of candidate solutions of a problem and evolve these populations by selection based on an objective function which emulates natural selection (Bäck 2003).

The key to making this work relates to how the problem is presented to the solving algorithm. Kinghorn and Shepherd (1999) show how a vector of numbers can be transformed to derive a pattern of animal selection and mate allocation – Mate Selection. This makes the optimisation problem simple – the DE algorithm has only to find the vector of numbers that maximises the objective function.

For each mating set tested, the component outcomes evaluated constitute the overall Mate Selection Index (MSI). Each component should be evaluated on the same scale, typically the scale of the breeding objective in units of, for example, dollars profit per breeding female per year. The MSI is a function of the selections and mate allocations chosen across all mating individuals in the population – it is used to form a mating set for the specified number of matings in the breeding program. The MSI can include predicted genetic merit in progeny, long-term inbreeding as reflected by parental coancestry, progeny inbreeding, heterosis (if relevant) and various costs and logistical constraints associated with the breeding program (Banks 2000; Kinghorn 2000). The objective function can be changed (variables added or deleted) to suit the desired outcomes of the breeder if the breeding objective changes or constraints arise. An example objective function is:

$$\mathrm{MSI} = \frac{x'G}{2M} + \lambda\frac{x'Ax}{4M^2} + \phi F + \chi C - \mathrm{cost} \tag{2}$$

Components of this MSI are described in Table 2

Comprehensive use of a *Mate Selection Tool* (MST) involves decision making at various stages of the life cycle, including selection/culling at various juvenile and adult stages, selections for main round matings and, in some cases, selection of sires for backup matings. Ideally, the MST will accommodate all members of the population, across different groups, including putative embryos *in utero*, immature juveniles, virgin females and pregnant females, as well as candidates for active matings. This helps to accommodate overlapping generations, and to account for contributions already made as reflected by juveniles and embryos in the analysis – sires that have already been used extensively will be somewhat inhibited from further use.

To implement such a scheme, there is a requirement to constrain the solution such that animals can only be allocated to members of the opposite sex that are in an appropriate group, with the two key groups being the male and female active mating groups (although these may be sub-grouped in turn, for example by location). One way to handle such complex grouping constraints is to penalise non-conforming solutions in the objective function. However, this has proved to result in slow computation speed, as most evolutionary pressure is used up maintaining legality of solutions. A method has now been developed to translate a vector of numbers to a legal mate selection solution that satisfies complex constraints, not previously handled by Kinghorn and Shepherd (1999), related to grouping and mate

Table 2 Component definitions of a sample mate selection index (MSI)

Item	Definition
x	A vector of number of matings to be made for each candidate over both sexes
G	A vector of genetic merit containing selection index values for candidates
M	Total number of matings to be made – usually the number of breeding females
λ	Weighting factor (typically negative) to discourage low effective population sizes
A	The numerator relationship matrix for candidates
ϕ	Weighting on predicted progeny inbreeding
F	Predicted progeny mean inbreeding
χ	Weighting on predicted progeny mean crossbred value
C	Predicted progeny crossbreeding value
cost	Cost of the mating policy implied by x

allocation among groups. This has resulted in considerably higher speed, giving much increased flexibility.

Part of the need for high speed in such MSTs relates to the need to explore the solution space to find a result that gives a satisfactory balance among the wide range of issues that can be represented in the objective function. It is difficult to be economically rational across issues such as level and cost of migration, pattern of trait distributions, marker genotype frequencies and number of mating paddocks required. Experience shows that iterations of quickly finding an optimal solution for the prevailing objective function followed by adjustments to weighting factors and/or thresholds set in the objective function gives the practitioner a good feel for the trade-offs involved in putting differential emphasis on component objectives. In decision-making theory these solutions are sometimes referred to as *satisficing* – they are optimal given all of the information at hand. This leads to some understanding of what can be achieved, and acceptance of a final mate selection solution. In practice, there is a trade-off between what rigorous scientific effort might determine to be the "best" overall direction, and the direction that is chosen by the practitioner. The latter has the advantage that it is in fact implemented, and not left hanging as a mere recommendation. In addition, of course, scientific method gets well implemented "around" the constraints and weightings applied by the practitioner – for example, the theory of optimal contributions (Meuwissen 1997) is invoked, given the prevailing constraints.

Breeding decisions are made on a frequent basis in the pig and poultry industries and so this discovery process needs to be captured in a manner permitting essentially automatic running of the MST with little or no user intervention. This can be achieved using algorithms to maintain certain key outcomes, such as keeping the relationship between parental coancestry and predicted progeny merit within a narrow trajectory, so that other issues cannot displace the final solution from that path. This is analogous to a desired gains selection index (e.g. Brascamp 1984), but where the ratio of gains between two traits (*issues* in the MST case) is softly constrained within certain bounds. Overall response surfaces can be very flat – giving good opportunity to move one issue in a favourable direction, with little compromise in other key issues.

MSTs that are available include Ani-Mate (Amer, www.abacusbio.co.nz/products.html), TGRM and X'Aim (X'Prime, www.xprime.com.au), GENCONT/OCSELECT (Meuwissen 2002; Hinrichs et al. 2006) and EVA (Berg et al. 2006). Tools such as these are now being used in breeding programs for sheep, beef and dairy cattle, pigs, poultry and other species.

Implementation level ranges from providing target numbers of matings for each candidate animal, with the aim of maximising genetic gain at a defined rate of inbreeding, through to dictation of all stages of selection, culling, semen collection, mate allocation and backup matings, with comprehensive attention to the full prevailing range of technical and operational constraint issues. More complete implementations take a considerable amount of setting-up. However, once established they provide for routine optimal implementation of progressive breeding programs. They also constitute an appropriately competitive framework for implementation of new technical opportunities, such as reproductive boosting, targeting pools of opposing homozygotes for nominated genetic markers, or engineering extra genetic variation in key traits well before the physical splitting of a breeding line.

Look ahead mating strategies (Hayes et al. 2002; Shepherd and Woolliams 2004) have been proposed, aiming to exploit non-additive components some generations ahead. This includes the setting up of "investment matings" (such as F1 crosses) in order to capitalise on "realisation matings" (such as 3-breed cross using F1 females). This constitutes short-term management of genetic variation for longer term gains. A simulated example has been given by Li et al. (2006) where an MST operating on cohorts rather than individuals was implemented for simultaneous decision making over generations, resulting in the setting up of divergent lines to exploit dominance at known loci. This approach has the potential to give tactical targeting of a suitable route through genotypic changes across multiple interacting loci, should we ever have sufficient understanding of the genotypic effects involved.

4.1 Maintaining Genetic Variation Across-Populations

Somewhat similar approaches have been developed to help make decisions about which breeds/populations to place emphasis on in programs that have maintenance of genetic diversity as their objective (Eding et al. 2002; Simianer et al. 2003), or part of their objective (Piyasatian and Kinghorn 2003). Genetic marker information is more relevant for maintaining variation between populations, as little or no migration between populations results in considerable disequilibrium – it is easier to maintain or improve this state than it is to engineer maintained or increased variation within a population, which almost invariably will have a very much higher degree of equilibrium.

However, when considering whole populations, an increased opportunity exists to use information other than genetic marker data to describe genetic variation. In particular, variation in trait merit deserves some attention, even although there may be challenges in relation to confounding with environment. This is particularly true for traits such as disease resistance and distinct morphological traits, but there is potential future value across traits (Smith 1985).

References

Bäck, T., Ed. 2003. *Handbook of Evolutionary Computation*, Institute of Physics Publishing, Bristol.

Banks, R. 2000. Genetic improvement in the animal breeding industries. In *Animal Breeding – Use of New Technologies* (B.P. Kinghorn, J.H.J. van der Werf, and M. Ryan, eds.), The Post Graduate Foundation in Veterinarian Science of the University of Sydney, Chapter 21.

Barker, J.S.F. 2001. Conservation and management of genetic diversity: a domestic animal perspective. *Can. J. For. Res.* **31**: 588–595.

Barton, N.H. and Keightley, P.D. 2002 Understanding quantitative genetic variation. *Nat. Rev. Genet.* **3**: 11–21.

Berg, P., Nielsen, J. and Sørensen, M.K. 2006. EVA: Realized and predicted optimal genetic contributions. *Proc. 8th World Congr. Genet. App. Livest. Prod.* Comm. No. 27-09.

Borevitz, J.O., Hazen, S.P., Michael, T.P., Morris, G.P., Baxter, I.R., Hu, T.T, Chen, H., Werner, J.D., Nordborg, M., Salt, D.E., Kay, S.A., Chory, J., Weigel, D., Jones, J.D.G. and Ecker, J.R. 2007. Genome-wide patterns of single-feature polymorphism in *Arabidopsis thaliana*. *Proc. Nat. Acad. Sci.* **104**: 12057–12062.

Brascamp, E.W. 1984. Selection indices with constraints. *Animal Breeding Abstracts* **52**: 645–654.

Brisbane, J.R. and Gibson, J.P. 1994. Balancing selection response and rate of inbreeding by including genetic relationships in selection decisions. *Proc. 5th World Congr. Genet. App. Livest. Prod.* **19**: 135–138.

Eding, J.H., Crooijmans, R.P.M.A., Groenen, M.A.M. and Meuwissen, T.H.E. 2002. Assessing the contribution of breeds to genetic diversity in conservation programs. *Genet. Sel. Evol.* **34**: 613–633.

Falconer, D.S. and Mackay, T.F.C. 1996. *Introduction to Quantitative Genetics Fourth Edition*, Longman Inc., New York.

Grundy, B., Villanueva, B. and Woolliams, J.A. 1998. Dynamic selection procedures for constrained inbreeding and their consequences for pedigree development. *Genet. Res.* **72**: 159–168.

Hayes, B.J., Shepherd, R.K and Newman, S. 2002. Look ahead mate selection strategies for multibreed populations. *Anim. Sci.* **74**: 13–24.

Hinrichs, D., Wetten, M. and Meuwissen, T.H.E. 2006. An algorithm to compute optimal genetic contributions in selection programs with large numbers of candidates. *J. Anim. Sci.* **84**: 3212–3218.

Jannink, J.-L. 2003. Selection dynamics and limits under additive x additive epistatic gene action. *Crop Sci.* **43**: 489–497.

Kinghorn, B.P. 2000. The tactical approach to implementing breeding programs. In *Animal Breeding – Use of New Technologies* B.P. Kinghorn, J.H.J. van der Werf and M. Ryan, (eds.), The Post Graduate Foundation in Veterinarian Science of the University of Sydney, pp 291–308.

Kinghorn, B.P. and Shepherd, R.K. 1999. Mate selection for the tactical implementation of breeding programs. *Assoc. Advmt. Anim. Breed. Genet.* **13**: 130–133.

Kinghorn, B.P., Meszaros S.A. and Vagg, R.D. 2002. Dynamic tactical decision systems for animal breeding. *Proc. 7th World Congr. Genet. Appl. Livest. Prod.* **33**: 179–186.

Li, Y., van der Werf, J.H.J. and Kinghorn, B.P. 2006. Optimisation of a crossing system using mate selection. *Genet. Sel. Evol.* **38**: 147–165.

Meuwissen, T.H.E. 1997. Maximizing the response of selection with a predefined rate of inbreeding. *J. Anim. Sci.* **75**: 934–940.

Meuwissen, T.H.E. 2002. GENCONT: An operational tool for controlling inbreeding in selection and conservation schemes. *Proc. 7th World Congr. Genet. Appl. Livest. Prod.* **28**: 769–770.

Meuwissen, T.H.E. and Sonesson, A.K. 1998. Maximizing the response of selection with predefined rate of inbreeding: Overlapping generations. *J. Anim. Sci.* **76**: 2575–2583.

Piyasatian, N. and Kinghorn, B.P. 2003. Balancing genetic diversity, genetic merit and population viability in conservation programs. *J. Anim. Breed. Genet.* **120**: 137–149.

Pong-Wong, R. and Woolliams, J.A. 2007. Optimisation of contribution of candidate parents to maximize genetic gain and restricting inbreeding using semi-definite programming. *Genet. Sel. Evol.* **39**: 3–25.

Shepherd, R.K. and Woolliams, J.A. 2004. Minimising inbreeding in small populations by rotational mating with frozen semen. *Genet. Res. Camb.* **84**: 87–93.

Simianer, H., Marti, S.B., Gibson, J.P., Hanotte, O. and Rege, J.E.O. 2003. An approach to the optimal allocation of conservation funds to minimize losses of genetic diversity between livestock breeds. *Ecol. Econ.* **45**: 377–392.

Smith, C. 1985. Scope for selecting many breeding stocks of possible economic value in the future. *Anim. Prod.* **41**: 403–412.

Storn, R. and Price, K. 1997. Differential evolution – a simple and efficient heuristic for global optimization over continuous spaces. *J. Global Optimization* **11**: 341–359.

Wray, N.R. and Goddard, M.E. 1994. Increasing long term response to selection. *Genet. Sel. Evol.* **26**: 431–459.

Managing Genetic Diversity, Fitness and Adaptation of Farm Animal Genetic Resources

Louis Ollivier and Jean-Louis Foulley

Abstract In this review, we first analyse the objectives to consider in preserving diversity, fitness and adaptability of farm animal genetic resources (AnGR), given the links between genetic diversity and fitness-adaptedness (FA) traits. Ways to measure diversity are then presented, and tools available for managing genetic diversity within given economic constraints are described, under a theoretical framework proposed by Martin Weitzman. The emphasis is on molecular diversity and the exploitation of genetic marker information for use in conservation management of AnGR, through the use of a diversity function which may be applied either to gene diversity (based on allele frequencies) or to allelic diversity (based on allele counts). The question is raised of the possible relationships between molecular diversity of AnGR and FA traits. Methods for investigating those relationships based on within-breed heterozygosity or between-breed molecular divergence are reviewed. The concepts are illustrated on data from a European study of pig diversity.

Keywords Animal genetic resources · biodiversity · conservation · diversity theory · genetic markers

1 Introduction

The present diversity of farm animal species reflects their adaptation to various environmental challenges as well as to the diversity of human needs in various parts of the world. Diversity is, on the other hand, recognised as an absolute prerequisite for the future evolution of farm animal species. In fact, conserving farm animal genetic resources (AnGR) presents some analogies with the more general problem of preserving biological diversity. Both aspects are indeed covered by the Convention on Biological Diversity, now ratified by most countries, following the Rio Conference of 1992. As noted by Reed and Frankham (2003), conservation of genetic variation in natural populations is needed for two essential reasons: (1) to

L. Ollivier (✉)
INRA-SGQA, 78350 Jouy-en-Josas, France
e-mail: louis.ollivier@free.fr

J. van der Werf et al. (eds.), *Adaptation and Fitness in Animal Populations*,

allow *evolution* in response to environmental changes, and (2) to maintain population *fitness* against inbreeding depression. So in a sense managing genetic diversity within species should amount also to managing their fitness-adaptedness (FA) traits. When it comes to domestic animal species, many other reasons justify conserving their genetic diversity. Various benefits to be derived from conserving AnGR have been detailed in a special issue of *Ecological Economics* (Drucker and Scarpa 2003).

The issue of farm animal diversity conservation, and its possible enhancement, has been receiving increased attention for the last 25 years. Various national and international bodies have been involved, and a multidisciplinary approach has generally been advocated. A number of scientific disciplines, such as genetics, economics, sociology, system analysis and ecology are recognised as necessary for an adequate analysis of AnGR situations. Whilst management of farm AnGR covers their characterisation, utilisation and conservation, this paper focuses on methodology for measuring diversity and application to conservation decision-making. We shall first analyse the *objectives* to be considered in the preservation of diversity, fitness and adaptability of farm AnGR. The way to *measure* diversity will then be presented, and *tools* available for managing genetic diversity within given economic constraints will be described. The emphasis will be on molecular diversity and the exploitation of genetic marker information. Finally, the possible relationships between *molecular and quantitative variation* will be discussed.

2 Objectives in Diversity Preservation

Diversity of the genetic stock, understood as genetic variation, is a prerequisite for animal improvement, since "variation is the raw material on which the breeder works" (Lush 1945). Genetic diversity may thus in a sense be considered as the primary value of genetic resources. In farm animal populations, both within-breed and between-breed variation have a role to play. The essential features of those two types of diversity are summarised in Table 1. In terms of FA, roughly speaking, *fitness* is a matter of within-breed variation whereas *adaptability* has more to do with between-breed variation.

It is to be noted that animal breeders have traditionally given much more attention to the maintenance of within-breed genetic variability than to the preservation of between-breed variability. Breed diversity, though important in the overall variation of each species, does indeed cover areas of variation which are often well outside the range of foreseeable production objectives. A clear perception by animal breeders of the economic dimension of breed conservation has for a long time been lacking. The economic relevance of AnGR conservation and its integration into mainstream animal breeding came progressively in the forefront in the early 1980s (Ollivier 2005).

As shown by Charles Smith in his pioneering paper (Smith 1985), when uncertainty and risk are taken into account in the benefits expected from selection, breeding for diversity may become an economically sound alternative. The role of between-breed diversity however needs a careful evaluation, if one wants to

Table 1 Comparative features of within- and between-breed genetic diversity

Feature	Within-breed diversity	Between-breed diversity
Utility	Fitness[1]	Adaptability[2]
Appraisal	Genetic parameters (heritabilities and genetic correlations)	Crossbreeding parameters (direct and maternal effects, heterosis)
	Heterozygosities	Allele frequencies (genetic distances)
Use in animal breeding	Genetic improvement	Choice of breeds or crossbreeding systems
Evolution in the medium/long term	Gradual changes of genetic variance	Abrupt changes through breed extinction
Maintenance	Management of inbreeding	*In situ* and *ex situ* conservation

[1]Concept 1 in Barker (2009)
[2]Concept 4 in Barker (2009)

give an objective basis to conservation policies required to maintain such diversity. It has been known for a long time that between-breed diversity plays an essential role in heterosis and complementarities between breeds exploited in current crossbreeding systems (Smith 1964). Compared to these relatively short-term advantages derived from breed (or strain) diversity possible longer-term targets for breed diversity in animal production have also been pointed out. These include: (i) *insurance* against changes in economic and environmental conditions, through the advantage of being able to cope with a range of production-marketing systems, as stressed by Smith (1985, 1986); (ii) increased *flexibility* of the genetic systems, as for instance in situations where breed replacement offers a meaningful alternative to within-breed improvement (e.g. Thiessen et al. 1985); (iii) preservation of *potentially useful characteristics*, not so far exploited, and which may even be presently unknown.

The advantages derived from genetic stock diversity are direct use values. They are associated with production and as such may be seen as a "producer good", as opposed to indirect uses and existence value which may be seen as a "consumer good" (Gollin and Evenson 2003). For example, environment/landscape arguments in favour of AnGR preservation should also be considered, as they involve local or national benefits. Mendelsohn (2003) gives the example of the dairy farms of North Eastern USA in a landscape of rolling green valleys surrounded by forested hills. Similar arguments hold for maintaining adapted breeds of cattle in the Alps. However, one should be aware that landscape/environment arguments may in some cases work against protecting domesticated animals. There may be more interest in maintaining wildlife in some areas, as for instance in East Africa.

Cultural and lifestyle value, as well as existence value, associated with farm animals are also strong arguments for diversity conservation. The cultural values of livestock populations, as historical witnesses and custodian of local traditions, have been analysed by Gandini and Villa (2003). These values are sometimes difficult to dissociate from the landscape/environment argument.

3 Measuring Genetic Diversity

3.1 Levels of Characterisation

A proper characterisation of the available genetic resources is obviously needed for their management, since the better the available resources are characterised the better they can be used (FAO 2007). Several levels of characterisation exist. First, breed inventories are being made to collect information aimed at a wide coverage of the genetic resources of each species. Extensive breed inventories, launched in Europe in the early 1980s, have been available for several years (Simon and Buchenauer 1993; Loftus and Scherf 1993) and are now stored in databases which may be accessed through various websites, at country, region or world level.

An important use of the breed information is for assessing the degree to which each breed is exposed to the risk of becoming extinct, i.e. its degree of endangerment (DE). Assessing DE properly, however, is a difficult task because it requires integrating genetic and demographic factors which interact in various ways to determine the survival prospect of any breed, as stressed by Gandini et al. (2004). The latter discuss various systems used for classifying breeds on their DE, in Europe as well as at the world level. Table 2 shows various criteria being presently taken into consideration.

Through regular updating the number of breeds recorded has continuously increased. The number of Mammalian breeds recorded worldwide has thus been about doubled between 1993 and 2000 (Scherf 2000). Though the number of breeds known to exist in a country, or in a region of the world, is a valid indicator of

Table 2 Criteria for evaluating the degree of endangerment of animal breeds (adapted from Gandini et al. 2004)

CRITERIA	Europe EU	EAAP[1]	World FAO[2]	RBI[3]
Census	EU countries pooled	Individual countries	Individual countries	Individual countries
Overall population size			X	
Population trend	X	X	X	
Number of breeding males (M)			X	
Number of breeding females (F)			X	
Effective size[4] (N_e)		X		
Annual registration of females				X
Generation interval	X	X		X
Additional criteria		X		X

[1]European Association for Animal Production
[2]Food and Agriculture Organisation of the United Nations
[3]Rare Breed International (Alderson 2003)
[4]$N_e = 4MF/(M+F)$ as defined by Wright (1931): a simple measure linked to the rate of inbreeding under the assumption of no selection (see Kinghorn et al. 2009)

diversity, it remains of limited value when comparisons are being made between different countries or regions, because the data presented very much reflect the information provided by each country. In spite of the efforts made to co-ordinate the inventories across different countries, through the FAO-driven Domestic Animal Diversity Information System (DAD-IS at http://dad.fao.org/), the information displayed should not be considered as necessarily representing a comprehensive and comparable coverage of all situations. Breed inventories may also include succinct information on the production characteristics of the various breeds, but this is mostly confounded with breed environment and therefore does not tell us much about real genetic diversity. Within country, however, confounding may to some extent be avoided by expressing breed performance as a deviation from a standard breed, as suggested by Simon and Buchenauer (1993). Well-designed experiments, such as diallel crosses, would be costly to implement in farm animals, because of the testing facilities required as soon as the number of breeds exceeds three or four. When the number of breeds to compare is large, multibreed designs, able to test a large number of breeds with a few animals per breed (Taylor 1976), offer a suitable approach, well adapted for instance to the testing of diversity among a large number of rare breeds.

Genetic markers offer the great advantage of a direct assessment of genetic diversity, by definition independent from environmental effects. Molecular markers have a set of properties that make them a tool of choice in the evaluation of genetic variability. Let us mention their polymorphism, their ubiquity over the genome and the possibility of automating their identification. An important question is to know whether the markers' variability well reflects variability at quantitative trait loci, for instance variability of production or FA traits. This aspect has been and continues to be widely discussed. Several studies, in animals as well as in plants, show that molecular and quantitative variabilities are often connected, even though the degree of those connections and their generality remain a matter of debate (Reed and Frankham 2001; Barker 2001). This aspect will be further discussed in the last section of the paper.

3.2 Marker-Based Measures of Diversity

3.2.1 Gene Diversity

One classical method of measuring genetic diversity is the calculation of *fixation indices* (F) as proposed by Wright in 1943 (see Nei 1977). F_{ST} in particular measure the degree of genetic differentiation among a set of populations, based on analyses of variance of allele frequencies (Weir and Cockerham 1984). A different approach is that of Nei (1973), who defines a concept of *gene diversity* and shows that the total gene diversity G_{ST} can be partitioned into intrapopulational and interpopulational gene diversities, based on the actual heterozygosities observed at several loci. Wright and Nei indices both derive from the concepts of identity and genetic distance between populations, which are the grounds of most methods of genetic diversity measurement. Here we shall retain the method advocated by an economist (Weitzman 1992, 1993), which is based on the more general concept of *distance*.

The basic problem tackled by Weitzman (1992) is to measure the diversity of a set S, denoted V(S), out of given distance measures between all pairs of members of the set. Members may represent species, subspecies, breeds, varieties, or any other taxonomical unit. Weitzman chooses to extend the notion of distance *between two members* by defining the distance *between a member and the remaining set* as the distance between the member considered and its closest neighbour in the set. The diversity function of Weitzman can be defined in words as the maximum, over all members of the set, of the distance of the member from its closest neighbour in the set augmented by the diversity of the set without that member. Taking $S\backslash i$ to define the set without member i, the formal definition of V is:

$$V(S) = \max_{i \in S}[d(i, S\backslash i) + V(S\backslash i)] \tag{1}$$

The above function is a recursion, which makes its computation quite challenging when the number n of members in set S becomes large. Fortunately, a "clever" algorithm, detailed in Weitzman (1992) and illustrated by an example in Thaon d'Arnoldi et al. (1998), allows considerable reduction in calculation complexity. The procedure starts by finding the two closest neighbours in the whole set, and the distance between them is taken as the initial value of V. One of the two is then defined as a "link" and the other as a "representative". Then the link is dropped and the same procedure is repeated on the remaining set of $n-1$ members. The distance between the two closest neighbours in this set is added to V and a new link is determined. The procedure goes on until only one member remains, and the final V is then a sum of $n-1$ distances.

Function (1) possesses a number of "nice" properties, two of which deserve being highlighted, namely (i) monotonicity in member, which means that the change in diversity resulting from the loss of a member of the set can never be positive, a property which does not apply to variance-based diversities, and (ii) continuity in distances, meaning that slight changes in distances also induce slight changes of diversity.

Formula (1) can obviously be applied to any subset of S. In particular, the decrease of diversity resulting from the loss of any given member i may be expressed as:

$$CB_i = 1 - V(S\backslash i)/V(S) \tag{2}$$

CB_i may be seen as the contribution of member i to between-member diversity, and considered as an estimate of its genetic uniqueness, thus leading to a partition of the total diversity among the individual members of the set. In this context it should be mentioned that the solution of (1) generates a tree representing a "perfect" taxonomy corresponding to a matrix of ultrametric distances.[1] The diversity function is the length of the tree and the algorithm explicitly determines the position of each member-leaf. Consequently, diversity can be "read" on the resulting tree, since each branch length represents (though approximately when the distances are not ultrametric) the contribution CB_i of the corresponding member (as defined above). The diversity decrease due to the loss of a group of breeds can also be read either by adding the ordinates of the "ancestors" disappearing from the tree, or as the ordinate of the group's common ancestor. This helps in understanding wh[illegible]vidual contributions may or may not add up, depending on their relative p[illegible]e tree. A software implementation of the Weitzman method, down t[illegible]wing, is available (Derban et al. 2002, 2005).

The diversity function V, developed in a framework of spec[illegible]vation, can also be applied to the situation of livestock breed diversity, as [illegible]ed by Thaon d'Arnoldi et al. (1998). Though the method has only recently b[illegible]n extended to the world of livestock species, it has already been applied to several farm animal species, such as cattle (Thaon d'Arnoldi et al. 1998; Canon et al. 2001; Simianer et al. 2003; Tapio et al. 2006; European cattle diversity consortium 2006), sheep and goats (Barker et al. 2001; Fadlaoui 2006), pigs (Laval et al. 2000; Ollivier et al. 2005), horse (Plante et al. 2007), poultry (Pinent et al. 2005) and fish (Rutten et al. 2004). In all studies, highly unequal breed contributions to diversity were evidenced. Breed diversity in those studies was based on genetic distances. The method, however, is quite general, and may be applied to other types of diversity, such as cultural diversity considered by Gandini and Villa (2003). As an illustration, a partitioning of microsatellite diversity among 22 breeds of pigs is presented in Fig. 1. The data are drawn from a European investigation described in Ollivier et al. (2003), and the panel of 50 microsatellites markers is detailed in Groenen et al. (2003). In this sub-sample of 22 breeds (as in the complete study involving 68 European domestic breeds), more than half of the breed diversity could be assigned to the 11 local breeds. Similar examples showing large contributions of "native" breeds have been found in cattle and sheep (Tapio et al. 2006). Figure 1 shows extremely variable breed contributions and also, perhaps surprisingly, that the highest European contribution (11.4% for the *Tamworth*) is not far behind the contribution of a highly "exotic" breed like the Chinese *Meishan* (13.3%). The corresponding Weitzman tree exemplifies the non-additivity of individual contributions, since in general the joint contribution of two breeds will be different from the sum of their corresponding individual contributions, i.e. $CB_{ij} \neq CB_i + CB_j$. Here the joint contribution of the two *Duroc* strains considerably exceeds the sum of their individual contributions (9.9% *vs* 4.2%), the reverse being seen for the *Gloucester Old Spots* and *Tamworth* breeds (9.9% *vs* 19.1%).

Other methods for analysing diversity have been proposed, either based on the gene diversity parameters of Nei (Petit et al. 1998), or on the concept of coancestry (Eding et al. 2002; Caballero and Toro 2002). These alternative methods, based on

[1] Distances are said to be ultrametric when the larger two distances of any triplet are equal

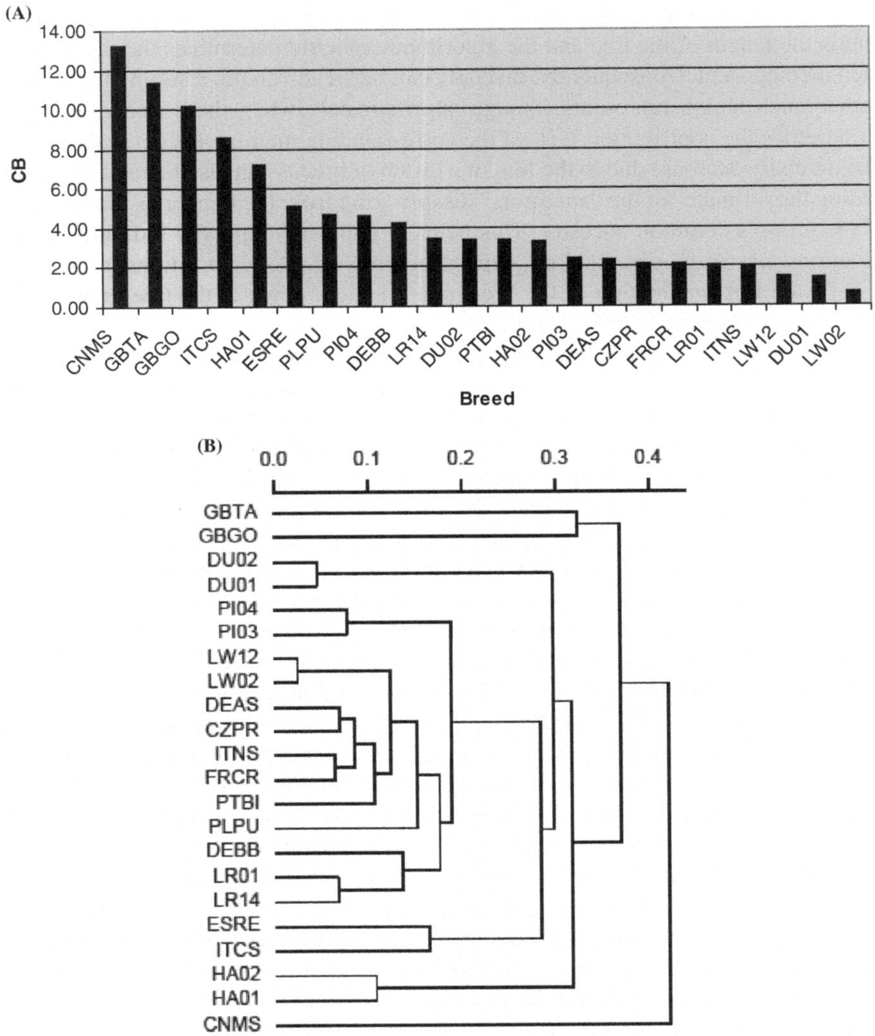

Fig. 1 Diversity partitioning among 22 breeds of pigs and corresponding Weitzman tree, based on 50-microsatellite Reynolds genetic distances (**A**) Between-breed diversity contributions (CB), in percent of total diversity (**B**) Corresponding Weitzman phylogenetic tree. Breed codes: CZPR: Presticke from Czech Republic; DEAS: Angler Sattelschwein from Germany; DEBB: Bunte Benheimer from Germany; HA02: Hampshire line from Germany; LR14: Landrace line from Germany; LW02: Large White line from Germany; PI03: Piétrain line from Germany; ESRE: Retinto from Spain; FRCR: Créole from France; LR01: Landrace line from France; LW12: Large White line from France; DU02: Duroc line from UK; GBGO: Gloucester Old Spots from UK; HA01: Hampshire line from UK; PI04: Piétrain line from UK; GBTA: Tamworth from UK; ITCS: Cinta Senese from Italy; DU01: Duroc line from Italy; ITNS: Nera Siciliana from Italy; PLPU: Pulawska from Poland; PTBI: Bisaro from Portugal. More details on breeds and markers are available at http://www.projects.roslin.ac.uk/pigbiodiv/publications.html

genetic variance and therefore restricted to genetic diversity, have been shown to give in practice results very similar to the Weitzman method in terms of *CB*, though they may yield negative contributions (Ollivier and Foulley 2005). In addition, the methods based on coancestry differ by allowing the contribution of each breed to a "core set" to be adjusted in order to maximise diversity.

The Weitzman approach has been criticised on the grounds that the "method or any other phylogenetic method that ignores within-breed variability should not be used when applied to within-species diversity" (Caballero and Toro 2002). This argument, however, may easily be refuted since one can combine the breed *CB* contribution defined above with the breed *CW* contribution to within-breed variability. By the way, this may be seen as an equivalent for farm animals of the "principle of complementarity" invoked in species conservation, whereby species are not considered only according to their dissimilarity from other species but also according to their intraspecific diversity (Bonin et al. 2007). The within-breed contribution *CW* may be defined, in a way similar to *CB*, as $\mathrm{CW}_i = 1 - \mathrm{H}(\mathrm{S}\backslash i)/\mathrm{H}(\mathrm{S})$, where H(S) is the mean heterozygosity of the set and $\mathrm{H}(\mathrm{S}\backslash i)$ the same quantity measured on the set without *i*, or alternatively by using a diversity function based on genetic distances between individuals. A global diversity *D* can then be obtained by applying appropriate weights α and β to *CB* and *CW*, such as $D = \alpha CB + \beta CW$ (Ollivier and Foulley 2005). Weights should be chosen according to the objective pursued and these may differ considerably depending on the use to be made of genetic diversity (see Table 1). Another criticism of the method is that it would tend to favour the most inbred breeds (European cattle diversity consortium 2006). It should however be noted that such a situation requires that *CB* and *CW* be negatively correlated, which is not a general rule (but see Section 5.2 below), and that within-breed diversity be ignored. In general, global diversity can be taken into account by combining *CB* and *CW* as indicated above.

3.2.2 Allele Diversity

Allelic richness, defined as the number of alleles per locus, is a diversity measure of obvious interest in conservation genetics. The observed allelic richness, however, needs correction for sample size, since the chances to discover a new allele increase each time a new individual is observed. This is a general problem in ecology, and several methods have been proposed for taking sample size into account (Foulley and Ollivier 2006a).

El Mousadik and Petit (1996) proposed to estimate allelic richness (*r*) by using the *rarefaction* methodology, based on the idea of taking the expected number of species in a sample of *g* individuals as a measure of species diversity (Hurlbert 1971), and replacing "species" by alleles and "individuals" by genes. The method is based on the probability $P_{ik}^{(g)}$ for an allele *k*, though observed in population *i*, to be absent in a sample of smaller size *g*, taken as a reference for comparing all populations. Allelic richness at any given locus is obtained by adding over the K_i alleles observed:

$$r_i = \sum_{k=1}^{K_i} \left(1 - P_{ik}{}^{(g)}\right) = K_i - \sum_{k=1}^{K_i} P_{ik}{}^{(g)} \quad (3)$$

where $P_{ik}{}^{(g)} = C(N_i - N_{ik}, g)/C(N_i, g)$, N_{ik} being the number of k alleles in the sample of size N_i.

Allelic richness (R) may otherwise be defined by *extrapolation*. Here one adds to the number of alleles actually observed in each population the number of alleles expected to be missing in that population, given its sample size and assuming the populations to be drawn at random from a "founder" population made of the whole set of populations (Foulley and Ollivier 2006a). Hence the following estimation of allelic richness R of population i based on a sample of size N_i in which K_i alleles are observed:

$$R_i = K_i + \Sigma_{km}(1 - \pi_{km})^{N_i} \quad (4)$$

where $\pi_1, \pi_2, \ldots, \pi_k, \ldots$ are the overall allelic frequencies in the sample considered and km means $k \in A^{(m)}$, the set of alleles actually missing in the i sample.

The concept of allelic richness quite naturally leads to that of *allele diversity*, which depends on the existence of alleles specific to each breed. An allelic dissimilarity (or "distance") may be defined as the number *d(i, j)* of alleles present in population i and absent in population j. It should be noted that allele dissimilarities are not symmetric since in general $d(i, j) \neq d(j, i)$. In order to account for sample size, Foulley and Ollivier (2006b) propose either a rarefaction-based distance using the $P_{ik}^{(g)}$ defined above:

$$d_r(i, j) = \Sigma_k(1 - P_{ik}{}^{(g)})P_{jk}{}^{(g)} \quad (5)$$

where the summation is over all observed alleles of i and j, or an extrapolation-based distance:

$$d_e(i, j) = \Sigma_{kom}[1 - (1 - \pi_{kom})^{N_j}] + \Sigma_{kmm}(1 - \pi_{kmm})^{N_i}[1 - (1 - \pi_{kmm})^{N_j}] \quad (6)$$

where the first summation is on the alleles $k \in A^{(om)}$, the set of alleles observed in i and missing in j, called *kom*, and the second summation is on the alleles $k \in A^{(mm)}$, the set of alleles missing in both populations, called *kmm*.

A parallel may be drawn with the Weitzman approach previously described. Given a set of n breeds, the diversity function V(S) applying here is the total allelic richness of the set S. The "species/library" model of Weitzman (1998), whereby species gene pools are compared to books in libraries, can be transposed to alleles in breeds, and the allelic diversity function is equivalent to the number of different books in a set of libraries. The diversity functions based on rarefaction and extrapolation will usually give similar results, at least in terms of relative diversity. Extrapolation, however, is recommended when sample sizes are low on average or highly unbalanced among populations.

Equivalently to the CB_i defined for gene diversity in (2), the contribution of breed i to allele diversity is the number of alleles present in population i and absent in all others. This is a measure of the allelic uniqueness of breed i, also called its "private" allelic richness. Here again simple counts, which ignore the number of copies of each private allele, should preferably be corrected for sample size. Kalinowski (2004) has proposed a rarefaction-based estimation, $PR_i = \Sigma_k(1 - P_{ik}^{(g)})\Pi_{j\neq i} P_{jk}^{(g)}$. In this approach, the probability of allele k being present in two populations i and j – easily extended to any number of populations – is defined as $1 - P_{ik}^{(g)} P_{jk}^{(g)}$, whereas the breed contribution of El Mousadik and Petit (1996) is based on a $P_{ijk}^{(g)}$ obtained by applying rarefaction to the joint sample of size $N_i + N_j$. The extrapolation-based estimate (PE_i) is the first term of equation (6) in which N_j is the total sample size of all populations other than i (Foulley and Ollivier 2006b). An example illustrating the two methods, based on the same breeds and microsatellites as in Fig. 1 is presented in Table 3. Due to the correction applied, the two methods widely diverge in their estimates of individual breed allelic richness, though they do not much differ in private allele estimates ($r = 0.99^{**}$). It should be noted that 5 populations devoid of any observed private alleles yield rarefaction-based positive numbers of private alleles. Figure 2 shows that with both methods the allelic richness of a set of breeds tends to equal the observed richness as the number of breeds increases, with rarefaction below and extrapolation above observed values. Figure 2 however shows a large difference in rarefaction-based allelic richness whether cumulated according to Kalinowski (2004) or El Mousadik and Petit (1996).

A note of caution is appropriate on the parallel which could be made between the partitioning of gene diversity and that of allele diversity into between-breed

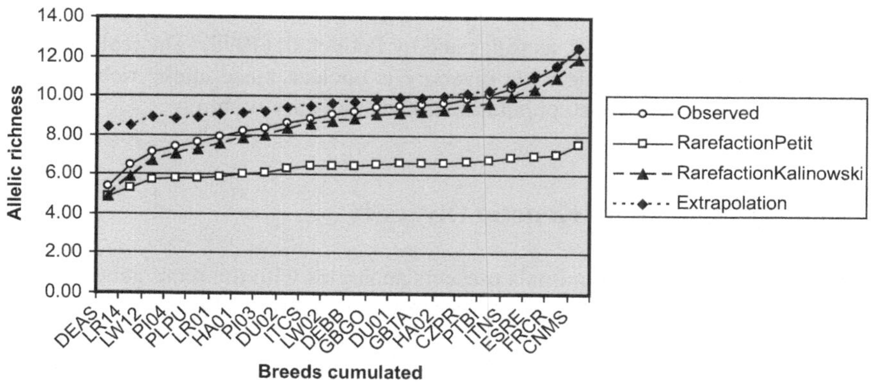

Fig. 2 Cumulated allelic richness over 22 breeds of pigs, in average number of alleles over 50 loci: observed and corrected by rarefaction – "RarefactionPetit" according to El Mousadik and Petit (1996) or "RarefactionKalinowski" according to Kalinowski (2004), as explained in text – or by extrapolation. The breeds are ranked in increasing order of private allelic richness from German Angler Sattelschwein to Chinese Meishan

Table 3 Estimation of allelic richness per locus and total number of private alleles for 50 loci in 22 breeds of pigs: rarefaction(RAR)- or extrapolation (EXT)-derived estimates. Breed codes are given in Fig. 1

Breed Status	Breed	Allelic Richness		Private Alleles	
		RAR	EXT	RAR	EXT
Local	CZPR	5.27	9.05	4.36	3.51
	DEAS	4.85	8.43	0.85	0.00
	DEBB	4.18	7.93	1.38	2.08
	ESRE	5.04	8.46	11.29	12.92
	FRCR	6.46	9.96	22.51	18.20
	GBGO	3.00	6.55	3.39	2.44
	GBTA	2.85	6.63	2.45	2.62
	ITCS	3.96	8.13	3.24	1.25
	ITNS	5.90	9.52	10.81	10.00
	PLPU	3.95	7.67	0.57	0.00
	PTBI	5.09	8.84	3.18	4.06
	DU01	3.79	7.36	2.79	2.50
International	DU02	3.09	6.63	1.29	1.00
	HA01	3.01	6.58	1.42	0.98
	HA02	4.19	8.13	3.33	3.22
	LR01	4.94	8.85	0.90	0.62
	LR14	4.04	7.68	0.33	0.00
	LW02	4.90	8.68	1.92	1.56
	LW12	4.56	8.93	0.64	0.00
	PI03	4.53	8.26	1.60	1.00
	PI04	3.82	7.40	0.32	0.00
Exotic	CNMS	4.35	7.31	46.60	43.95
	MEAN	**4.35**	**8.04**	**5.69**	**5.09**

and within-breed components, as suggested by Petit et al. (1998). The real meaning of within-breed variability in allele diversity is unclear, since allelic richness is a characteristic attached to each population and cannot vary within it.

4 Tools for Managing Genetic Diversity

Conserving breeds of farm animals presents analogies with the more general question of preserving biological diversity. In both cases, owing to the limited resources which can be devoted to conservation, the central question is "what to protect" (see Vane-Wright et al. 1991). Once diversity is analysed, additional information is needed on vulnerability and costs of conservation for each taxonomical unit. The choices are difficult and an operational framework is needed. An economic theory for guiding actual conservation policies in a diversity-improving direction has been provided in a series of papers by Weitzman (1992, 1993, 1998). The problem of optimum diversity preservation under a budget constraint has been formalised as the "Noah's Ark Problem" (Weitzman 1998). Essentially, boarding the Ark is a metaphor for investing in a conservation project to improve survivability of a set of n elements (e.g. species or breeds) subject to an overall budget constraint.

4.1 The Information Needed

The four fundamental ingredients of an optimum policy are *utility, distinctiveness, probability of survival* and *cost* of improving survivability. These quantities are defined for each breed i as U_i, D_i, P_i and C_i, respectively. Utility is a measure which can be obtained by various procedures detailed elsewhere (see Ollivier 2005). We shall concentrate here on how the three remaining factors can be combined.

4.1.1 Distinctiveness and Probability of Survival

The approach to distinctiveness advocated by Weitzman is based on the concept of distance, as we have seen above. More generality may be given to the diversity function by taking into account the risks of extinction of each breed. Based on the survival probability P_i of any breed i over a period of time t, 2^n extinction-survival patterns may occur with given probabilities, and assuming those events to be independent an *expected present discounted diversity* can be defined (Weitzman 1993):

$$\mathrm{E(V)} = \sum_t \sum_{\mathrm{Q}} \mathrm{P(Q}, t)\mathrm{V(Q)}(1 + r)^{-t} \tag{7}$$

where V(Q) is the diversity of subset Q, P(Q,t) its survival probability at time t, and r the discount rate. Risk of extinction or degree of endangerment can be evaluated using the criteria defined in Table 1, examples of which can be found for African cattle in Reist-Marti et al. (2003) or for Europe in Gandini et al. (2004).

According to whether gene or allele diversity is being considered (see Section 3.2), V(Q) will be the diversity function – equation (1) – or the allelic richness – equations (3) or (4) – of the subset considered. The *distinctiveness* D_i of breed i, which may also be called its *marginal diversity*, is then obtained as the partial derivative of expected diversity with respect to P_i. D_i expresses the increase in expected diversity with respect to the increase in survival probability of breed i (Weitzman 1993). Equation (7) may easily accommodate the case when a subset of breeds is safe, and distinctiveness then only applies to endangered breeds. An application to the endangered breeds of pigs considered in Fig. 1 is presented in Table 4, ignoring the discount rate r in equation (7). Distinctiveness is here limited to the 11 local breeds, assuming the remaining set of 11 international breeds is safe, though the diversity evaluations, in all situations, include the 22 breeds considered.

Gene distinctiveness and allelic distinctiveness are negatively correlated ($r = -0.32$ ns). Estimates of allelic distinctiveness with rarefaction and extrapolation are in general agreement ($r = 0.97^{**}$), though markedly lower values are obtained with extrapolation compared to rarefaction for the two breeds devoid of any private allele,

Table 4 Gene and allelic distinctiveness of 11 endangered breeds of pigs and BREED codes defined in Fig. 1

BREED	CB	PR	PE	GD	ADR	ADE
CZPR	3.81	3.51	0.06	4.41	5.42	4.16
DEAS	4.19	0.00	0.42	4.37	1.66	0.55
DEBB	7.59	2.08	0.54	7.48	2.09	2.31
ESRE	9.13	12.92	0.06	8.74	12.46	13.91
FRCR	3.81	18.20	0.15	3.75	23.87	18.91
GBGO	18.12	2.44	0.29	17.97	3.84	2.79
GBTA	20.13	2.62	0.10	20.12	3.19	2.74
ITCS	15.33	1.25	0.09	15.16	3.72	1.56
ITNS	3.57	10.00	0.10	3.78	11.64	10.70
PLPU	8.33	0.00	0.38	8.20	0.94	-0.07
PTBI	5.98	3.51	0.27	6.04	3.48	4.09

CB: breed contribution (%) to genetic diversity
PR: breed contribution (in number of private alleles for 50 loci) to allelic diversity based on extrapolation, from Table 3
PE: probability of extinction as defined in Ollivier et al. (2005)
GD: gene distinctiveness standardised to sum 100
ADR: rarefaction-based allelic distinctiveness (total number of alleles for 50 loci)
ADE: extrapolation-based allelic distinctiveness (total number of alleles for 50 loci)

Angier and *Pulawska*. The latter breed shows a low negative value with extrapolation, indicating that this correction does not guarantee the monotonicity in breed of the diversity function used. Monotonicity in breed is not even guaranteed when distinctiveness is based on the El Mousadik and Petit (1996) breed contributions (results not shown).

4.1.2 Costs of Improving Survivability

The costs C_i of improving survivability of breed i will in general depend on the desired increase in survival probability of the breed considered, ΔP_i, and vary according to the situation of each breed and to the various techniques of conservation.

4.2 Finding an Optimum Solution

When we know for each breed its utility, distinctiveness, the chosen increases in survival probability, ΔP, and the corresponding costs C, the formal solution of the Noah's Ark Problem, as demonstrated by Weitzman (1998), is an "extreme policy", which means that each breed should be made either totally safe or left to become extinct, within the budget allowed. Ranking the breeds is a linear programming problem whose solution is to rank the breed projects according to the following criterion:

$$R_i = (U_i + D_i)\Delta P_i / C_i \tag{8}$$

This shows the need to link survivability to conservation expenditures, and to include the relevant information on preservation costs, which in general may vary according to the situation of each breed. Two particularly simple situations, however, are worth considering. The first situation is when the costs C are proportional to the survivability improvements achieved ΔP. One can see from equation (8) that $\Delta P_i/C_i$ is then a constant and the diversity ranking should be based on the breeds' distinctiveness. In some cases the more endangered breed may appear as less distinctive, as the crowned-crane paradox of Weitzman (1993) shows. It will then be optimal to make the safe breed even safer at the expense of making the endangered one even more endangered. A similar situation, though not so extreme, is also illustrated for gene distinctiveness in Table 4, where the more endangered *Angler* appears to be less distinctive than the *Presticke*, though the latter has a lesser contribution to diversity. The second simple situation is when the costs C_i may be assumed to be the same for all breeds. Then if ΔP_i is taken at its maximum value, i.e. equal to the breed extinction probability E_i, ($E_i = 1 - P_i$), the optimal diversity ranking should be based on the product $D_i E_i$, named *conservation potential* (Weitzman 1993). This parameter, considered by Weitzman as the "single most useful indicator" in the absence of specific knowledge about preservation costs, would apply to livestock species conservation, where cryopreservation is available and can be assumed to make a breed safe. In line with equation (8), and assuming cryogenic preservation costs to be equal over different breeds of the same species, cryopreservation priorities could then be based on a *cryopreservation potential* analogous to the conservation potential of Weitzman.

The logic underlying equation (8) in those simple situations is illustrated in Table 5 on a set of 4 breeds, which combines extreme cases of breed uniqueness and degree of endangerment, utility being not considered. In this example, it can be seen that three different top priorities can be defined: either breed US based on distinctiveness, or breed UE based on conservation potential, or breed CE under a budget constraint. It can be seen that priority does not always go to the most endangered nor to the most unique breed. In practice, the relationship between costs of conservation and survival improvement will lie somewhere between constant costs and costs proportional to survival improvement. A wide spectrum of situations particularly applies to systems of live animal conservation, making it difficult to link expenditures and survival improvement. The possibility to develop iterative algorithms for ranking breeds on the basis of conservation costs is then an alternative to consider, as applied to breeds of cattle in Africa by Simianer et al. (2003).

It must be emphasised that equation (8) should be seen as a general framework, and may not be easy to apply in the real world, due to the difficulty to evaluate the four necessary ingredients, U_i, D_i, P_i and C_i, with sufficient precision. As to the diversity to consider, we should keep in mind, as stressed previously, that both between-breed and within breed components of diversity are to be considered. We have also seen that gene diversity and allele diversity contributions may not be similarly distributed over breeds (Table 4). Equation (8) in addition implies that proper weights should be given to distinctiveness and utility. Here again choices have to be made. A method to balance genetic diversity, perceived merit and population

Table 5 An example of diversity conservation diagnostics and priorities (top values in bold)

	Diversity conservation diagnostics				Diversity conservation under a budget constraint	
Breed[1] (i)	Contribution to diversity[2] (V_i)	Probability of extinction (E_i)	Distinctiveness (D_i)	Conservation potential ($D_i \times E_i$)	Cost of conservation (C_i)	Ranking criterion ($R_i = D_i \times E_i/C_i$)
CE	10	**0.90**	15	14	1	**14**
CS	10	0.10	55	6	2	3
UE	50	0.80	51	**41**	4	10
US	**70**	0.10	**65**	6	6	1

[1]CE: common endangered; CS: common safe; UE: unique endangered; US: unique safe

[2]$V_i = V(S) - V(S\backslash i)$ as defined in equation (2) assuming the matrix of ultrametric distances as follows.

	CE	CS	UE	US
CE	0	10	50	70
CS	10	0	50	70
UE	50	50	0	70
US	70	70	70	0

viability has been proposed by Piyasatian and Kinghorn (2003). Depending on the emphasis on diversity, the solutions may give priority to the conservation of pure breeds or of crosses between closely related breeds more adapted to enhancing populations' viability

5 Relationships Between Molecular and Quantitative Variation

There has been an explosion of projects in recent years aimed at calculating genetic distances between domestic breeds, mostly based on microsatellite marker loci. Studies of DNA markers, in addition to the information they provide on the history of domestication (Bruford et al. 2003), offer objective measures of diversity which can help decision-makers to identify the most unique breeds that may then be prioritised for conservation, as we have seen above. The relevance of molecular markers in studies of diversity, however, has been questioned on the grounds that neutral loci are being used and differences between populations at loci under selection are consequently not measurable (Ruane 1999). We know, on the other hand, that selection acts on the whole genome and the dynamics of multilocus systems should therefore be considered, particularly the interactions between selected and linked neutral loci. The so-called hitchhiking effect of selected genes is known to be determined by the relative importance of the recombination fraction between the neutral and selected loci and of the selection pressure applied to the latter (Maynard Smith and Haig 1974; Thompson 1977). Selection may thus affect diversity at both levels of within-breed and between-breed diversity of molecular markers. The *anonymous* (or *tautological*[2]) diversity, however not necessarily neutral, that genetic markers supposedly reflect is often opposed to *functional* diversity, when genes of interest are identified. Both are obviously of value, given the uncertainties on the future animal functions of interest. Marker diversity may indeed include some *cryptic* diversity of potential usefulness in the future, though escaping present detection, since molecular markers can only reveal, or "sign", past selection (or adaptation) episodes.

A priori some connection is to be expected between population FA traits on the one hand and within-breed and between-breed molecular diversity on the other hand. We shall consider in turn these two facets of the molecular-quantitative trait correlations, keeping in mind they are by nature intimately connected. The approach which will be discussed below, also called "hitchhiking mapping" (Schlötterer 2003), is marker-based and does not require measuring any particular trait. Alternatively, quantitative trait locus (QTL) mapping or candidate gene mapping, generally based on experimental crosses, requires measured traits in order to identify genome-wide targets of past selection. Quite extensive QTL maps are now available for most farm animal species. They show that markers linked to some traits can be found on all parts of the genome (see the example of Fig. 3). Adequate genome coverage

---

[2] As used by Arnason and Barker (2000) to qualify *tautological vs functional* fitnesses

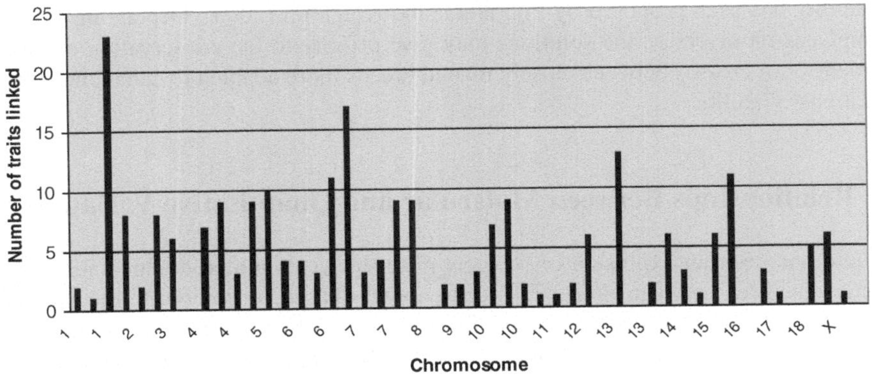

Fig. 3 Distribution of the number of traits reported in the literature as linked to the 50 microsatellites selected in a diversity study in pigs (data on QTL flanking markers from http://www.animalgenome.org/QTLdb/ as of 7 May 2007). The 50 loci are ranked according to their position on chromosomes 1 to 18 and chromosome X, with 2 to 4 loci per chromosome according to size (panel described in Groenen et al. 2003)

by markers is therefore necessary to generate evidence of molecular-quantitative relationships.

5.1 Fitness and Heterozygosity

It is currently accepted that heterozygous individuals possess fitness advantages and so departure from neutrality is expected to be common. In a classic paper, Watterson (1977) developed a test of neutrality based on a model of expected homozygosity F (sum of squared allelic frequencies) where a given number of alleles are subject to mutation and selection. The test criterion was the deviation of F from its expectation under neutrality, negative values indicating selection in favour of heterozygotes. Though the mutation model underlying the test is more likely to apply to allozymes than to microsatellites, the mostly negative (though non-significant) values observed for the microsatellite loci considered in Fig. 1 (47 negative locus average values out of 50, results not shown) may be indicative of a slight heterozygote superiority.

Another path, widely investigated in natural populations, is to study marker heterozygosity and fitness trait correlations (HFC). The theory behind HFC is rather complex and does not offer unequivocal explanatory mechanisms. Three possible explanations for positive HFC are generally invoked, namely (i) *direct effect* of the marker locus, not likely to apply to microsatellites though their intrinsic neutrality has been questioned (Kashi et al. 1997; Zhang and Hewitt 2003; Kashi and King 2006), (ii) *general effect* reflecting genome wide heterozygosity equally expressed at marker and fitness trait loci, and (iii) *local effect* for marker loci being in linkage disequilibrium with (over)dominant fitness trait loci. The latter effect can also explain why the correlation predicted under (ii), and approximated by $\sqrt{r/n}$ where r and n are, respectively, numbers of loci assayed and in the genome, should underestimate the true correlation (Aparicio et al. 2007).

The strength of HFC is a matter of considerable debate. Negative as well as positive correlations have been reported in natural populations (see references in Da Silva et al. 2006). In domesticated plants, attempts have also been made, with variable success, to correlate heterosis and marker heterozygosity derived from genetic distances. Very few such studies have been done in farm animals, though significant associations have been reported in pigs (Glodek 1974) and cattle (Goddard and Ahmed 1982). More recently, a highly significant correlation was reported by Gavora et al. (1996) between heterosis for egg production traits and heterozygosity for DNA fingerprints in poultry. Several studies have approached HFC by considering marker heterozygosity as a surrogate for inbreeding in order to evaluate inbreeding effects on various traits. Slate et al. (2004) showed that the correlation between individual heterozygosity H and inbreeding coefficient f is a rather simple function of the mean and variance of f and of the variance of H. Low H-f correlations are to be expected unless inbreeding is high and highly variable, and unless a sufficient number of markers are assayed. In the data they reviewed, the average predicted correlation was low (−0.21), and ranged from −0.08 (for a lowly inbred flycatcher population typed for 3 microsatellites) to −0.71 (for a highly inbred captive wolf population typed for 29 microsatellites). The low correlation between H and inbreeding depression on life-history traits at the individual level (see the meta-analysis of Coltman and Slate 2003) is in keeping with the above theoretical prediction. One may thus expect that heterozygosity and fitness at the population level will also be weakly correlated. This issue has been explored in a meta-analysis of 34 studies in which fitness was measured in three or more populations together with, effective size, heritability and heterozygosity, taken as three possible surrogates for genetic diversity (Reed and Frankham 2003). They found a highly significant correlation (r=0.43) between genetic diversity and population fitness, and all three diversity surrogates were also positively and significantly correlated with population fitness (correlations of 0.35 0.51 and 0.45 respectively), suggesting that marker heterozygosity may have a better predictive value for comparing populations than for comparing individuals. In contrast, Pfrender et al. (2000) demonstrated that molecular and quantitative genetic diversity were essentially uncorrelated within natural populations of *Daphnia*, while suggesting that molecular markers may be valid indicators of population subdivision for quantitative traits.

5.2 Adaptive and Molecular Divergence

Since Kimura (1968), the testing of neutrality has been a prime objective of population genetics. By 1986, 10 different methods were under discussion (in Endler's *Natural Selection in the Wild* quoted by Arnason and Barker 2000). The field has since been extensively developed and it is only very briefly reviewed here as it may be of interest for livestock species. The adaptation of domestic breeds to local conditions or to specific production objectives results in an increase in the between population diversity at loci underlying the traits under selection, and a reduction of within-population diversity at those loci. Thus, the methods for detecting diversifying selection can be divided in two classes, those based on the levels of differentiation between populations and those based on levels of diversity within-population. Such a dichotomy has been retained in the detailed review by Storz (2005) that we follow below.

5.2.1 Tests Based on Differentiation Between Populations

Differentiation between populations as measured by F_{ST} is the basis of the early test of selective neutrality of Lewontin and Krakauer (1973). The basic argument behind this LK test is that, in the absence of selection, differentiation at all loci should be the same. In such a situation, under the null hypothesis of neutrality, the expected variance of F_{ST} is $\leq 2\overline{F}^2/(n-1)$, where n is the number of populations, against which the observed variance of F_{ST} across loci can be tested. This variance, however, depends strongly on the pattern of relationship among populations, and any "structured" relationship will tend to increase its expectation (Robertson 1975). One would then expect a lower variance of F_{ST} when breeds exhibit a star-like pattern of phylogeny making them more independent compared to a clustering structure. This effect has indeed been confirmed in a study of AFLP diversity in pigs, in which a lesser departure from neutrality was observed by removing half of the breeds in order to approximate a star-like pattern of phylogeny (Foulley et al. 2006). The LK test, however, remained highly significant, in keeping with results obtained by other methods in the same species (Cameron et al. 2003) or in other species (e.g. Campbell and Bernatchez 2004; Cavagnaro et al. 2006). Several improvements of the LK test have been proposed. Robertson (1975) suggested adding to the expectation of F_{ST} a correction for the phylogenetic pattern equal to the squared coefficient of variation of the distances. This would amount to taking the expected variance of F_{ST} as equal to twice the variance of Reynolds distances when n becomes large, since $\overline{F}$ equals the mean of the multilocus Reynolds distances. Another proposed modification of the LK test is to replace the expected theoretical variance by the observed variance for a set of markers that can be assumed to be neutral (Pogson et al. 1997; Arnason and Barker 2000; Barker et al. 2001). The above modifications of the LK test preserve its advantage of being a model-free approach based on detecting outlier loci. This type of approach is likely to prevail when a large number of loci will be available, as with the single nucleotide polymorphism (SNP) study by Akey et al. (2002) in human populations.

Other improvements of the LK test have used simulations to generate multilocus distributions of F_{ST} under neutral models implying various assumptions on the populations' demographic history (reviewed in Storz 2005). This opens other possibilities for detecting outlier and presumably adaptive loci and thus for eventually defining an adaptive molecular diversity (Bonin et al. 2007). This class of tests, however, offers model-based tests of selection whose applicability to farm animals depends on finding an adequate population genetic model, which appears to be difficult due to the complex demographic history of livestock breeds.

5.2.2 Tests Based on Diversity Within Populations

The neutrality tests based on relative levels of diversity within populations exploit the reduction of variability around a selected locus due to hitchhiking, a phenomenon called "selective sweep". Schlötterer et al. (1997) proposed a test based on the variance of repeat number at microsatellite loci. In the absence of selection, the logarithm of the variance in repeat number is expected to follow an additive model over any set of populations-loci, and selective sweeps will be detected by population × locus interactions.

Unusually low variances for particular locus × population combinations reveal directional selection whereas unusually high variances indicate within-breed balanced polymorphism. Such a test was applied to the 42 (populations) × 50 (loci) subset from the pig data of Ollivier et al. (2005) and did not reveal significant overall population × locus interaction, though several locus × population combinations showed large deviations from their expectation (results not shown). This test, however, may be sensitive to the demography-mutation model assumed. In a search for more robust tests, Schlötterer (2002) suggested comparing a large number of microsatellite loci in two groups of populations. The test statistic is the log of the ratio of variance (ln RV) in repeat number in the two groups, which was shown to follow a normal distribution under neutrality. Computer simulations also showed ln RV distribution to be quite robust to various demographic situations. An application of this test to pig microsatellite data is presented in Fig. 4, which clearly shows the existence of two "outlier" loci with contrasting allele size distributions between groups of breeds.

6 Conclusion

Our task in preparing this review was to consider approaches to analysing domestic animal genetic resources in order to assist management decision-making on which resources to preserve and possibly improve for fitness-adaptability (FA) traits. Until now decision-makers have essentially been assuming that a sound management of AnGR will be beneficial to FA. We have also emphasised that, whatever the type of diversity considered, there is a need to measure this properly and to use appropriate management tools. The economic framework proposed by Martin Weitzman can be recommended, since it integrates diversity and vulnerability of AnGR as well as cost-effectiveness of the conservation techniques available. Nevertheless, the evidence on the benefits to FA from preserving genetic diversity in farm animals needs

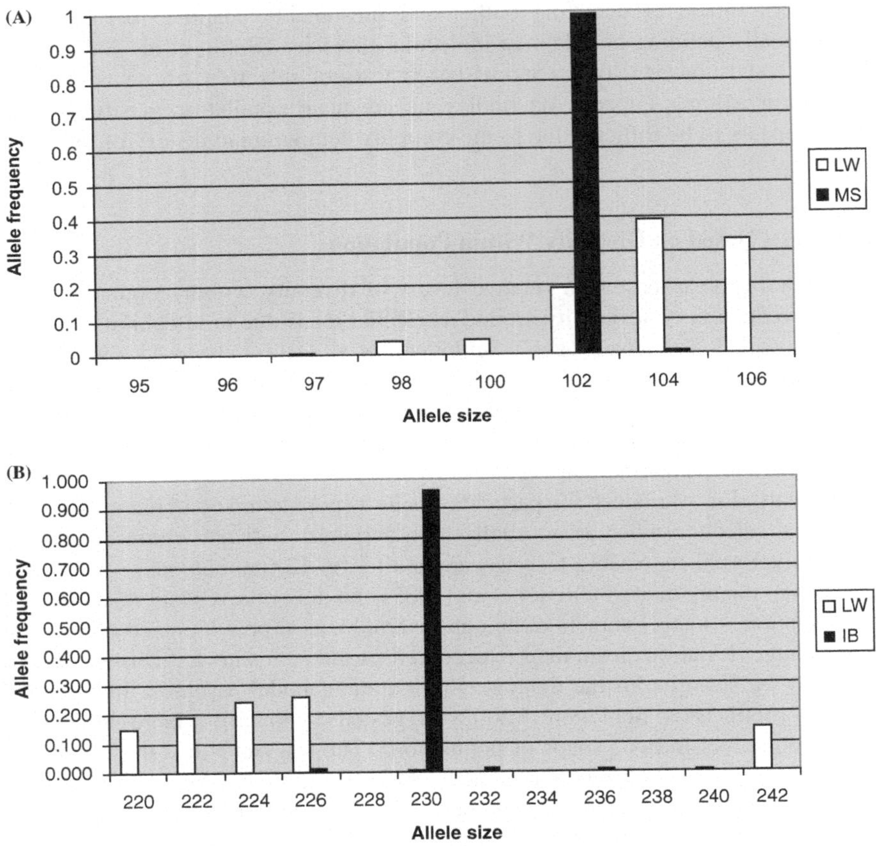

Fig. 4 Two examples of contrasting allele size distributions of microsatellites in groups of pig breeds: (**A**) Microsatellite SW2008 in a set of Large White breeds (LW: 820 alleles) compared to the Chinese Meishan breed (MS: 110 alleles); (**B**) Microsatellite S0228 in a set of Large White breeds (LW: 862 alleles) compared to Iberian breeds (IB: 228 alleles)

further substantiation. The situation is made difficult because of the large numbers of breeds available in each domestic species and our lack of objective assessments of FA traits over a sufficiently wide panel of germplasm. In emphasising the indirect means of molecular characterisation for assessing FA diversity, we have implicitly assumed that molecular reliability might compensate for the indirectness of the approach. Though more research is certainly needed on the controversial issues around molecular and quantitative trait diversity, molecular characterisation is now generally recognised as a valuable tool for exploring genetic diversity. Molecular technology for characterising AnGR has indeed been given priority in the recently agreed Global Plan of Action on Animal Genetic Resources (FAO 2007 and DAD-IS library at http://dad.fao.org/). Molecular characterisation of diverse AnGR is otherwise a tool for elucidating the genetic basis of FA traits. One may hope that with increasing knowledge of farm animal genomes more insight will be gained on those issues. The biological samples collected in diversity projects can in future be expected to gain in scientific value and to contribute to the devising of more efficient conservation programmes.

Acknowledgments The pig investigation used as an example in this paper has received support from several programmes of the European Union (contracts BI02-CT94-3044, RESGEN-CT95-012 and BIO4-CT98-0188), which are gratefully acknowledged. Special thanks should go to A. Archibald, C.S. Haley and A. Law (Roslin, UK) for the maintenance of the "pigbiodiv" database. Comments made by two anonymous referees are also gratefully acknowledged.

References

Akey, M., Zhang, G., Zhang, K., Li, J. and Shriver, M.D. 2002. Interrogating a high-density SNP map for signatures of natural selection. *Genome Res.* **12**: 1805–1814.

Alderson, L. 2003. Criteria for the recognition and prioritisation of breeds of special genetic importance. *Anim. Genet. Res. Information* **33**: 1–9.

Aparicio, J.M., Ortego, J. and Cordero, P.J. 2007. Can a simple algebraic analysis predict markers-genome heterozygosity correlation? *J. Hered.* **98**: 93–96.

Arnason, E. and Barker, J.S.F. 2000. Analysis of selection in laboratory and field populations. In *Evolutionary Genetics: From Molecules to Morphology* (R.S. Singh and C.B. Krimbas, eds.), Cambridge University Press, Cambridge, UK, pp. 182–203.

Barker, J.S.F. 2001. Conservation and management of genetic diversity: a domestic animal perspective. *Can. J. For. Res.* **31**: 588–595.

Barker, J.S.F. 2009. Defining fitness in natural and domesticated populations. In Adaptation and Fitness in Animal Populations. Evolutionary and Breeding Perspectives on Genetic Resource Management (J.H.J. van der Werf, H.-U. Graser, R. Frankham and C. Gondro, eds.), Springer, The Netherlands.

Barker, J.S.F., Tan, S.G., Moore, S.S., Mukherjee, T.K., Matheson, J.L. and Selvaraj, O.S. 2001. Genetic variation within and relationships among populations of Asian goats. *J. Anim. Breed. Genet.* **118**: 213–233.

Bonin, A., Nicole, F., Pompanon F., Miaud C. and Taberlet, P. 2007. Population adaptive index: a new method to help measure intraspecific genetic diversity and prioritize populations for conservation. *Conserv. Biol.* **21**: 697–708.

Bruford, M.W., Bradley, D.G. and Luikart, G. 2003. DNA markers reveal the complexity of livestock domestication. *Nat. Rev. Genet.* **4**: 900–910.

Caballero, A. and Toro, M.A. 2002. Analysis of genetic diversity for the management of conserved subdivided populations. *Conserv. Genet.* **3**: 289–299.

Cameron, N.D., van Eijk, M.J.T., Brugmans, B. and Peleman J. 2003. Discrimination between selected lines of pigs using AFLP® markers. *Heredity* **91**: 494–501.

Campbell, D. and Bernatchez, L. 2004. Generic scans using AFLP markers as a means to assess the role of directional selection in the divergence of sympatric whitefish ecotypes. *Mol. Biol. Evol.* **21**: 945–956.

Canon, J., Alexandrino, P., Bessa, I., Carlos, C., Carretero, Y., Dunner, S., Ferran, N., Garcia, D., Jordana, D., Laloë, D., Pereira, A., Sanchez, A. and Moazami-Goudarzi, K. 2001. Genetic diversity measures of local European beef cattle breeds for conservation purposes. *Genet. Sel. Evol.* **33**: 311–322.

Cavagnaro, P.F., Cavagnaro, J.B., Lerner, J.B., Masuello, R.W. and Passera, C.B. 2006. Genetic diversity among varieties of the native forage grass *Trichloris crinata* based on AFLP markers, morphological markers, and quantitative agronomic traits. *Genome* **49**: 906–918.

Coltman, D.W. and Slate, J. 2003. Microsatellite measure of inbreeding: a meta-analysis. *Evolution* **57**: 971–983.

Da Silva, A., Luikart, G., Yoccoz, N.G., Cohas, A. and Allainé, D. 2006. Genetic diversity-fitness correlation revealed by microsatellite analyses in European alpine marmots (*Marmota marmota*). *Conserv. Genet.* **7**: 371–382.

Derban, S., Foulley, J.-L. and Ollivier, L. 2002, 2005. *WEITZPRO: a software for analysing genetic diversity*. INRA, Paris. (http://www-sgqa.jouy.inra.fr/rubrique.php3?id_rubrique=5)

Drucker, A. and Scarpa, R. 2003. Introduction and overview of the Special Issue on animal genetic resources. *Ecol. Econ.* **45**: 315–317.

Eding, H., Crooijmans, R.P.M.A., Groenen, M.A.M. and Meuwissen, T.H.E. 2002. Assessing the contribution of breeds to genetic diversity in a conservation scheme. *Genet. Sel. Evol.* **34**: 613–633.

El Mousadik, A. and Petit, R.J. 1996. High level of genetic differentiation for allelic richness among populations of the argan tree (Argania spinosa L. Skeels) endemic to Morocco. *Theor. Appl. Genet.* **92**: 832–836.

European Cattle Diversity Consortium, 2006. Marker assisted conservation of European cattle breeds: an evaluation. *Anim. Genet.* **37**: 475–481.

Fadlaoui, A. 2006. *Modélisation bioéconomique de la conservation des ressources génétiques animales*. Thèse de doctorat en sciences agronomiques et ingénierie biologique, Université catholique de Louvain, Louvain-la-Neuve, 250 pages.

FAO, 2007. *The state of the world's animal genetic resources for food and agriculture*, Rome.

Foulley, J.-L., van Schriek, M.G.M., Alderson, L., Amigues, Y., Bagga, M., Boscher, M.Y., Brugmans, B., Cardellino, R., Davoli, J.V., Fimland, E., Gandini, G.C., Glodek, P., Groenen, M.A., Hammond, K., Harlizius, B., Heuven, H., Joosten, R., Martinez, A.M., Matassino, D., Meyer, J.N., Peleman, J., Ramos, A.M., Rattink, A.P., Russo, V., Siggens, K.W., Vega-Pla, J.L. and Ollivier, L. 2006. Genetic diversity analysis using lowly polymorphic dominant markers: the example of AFLP in pigs. *J. Hered.* **97**: 244–252.

Foulley, J.-L. and Ollivier, L. 2006a. Estimating allelic richness and its diversity. *Livest. Sci.* **101**: 150–158.

Foulley, J.-L. and Ollivier, L. 2006b. Allele diversity. *Proc. 8th World Congr. Genet. Appl.. Livest. Prod.* Comm. No. **33**–09.

Gandini, G.C. and Villa, E. 2003. Analysis of cultural value of local livestock breeds: a methodology. *J. Anim. Breed. Genet.* **120**: 1–72.

Gandini, G.C., Ollivier, L., Danell, B., Distl, O., Georgoudis, A., Groeneveld, E., Martyniuk, E., van Arendonk, J.A.M. and Woolliams, J. 2004. Criteria to assess the degree of endangerment of livestock breeds in Europe. *Livest. Prod. Sci.* **91**: 173–182.

Gavora, J.S., Fairfull, R.W., Benkel, B.F., Cantwell, W.J. and Chambers, J.R. 1996. Prediction of heterosis from DNA fingerprints in chicken. *Genetics* **144**: 777–784.

Glodek, P. 1974. Specific problems of breed evaluation and crossing in pigs. *Proceedings of the Working Symposium on Breed Evaluation and Crossing Experiments with Farm Animals*, Zeist, The Netherlands, pp. 267–281.

Goddard, M. E. and Ahmed, A.M. 1982. The use of genetic distance between cattle breeds to predict the heterosis in crosses. *Proc 2nd World Congr. Genet. Appl. Livest. Prod.* **8**: 377–382.

Gollin, D. and Evenson, R. 2003. Valuing animal genetic resources: lessons from plant genetic resources. *Ecol. Econ.* **45**: 353–363.

Groenen, M.A.M., Joesten, R., Boscher, M.Y., Amigues, Y., Rattink, A., Harlizius, B., van den Poel, J.J. and Crooijmans, R. 2003. The use of microsatellites genotyping for population studies in the pig with individual and pooled samples. *Arch. Zootec.* **52**: 145–155.

Hurlbert, S.H. 1971. The nonconcept of species diversity: a critique and alternative parameters. *Ecology* **52**: 577–586.

Kalinowski, S.T. 2004. Counting alleles with rarefaction: private alleles and hierarchical sampling designs. *Conserv. Genet.* **5**: 539–543.

Kashi, Y., King, D. and Soller, M. 1997. Simple sequence repeats as a source of quantitative genetic variation. *Trends Genet.* **13**: 74–78.

Kashi, Y. and King, D. 2006. Simple sequence repeats as advantageous mutators in evolution. *Trends Genet.* **22**: 253–259.

Kimura, M. 1968. Evolutionary rate at the molecular level. *Nature* **217**: 624–626.

Kinghorn, B.P., Banks, R., Gondro,C., Kremer, V.D., Meszaros, S.A., Newman, S., Shepherd, R.K., Vagg, R.D., van der Werf, J.H.J. 2009. Strategies to exploit genetic variation while maintaining diversity. In Adaptation and Fitness in Animal Populations. Evolutionary and Breeding Perspectives on Genetic Resource Management (J.H.J. van der Werf, H.-U. Graser, R. Frankham and C. Gondro, eds.), Springer, The Netherlands.

Laval, G., Iannuccelli, N., Legault, C., Milan, D., Groenen, M.A.M., Giuffra, E. andersson, L., Nissen, P.H., Joergensen, C.B., Beeckmann, P., Geldermann, H., Foulley, J.-L., Chevalet, C. and Ollivier, L. 2000. Genetic diversity of eleven European pig breeds. *Genet. Sel. Evol.* **32**: 187–203.

Lewontin, R.C. and Krakauer, J. 1973. Distribution of gene frequency as a test of the theory of the selective neutrality of polymorphisms. *Genetics* **74** 175–195.

Loftus, R.T. and Scherf, B. 1993. *World Watch List for Domestic Animal Diversity* (1st ed.), FAO, Rome, Italy.

Lush, J.L. 1945. *Animal Breeding Plans* (3rd ed), Iowa State College Press, Ames, Iowa.

Maynard Smith, J. and Haig, J. 1974. The hitch-hiking effect of a favourable gene. *Genet. Res.* **23**: 23–35.

Mendelsohn, R. 2003. The challenge of conserving indigenous domesticated animals. *Ecol. Econ.* **45**: 501–510.

Nei, M. 1973. Analysis of gene diversity in subdivided populations. *Proc. Natl. Acad. Sci. USA* **70**: 3321–3323.

Nei, M. 1977. Statistics and analysis of gene diversity in subdivided populations. *Ann. Hum. Genet.* **41**: 221–233.

Ollivier, L. 2005. Economic relevance of animal diversity conservation. In *Animal production and animal science worldwide - WAAP book of the year 2005* (A. Rosati, A. Teewolde and C. Mosconi, eds.), Wageningen Academic Publishers, Wageningen, The Netherlands, pp. 271–279.

Ollivier, L. Amigues Y. and Boscher M.-Y. 2003. An EC-funded project on characterisation of genetic variation in the European pig. Objectives, organisation, breed sampling, DNA preparation and circulation. *Arch. Zootec.* **52**: 137–144.

Ollivier, L. and Foulley J. L. 2005. Aggregate diversity: new approach combining within- and between-breed diversity. *Livest. Product. Sci.* **95**: 247–254.

Ollivier, L., Alderson, L., Gandini, G.C., Foulley, J.-L., Haley, C.S., Joosten, R., Rattink, A.P., Harlizius, B., Groenen, M.A.M., Amigues, Y., Boscher, M.-Y., Russell, G., Law, A., Davoli, R., Russo, V., Matassino, D., De'saute's, C., Fimland, E., Bagga, M., Delgado, J.-V., Vega-Pla, J.L., Martinez, A.M., Ramos, A.M., Glodek, P., Meyer, J.-N., Plastow, G.S., Siggens, K.W., Archibald, A.L., Milan, D., San Cristobal, M., Laval, G., Hammond, K., Cardellino, R., Chevalet, C. 2005. An assessment of the European pig diversity using molecular markers: partitioning of diversity among breeds. *Conserv. Genet.* **6**: 729–741.

Petit, R.J., El Mousadik, A. and Pons, O. 1998. Identifying populations for conservation on the basis of genetic markers. *Conserv. Biol.* **12**: 844–855.

Pfrender, M.E., Spitze K., Hicks J., Morgan K., Latta L. and Lynch M. 2000. Lack of concordance between genetic diversity estimation at the molecular and quantitative-trait levels. *Conserv. Genet.* **1**: 263–269.

Pinent, T., Simianer, H. and Weigend, S. 2005. Weitzman's approach and conservation of breed diversity: first application to German chicken breeds. *The Role of Biotechnology*, Vila Gualino, Turin, Italy, 5–7 March, 2005.

Piyasatian, N. and Kinghorn, B.P. 2003. Balancing genetic diversity, genetic merit and population viability in conservation programmes. *J. Anim. Breed. Genet.* **120**: 137–149.

Plante, Y., Vega-Pla, J.-L., Lucas, Z., Colling, D., de March, B. and Buchanan, F. 2007. Genetic diversity in a feral horse population from Sable Island. *Canad. J. Hered.* **98**: 594–602.

Pogson, G.H., Mesa, K.A. and Boutilier, R.C. 1995. Genetic population structure and gene flow in the Atlantic cod *Gadus morhua*: a comparison of allozyme and nuclear RFLP loci. *Genetics* **139**: 375–385.

Reed, D.H. and Frankham, R. 2001. How closely correlated are molecular and quantitative measures of genetic variation? A meta-analysis. *Evolution* **55**: 1095–1103.

Reed, D.H. and Frankham, R. 2003. Correlation between fitness and genetic diversity. *Conserv. Biol.* **17**: 230–237.

Reist-Marti, S.B., Simianer, H., Gibson, J., Hanotte, O. and Rege, J.E.O. 2003. Weitzman's approach and conservation of breed diversity: an application to African cattle breeds. *Conserv. Biol.* **17**: 1299–1301.

Robertson, A. 1975. Gene frequency distributions as a test of selective neutrality. *Genetics* **81**: 775–785.

Ruane J. 1999. A critical review of the value of genetic distance studies in breed conservation. *J. Anim. Breed. Genet.* **116**: 317–323.

Rutten, M.J.M., Komen, H., Deerenbeerg, R.M. and Siwek, M. 2004. Genetic characterization of four strains of Nile tilapia (*Oreochromis niloticus* L.). *Anim. Genet.* **35**: 93–97.

Scherf, B. 2000. *World Watch List for Domestic Animal Diversity* (3rd ed.), FAO, Rome, Italy.

Schlötterer, C. 2002. A microsatellite-based multilocus screen for the identification of local selective sweeps. *Genetics* **160**: 753–763.

Schlötterer, C. 2003. Hitchkiking mapping- functional genomics from the population genetics perspective. *Trends Genet.* **19**: 32–38.

Schlötterer, C., Vogl, C. and Tautz, D. 1997. Polymorphism and locus-specific effects on polymorphism at micosatelite loci in natural *Drosophila melanogaster* populations. *Genetics* **146**: 309–320.

Simianer, H., Marti, S.B., Gibson, J., Hanotte, O. and Rege, J.E.O. 2003. An approach to the optimal allocation of conservation funds to minimize loss of genetic diversity between livestock breeds. *Ecol. Econ.* **45**: 377–392.

Simon D.L. and Buchenauer D. 1993. *Genetic Diversity of European Livestock Breeds*, EAAP publication No. 66, Wageningen Pers, Wageningen, The Nethelands.

Slate, J., David, P., Dodds, K.G., Veenvliet, B.A., Glass, B.C., Broad T.E. and McEwan, J.C. 2004. Understanding the relationship between the inbreeding coefficient and multilocus heterozygosity: theoretical expectations and empirical data. *Heredity* **93**: 255–265.

Smith, C. 1964. The use of specialised sire and dam lines in selection for meat production. *Anim. Prod.* **6**: 337–344.

Smith, C. 1985. Scope for selecting many breeding stocks of possible economic value in the future. *Anim. Prod.* **41**: 403–412.

Smith, C. 1986. Variety of breeding stock for the production-marketing range, and for flexibility and uncertainty. *Proc. 3rd World Congr. Genet. Appl. Livest. Prod.* **10**: 14–18.

Storz, J.F. 2005. Using genome scans of DNA polymorphism to infer adaptive population divergence. *Mol. Ecol.* **14**: 671–688.

Tapio, I., Varv, S., Bennewitz, J., Maleviute, J., Fimland E., Grislis, Z., Meuwissen, T.H.E., Miceikine, I., Olsaker, I., Viinalas, H., Vilkki, J. and Kantanen, J. 2006. Prioritization for conservation of northern European cattle breeds based on analysis of microsatellite. *Conserv. Biol.* **20**: 1768–1779.

Taylor, St.C.S. 1976. Multibreed designs. I. Variation between breeds. *Anim. Prod.* **23**: 133–144.

Thaon d'Arnoldi, C., Foulley, J.-L. and Ollivier, L. 1998. An overview of the Weitzman approach to diversity. *Genet. Sel. Evol.* **30**: 149–161.

Thiessen, R.B., Taylor, St.C.S. and Murray, J. 1985. Multibreed comparisons of British cattle: variation in relative growth rate, relative food intake and food conversion efficiency. *Anim. Prod.* **41**: 193–199.

Thompson, G. 1977. The effect of a selected locus on linked neutral loci. *Genetics* **85**: 753–788.

Vane-Wright, R.I., Humphries, C.J. and Williams, P.H. 1991. What to protect? Systematics and the agony of choice. *Biol. Cons.* **55**: 235–254.
Watterson, G.A. 1977. Heterosis or neutrality? *Genetics* **85**: 89–814.
Weir, B.S. and Cockerham, C.C. 1984. Estimating F-statistics for the analysis of population structure. *Evolution* **38**: 1358–1370.
Weitzman, M.L. 1992. On diversity. *Quart. J. Econ.* **107**: 363–405.
Weitzman, M.L. 1993. What to preserve? An application of diversity theory to crane conservation. *Quart. J. Econ.* **108**: 157–183.
Weitzman, M.L. 1998. The Noah's ark problem. *Econometrica* **66**: 1279–1298.
Wright, S. 1931. Evolution in Mendelian populations. *Genetics* **16**: 97–159.
Zhang, D.-X. and Hewitt, G.M. 2003. Nuclear DNA analyses in genetic studies of populations: practice, problems and prospects. *Mol. Ecol.* **12**: 563–584.

Livestock Genetic Resources: Preserving Genetic Adaptations for Future Use

John P. Gibson

Abstract Since domestication, livestock have evolved and been actively selected for adaptation to a wide range of physical, climatic, disease challenge and production environments. Relatively free access to these farm animal genetic resources (FAnGR) has delivered large benefits to developed and developing countries, notwithstanding examples of inappropriate promotion and use of specialised breeds in the developing world. Rapid changes in economic and agricultural environments globally are creating major threats to livestock diversity. The problem is most acute in the developing world, where information about the characteristics and status of indigenous FAnGR is limited and there is little or no capacity to protect such genetic resources. Recent international fora have recognised the need for urgent action to conserve FAnGR of the developing world. Priority is given to conservation through sustainable use but an urgent need for cryo-conservation is now recognised, overturning the consensus arrived at last decade that cryo-conservation was neither required nor desirable. Implementing conservation of developing world FAnGR will require regional and international collaboration and support, mechanisms for which have yet to be developed. With the accelerating rates of change in the developing world, failure to implement conservation and sustainable utilisation programs in the developing world in the very near future will likely lead to the loss of a substantial proportion of the world's FAnGR.

Keywords Diversity · animal genetic resources · conservation

1 Extended Summary

The publication of the present book coincides with the publication of four papers that summarise the current state of knowledge about livestock genetic resources, their conservation and utilisation (Nimbkar et al. 2007; Sere et al. 2007;

J. P. Gibson (✉)
The Institute for Genetics and Bioinformatics, University of New England, Armidale NSW 2351, Australia
e-mail: jgibson5@une.edu.au

Tixier-Boichard et al. 2007; Woolliams et al. 2007). It also follows the recent publication of the first global assessment of the state of the world's livestock genetics resources, country by country (FAO 2007a). The presentation at the Symposium in honour of Professor Barker was based on these papers, and I here provide an extended summary while pointing the reader to the much more complete reviews referred to above.

The domestication of livestock, starting approximately 10,000 years ago, resulted in a prolonged period of intense natural and artificial selection for a wide variety of adaptation and performance traits. Livestock would initially have evolved to adapt to the new environment of living in association with and/or being controlled by humans. As livestock use spread around the world, livestock subsequently evolved adaptations to a much wider range of physical, climatic, disease and feed environments than their wild ancestors were exposed to within their original home range. Today's breeds of livestock are genetically very different from their wild ancestors and possess a greater range of genetic variation for adaptation and performance traits.

Agriculture systems continue to evolve, driving continued genetic change, most obviously through breed substitution and within breed artificial selection by man, but also through undocumented natural selection that will necessarily still be taking place for fitness within changing livestock production systems. In Europe as modern agriculture systems have evolved over the past few hundred years, it has become possible for livestock to have much higher production levels. Farmers naturally went through a prolonged period choosing among breeds or races those better suited to the new production systems. More recently, particularly during the past 50 years, farmers, breeding companies and cooperatives have practiced increasingly intense within-breed selection for higher production performance, focusing on the small number of breeds that remained economically competitive within modern farming systems. A substantial number of European breeds have gone extinct in the past 100 years, but many have remained as small remnant populations, that have increasingly found niche markets and renewed interest for cultural and social functions in recent times.

In many parts of the developing world, farmers continue to use their traditional indigenous breeds but there is also massive change, mostly through introduction of new breeds and crossbreeding. Rapid growth in consumption of animal products in the developing world has been supplied primarily by widespread introduction of intensive animal production systems. Such systems almost exclusively use high production breeds from the developed world. But there is also widespread genetic change within small-holder and pastoral production systems, where either high production developed country breeds or indigenous breeds with modest production potential have been introduced widely. In many cases such introductions of new germplasm have benefited poor farmers, but there are also many cases of failure and many more cases where the value of the change has not been demonstrated.

It seems likely that in most cases, genetic change is not optimal because choice of breeds is limited by the lack of good information about the fitness and production potential of the vast majority of breeds. Only a few breeds, mostly high production potential developed country breeds, are very well characterised and even for these breeds the fitness and production potential in developing world production systems is usually poorly characterised. Choice of new breeds to introduce is highly biased to the small proportion of breeds that have been characterised. Better characterisation of the majority of the world's breeds of livestock has been identified as an urgent priority. Along with this is an urgent need to capture all existing information on breed characterisation and make it globally available. This will allow more informed choices about which livestock genetic resources will provide better options, particularly to poor farmers in low-input systems.

The current rapid rate of genetic change is generating severe threats to the continued existence of a high proportion of the world's livestock breeds. Additional threats to genetic diversity are created by economic change leading to loss of viability of livestock production, social and political crises (e.g. the genocide in Rwanda and Burundi led to loss of most local livestock) and disease control (e.g. slaughter policies to control foot and mouth disease and avian influenza can eliminate local breeds or races). Recognising the increasing threats to livestock genetic resources, it has recently been recognised that there is urgent need for both *in situ* live animal conservation and *in vitro* conservation to ensure that the current livestock diversity is available to provide options for genetic change in future generations (Gibson et al. 2006). But recognition of the need is only the first step. It remains unclear who will pay for and organise such conservation programs (FAO 2007b). Without clear mechanisms for implementation of conservation where it is most needed, there remains severe risk that a substantial proportion of the existing livestock genetic diversity will be lost in the next ten to twenty years.

The diverse and intense natural and artificial selection of livestock over the nearly 10,000 years since domestication is a unique experiment on a grand scale that will never be repeated. Some of the products of that random experiment will undoubtedly have potential to provide better options in low input livestock systems, particularly those of resource poor farmers, and some may have genetic characteristics of value to higher input systems. They also represent a valuable source of study to understand genetic control of fitness, disease resistance and production traits. It would be a huge loss to lose even a small proportion of these products of this grand evolutionary experiment since domestication without exploring the opportunities for knowledge generation and improved livelihoods and food security.

References

Gibson, J.P., Gamage, S., Hanotte, O., Iñiguez, L., Maillard, J.C., Rischkowsky, B., Semambo, D., Toll, J. 2006 Options and Strategies for the Conservation of Farm Animal Genetic Resources, Report of an International Workshop (7–10 November 2005, Montpellier, France), CGIAR System-wide Genetic Resources Programme (SGRP)/Bioversity International, Rome, Italy, 53pp. ISBN-13: 978-92-9043-712-3, ISBN-10: 92-9043-712-X.

FAO. 2007a. *The State of the World's Animal Genetic Resources for Food and Agriculture*, (B. Rischkowsky and D. Pilling, eds.), Rome, ftp://ftp.fao.org/docrep/fao/010/a1250e/a1250e.pdf

FAO. 2007b. Report of the International Technical Conference on Animal Genetic Resources for Food and Agriculture, Interlaken, Switzerland, 3–7 September 2007. ITC-AnGR/07/REP, http://www.fao.org/ag/againfo/programmes/en/genetics/documents/Interlaken/Final_Report_en.pdf

Sere, C., van der Zijpp, A., Persle, G. and Rege, E. 2007. Dynamics of livestock production systems, drivers of change and prospects for animal genetic resources. Report of the Scientific Forum on Animal Genetic Resources during the International Technical Conference on Animal Genetic Resources for Food and Agriculture, Interlaken, Switzerland, 3 September 2007. Rome, 4–26.

Tixier-Boichard, M., Ayalew, W. and Jianlin, H. 2007 Inventory, characterization and monitoring. Report of the Scientific Forum on Animal Genetic Resources during the International Technical Conference on Animal Genetic Resources for Food and Agriculture, Interlaken, Switzerland, 3 September 2007. Rome, 27–43.

Nimbkar, C., Gibson, J.P., Okeyo, M. and Boettcher, P. 2007. Sustainable use and genetic improvement. Report of the Scientific Forum on Animal Genetic Resources during the International Technical Conference on Animal Genetic Resources for Food and Agriculture, Interlaken, Switzerland, 3 September 2007. Rome, 44–59.

Woolliams, J.A., Matika, O. and Pattison, J. 2007 Conservation of animal genetic resources: approaches and technologies for in situ and ex situ conservation. Report of the Scientific Forum on Animal Genetic Resources during the International Technical Conference on Animal Genetic Resources for Food and Agriculture, Interlaken, Switzerland, 3 September 2007. Rome, 60–73.

Discussion

Frank W. Nicholas

The following points were made in the general discussion:

- There are two documents just published by FAO that contain a wealth of information and highlight areas of profound ignorance. These are "The State of the World's Animal Genetic Resources for Food and Agriculture" edited by Barbara Rischkowsky and Dafydd Pilling (http://www.fao.org/docrep/010/a1250e/a1250e00.htm), and the "Global Plan Of Action for Animal Genetic Resources and the Interlaken Declaration" (http://www.fao.org/ag/againfo/programmes/en/genetics/documents/Interlaken/GPA_en.pdf) recently agreed at the International Technical Conference on Animal Genetic Resources for Food and Agriculture, Interlaken, Switzerland, 3–7 September 2007.
- The typical commercial short-term-outlook reality is such that fitness traits have been, and will continue to be, neglected in many practical selection programs.
- There are good examples where breed substitution has been a far better option than selection within existing breeds, e.g. the introduction of *indicus* cattle to northern Australia; and this type of major productivity enhancement will be possible in the future, only if a diverse range of breeds is conserved.
- By far the most important argument for conserving domestic animal diversity is as an insurance against the unknown and unpredictable future. The weakness of this argument is that the need for this type of insurance is deemed (by commercial short-term-outlook reality) as being so low as to warrant few or no funds being devoted to it, e.g. there are no present-day commercial animal-improvement companies or organisations that feel the need to invest in conservation as an insurance against the unknown and unpredictable future. Are the proponents of conservation, therefore, mainly driven by unquantifiable feel-good motives (that conservation is good because it is good)?
- Another weakness of the conservation-as-insurance argument is that even in countries where highly adapted breeds have evolved, typically these breeds are not perceived as having sufficient immediate utility to make them commercially

F.W. Nicholas (✉)
Reprogen, Faculty of Veterinary Science, University of Sydney, NSW 2006, Australia
e-mail: frankn@vetsci.usyd.edu.au

J. van der Werf et al. (eds.), *Adaptation and Fitness in Animal Populations*,

viable (hence we encounter the anomaly of governments of developing countries being asked to spend funds to conserve a breed because it is deemed to be not commercially viable, despite its adaptations). If anomalies such as this did not exist, there would be little need for funds to conserve locally-adapted breeds – they would automatically be conserved by utilisation.

- Between-breed variation is far more worthy of being conserved than within-breed variation, i.e. the main perceived insurance benefit of conservation is the conservation of adapted alleles (or sets of alleles) that are confined to one or a few breeds.
- Another important justification for conservation of breeds is the value of breed comparisons as the raw material for the increasingly powerful genomics analyses that have the potential to shed much light on the basic biology of adaptive traits.
- There is a serious need for comprehensive long-term breed characterisation studies, and for the public dissemination of existing knowledge of breed comparisons (often not readily available), to inform decisions as to the most appropriate breeding programs to be implemented right now, especially in developing countries. Two resources that are increasingly meeting this need are the Domestic Animal Diversity Information System (DADIS), available at http://dad.fao.org/; and the Domestic Animal Genetics Resources Information Service (DAGRIS), available at http://dagris.ilri.cgiar.org/. The former relies on input from individual countries; the latter is compiled centrally and, in particular, is capturing the "grey" literature on breed comparisons (i.e. reports of useful experiments that have not been published in the scientific literature); they are complementary and, increasingly, there is collaboration between the two groups responsible for them.
- If one accepts the conservation-as-insurance argument, a very strong case can be made for giving far higher priority to simple, cheap conservation of existing resources, and far lower priority to characterisation. Irrespective of how much money we spend now in breed characterisation, it is most unlikely that the traits we concentrate on today will be considered the most important in 20 or 50 years' time; indeed, it is quite possible that the most important issue in the future will be resistance to a disease that has not yet occurred! In addition, the molecular genomics toolbox likely to be available in the future will enable comparative studies the likes of which we cannot even imagine at present. And these same molecular/genomic tools will open up completely novel means of exploiting genetic resources (very likely bypassing traditional breeding practices). Moreover, in the just-published "State of the World's Animal Genetic Resources for Food and Agriculture", it is stated that breeds are becoming extinct at the rate of around one per month! For all these reasons, and in partial contradiction to the previous dot point, it makes very good sense to devote the majority of present conservation budgets to freezing (in liquid nitrogen) samples from existing breeds. This could include (or should even concentrate on) ova and sperm from abattoir material, and somatic cells, e.g. ear clips (the latter in anticipation of the increasing effectiveness of somatic cloning).

Concluding Summary

Fitness traits are like any other quantitative trait; they can be measured and improved by selection.

Short-term commercial imperatives often militate against inclusion of fitness traits in practical breeding programs. However, as illustrated by present trends in the dairy and poultry meat industries, if fitness traits become sufficiently important to livestock producers, the industry is increasingly willing to go to whatever extra trouble and expense is required to enable such traits to be included in practical breeding programs.

There is negligible evidence from well-designed animal breeding programs that quantitative traits run out of genetic variation during long term selection. Moreover, genetic resources generally exhibit new genetic variation when placed in a new environment. This variation may assist the animal breeder or prove to be detrimental, depending upon the (redefined) breeding objective.

Sustainable breeding programs need to avoid small population sizes, thereby controlling inbreeding, and avoiding the worst excess of reduction of genetic variation and inbreeding depression.

The appropriate strategy for avoiding problems due to genotype-environment interaction is to measure traits in the environment(s) in which they are to be utilised.

There is very good evidence that the breeders' equation ($R = h^2S$) works very well in the short and long term for predicting response to selection within populations; even intensely selected populations still retain adequate genetic variation, and hence show continued response to selection. Furthermore, even after decades of arguably the most intensive selection in any livestock species, the chicken industry perceives no need for conservation of any genetic resources (either within or between populations) that are not presently being utilized. Consequently, it can be strongly argued that there is no need to worry about conserving genetic variation that is not being utilized at present.

The conservation-as-insurance-against-the-unknown-future argument is often used to counter balance this. The latter argument has strong emotional appeal, and also has some practical precedents, mostly concerned with exploitation of between-breed variation by breed substitution or as a basis for crossing. However, if breeds

J. van der Werf et al. (eds.), *Adaptation and Fitness in Animal Populations*,

that are highly-adapted to local circumstances (often low-input circumstances in developing countries) are so valuable because of their adaptations to circumstances that could be increasingly common and widespread in the future, why are so many of them threatened with extinction by neglect in their own countries? One possible answer is that there is insufficient comparative information to highlight their merit (which is an argument for increased funding for characterisation). In any case, to the extent that the conservation-as-insurance argument has validity, and especially in the present global context of breeds becoming extinct at the rate of one per month, there is a strong argument that the majority of existing conservation funding should be devoted to the relatively low-technology (and relatively cheap) preservation in liquid nitrogen of ova and sperm from abattoir material, and somatic cells.

Stuart Barker's Contributions to Population Genetics and Animal Breeding: Exploring Fitness, Evolution and Animal Genetics

Frank W. Nicholas and Keith Hammond

Abstract Stuart Barker's initial contributions to genetics during the mid-1950s were down-to-earth instructive articles on genotype-by-environment interaction in cattle and random-sample tests for chicken strains, and pedigree analyses of Australian cattle breeds. Taking up a lectureship in animal genetics at the University of Sydney, Stuart rapidly became involved in experimental population genetics research with *Drosophila*. He also conducted some of the first Monte Carlo simulations of genetics, using SILLIAC – Australia's second automatic computer. During a year's sabbatical, Stuart performed some innovative quantitative genetic analyses of English dairy data (working with Alan Robertson) and was further stimulated in New York working with Theodosius Dobzhansky on the genetics of natural populations. Returning to Sydney, Stuart gathered a fine group of postgraduate students who used *Drosophila* to conduct comprehensive investigations of many aspects of artificial selection for quantitative traits. In 1970, Stuart was introduced to a powerful natural model (the cactus-yeast-*Drosophila* system) that has proven to be a particularly rich source of both field and laboratory experiments, aimed primarily at identifying the selective forces that maintain genetic polymorphisms in natural populations. On moving to the University of New England in 1979, Stuart became involved in important research on resistance and resilience to internal parasites in sheep. At about the same time, Stuart became heavily involved with the characterisation and management of domestic animal diversity, contributing to many FAO meetings and committees, and collaborating widely throughout Asia in measuring domestic animal diversity. Although now ten years into retirement, Stuart continues to conduct research and publish papers, primarily on the *Drosophila*-cactus-yeast system.

Keywords *Drosophila* · fitness · natural selection · artificial selection · domestic animal diversity · genetic resources

F.W. Nicholas (✉)
Reprogen, Faculty of Veterinary Science, University of Sydney, NSW 2006, Australia
e-mail: frankn@vetsci.usyd.edu.au

J. van der Werf et al. (eds.), *Adaptation and Fitness in Animal Populations*,

1 Introduction

Stuart Barker's work encompasses more than half a century of developing, understanding and communicating the science of genetics. In this paper, we review some of Stuart's contributions, with a particular emphasis on his work in population and quantitative genetics as it relates to animal breeding. We have made no attempt to provide a comprehensive account of Stuart's career. Instead, our aim has been to highlight important aspects of his research and teaching, to mention briefly some of his other research, and to record some of the background that may help to illuminate his many and varied achievements. Complementary to this present review is the summary of Stuart's achievements in McCann and Batterham's (1993) history of Australian genetics.

2 Casting the Die

2.1 A Genetics Honours Year in Sydney

The story of Stuart's contribution to genetics starts in the year 1953 – the same year as the structure of DNA was elucidated. The previous year, Stuart had completed the four coursework years of a BAgrSc degree at the University of Queensland, on a scholarship from the Queensland Department of Agriculture and Stock (now Primary Industries). Having told Stuart that he was to become a geneticist (he had no say in the matter!), the Department sent him to Sydney for an honours year under James Rendel, who, only two years previously, had arrived from England (having studied under J.B.S. Haldane) to take up the position of Officer in Charge of the newly-formed Animal Genetics Unit of the Commonwealth Scientific and Industrial Research Organisation (CSIRO) (Franklin et al. 2004). During its first 12 years of operation, the Animal Genetics Unit (which became the Division of Animal Genetics in 1959) was housed on the main (Camperdown) campus of the University of Sydney, in the then Department of Zoology.

Having very wide genetical interests, and being very keen on the use of *Drosophila* as a model species for genetics research, Rendel gave Stuart a *Drosophila* project for his honours year. This was Stuart's introduction to a genus that has been his passion throughout his entire scientific career. It also provided an introduction to Rendel's first laboratory assistant – a certain Miss Maureen Ferguson, who helped with the honours project, and who was soon to become Stuart's wife and life-long companion *extraordinaire*. The project itself concerned the mutagenic effects of formaldehyde. However, Rendel's broad interests, and those of Rendel's immediate CSIRO colleagues Alex Fraser and Bill Sobey, ensured that Stuart was given a wide-ranging and stimulating genetics education. And, of course, there was *that* 1953 discovery. In Stuart's own words, "At morning tea one day, Alex Fraser stormed into the tea room waving a copy of *Nature* and shouting 'DNA is the genetic material' or words to that effect. I think morning tea then continued to about lunch time."

2.2 Assistant Husbandry Officer, Yeerongpilly

Upon completion of his honours project, Stuart returned to Queensland, to be employed by his scholarship provider as Assistant Husbandry Officer at the new Animal Research Institute at Yeerongpilly (not unlike Sewall Wright's first post-doctoral appointment as a "Senior Animal Husbandman" with the United States Department of Agriculture; Provine 1986). Stuart remained in this job for almost two years, generating (among other things) publications on genotype × environment interaction (G × E) (Barker 1955) and random-sample tests (Barker 1956; Barker and Burton 1956). The GXE paper is an easy-to-read, no-nonsense summary of the latest research results from around the world, clearly showing that breeders should be selecting animals in the environment and under the circumstances in which their progeny are to be raised, rather than in circumstances aimed at enabling the best possible expression of traits. The other two papers arose from the first random-sample poultry test to be conducted in Queensland: Barker (1956) provides a very straightforward and clear explanation of why the old egg-laying competitions (comparing different breeds/strains of poultry via small, unrepresentative samples) were ineffective and had to be replaced by random-sample tests; and Barker and Burton (1956) present the results of the first test.

During his two years at Yeerongpilly, Stuart also commenced what was to become a series of pedigree analyses of Australian cattle breeds (Barker 1957a,b). Having learned from Rendel about Alan Robertson's development of relatively simple methods of pedigree analysis (Robertson and Asker 1951; Robertson 1953), Stuart realised that the same analyses could be employed fruitfully on Australian data, and (importantly) could be done without external financial support. There was also the first of two papers from his honours project (Barker and Davern 1956) – his first *Drosophila* paper.

2.3 Lectureship, University of Sydney

In April 1956, at the invitation of Professor Terry Robinson, and on the recommendation of Rendel (and without any selection committee!), Stuart commenced work as a part-time temporary lecturer in animal genetics in the recently-founded Department of Animal Husbandry at the University of Sydney. The Department had substantial responsibilities and involvement in both the Faculty of Agriculture and the Faculty of Veterinary Science. In Stuart's case, these were manifest as two genetics courses to veterinary students (26 lectures on basic and quantitative genetics in the third year and a further 13 lectures on animal breeding in the fifth year), and a course in population and quantitative genetics to agriculture students specialising in animal husbandry and in genetics (63 lectures and 26 practical classes!) For someone who himself had graduated only two years previously, this must have been a daunting challenge. In the case of the veterinary students, the material had to cover all domesticated animals (including companion animals), while for the agriculture students, the course concentrated on livestock. In both cases, Stuart's approach was to deal simultaneously rather than serially with all relevant species, utilising at each point the best examples from different species. This strategy added great strength to his teaching.

Stuart quickly developed into a very effective lecturer. Indeed, he is fondly remembered by many hundreds of undergraduates as one of the best lecturers they had. Part of his success was no doubt due to his willingness – even keenness – to receive feedback and to act upon it. At a time when there was still very strong opposition to the very idea of student evaluation of teaching, Stuart had designed a questionnaire, and every year handed it out to each of his undergraduate classes. Another reason for his success was that he could explain complex concepts very clearly. In addition, as his other responsibilities grew and grew, he still found time to mark assignments in a timely manner, and to provide constructive feedback.

In addition to his onerous teaching responsibilities, Stuart was required (not unreasonably) to enrol for a PhD. Indeed, this was the reason for his part-time appointment. Rendel was the obvious supervisor. However, apart from mentioning fitness in very general terms when Stuart re-established contact, Rendel stayed very much in the background, leaving Stuart to his own devices to design and conduct his PhD projects, which (as described below) turned out to be a very productive examination of a range of fitness issues in *Drosophila*. To conduct this challenging work, Stuart had to establish laboratory facilities from scratch and, in so doing, he improved the design of experimental cages (Barker 1960a) and added a set of feed bottles for oviposition, which could be replaced in sequence over time, compatible with the lifecycle (Barker 1960b).

During these early years, Stuart had the benefit of interactions not only with Rendel, Sobey and Fraser, but also with Helen Newton Turner, then in the CSIRO McMaster laboratory in the grounds of the Veterinary Faculty. She had helped him a little with statistics during his honours year, and now proved a valuable sounding board for some of the many lectures he was preparing. It was also through Helen that Stuart met and interacted with Sir Ronald Fisher during one of Fisher's visits to Sydney. There were also new arrivals in the Animal Genetics Unit in the Zoology Building, including Pamela Pennycuick, Berenice Kindred and Bruce Sheldon, who had done his honours year with Rendel in 1952, and who had arrived back also to pursue a *Drosophila* PhD under Rendel.

Given the enormous effort involved in working up and presenting the three courses, it is remarkable that Stuart's PhD thesis was submitted at the end of 1960, only four and a half years after he had commenced his lectureship. This time scale is all the more remarkable (though not surprising to anyone who has worked with him) when one realises that other work (unrelated to his PhD) was being published during this time; there were three more papers on pedigree analysis of Australian cattle (Barker 1959; Barker and Allingham 1959; Barker and Davey 1960) and, famously, there was his involvement in the pioneering adventure with the world's first genetic simulation on a computer.

2.4 Simulation of Genetic Systems on SILLIAC

The computing story begins in 1956, which is not only the year in which Stuart took up his lectureship, but is also the year in which SILLIAC was "born". The first automatic computer built within an Australian university (being a copy of a computer called ILLIAC built at the University of Illinois), SILLIAC was only the second computer in the country (the first, CSIR Mk1 (later known as CSIRAC), built by the then CSIR (which became CSIRO) – as it so happens, on the campus of the University of Sydney – had been operational since 1949). It was a young Peter Claringbold, then a senior lecturer in veterinary physiology, but later to become Chief of CSIRO's Division of Computing Research, who had seen the potential of the then very new digital computers to simulate genetic systems. Two "words" each of *n* binary digits (bits) could represent the two haplotypes of a diploid individual comprising *n* loci, each with two alleles. For example, the words 1010 and 1001 could represent an individual homozygous for the "1" allele at the first locus, homozygous for the "0" allele at the second locus, and heterozygous at the third and fourth loci. Utilising the logical arithmetic by which digital computers did their calculations, diploid genotypes and consequently multi-locus phenotypes could be deduced (including, if desired, the effect of interactions between loci, i.e. epistasis). Most importantly, the various stochastic events inherent in biological reproduction could be simulated by the use of random numbers that could so readily be created by the computer; so-called Monte Carlo simulation. Gametes could be created by "random walks" along a pair of words, i.e. along a pair of homologous chromosomes. Any degree of linkage could be simulated by specifying the recombination value between pairs of loci, and then using a random number to decide which particular haplotype to take for each section of a particular gamete. Thus was a whole new world of genetics research created. Stimulated and encouraged by Peter Claringbold, Rendel's colleague Alex Fraser took up the challenge, writing a program in machine language ("assembly language"), using it to simulate artificial directional selection for a quantitative trait whose variation was determined by six loci, and publishing two papers in 1957 (Fraser 1957a, b) and a brief report in *Nature* the following year (Fraser 1958). Mixing regularly with Claringbold and Fraser, Stuart was also encouraged to become involved. In contrast to Fraser, who concentrated on multi-locus simulation of quantitative traits in order to investigate the effects of selection on a quantitative trait, Stuart used SILLIAC to explore population-genetic issues at a single locus. Using a program he wrote himself, also in assembly language, Stuart showed in his first simulation paper (Barker 1958a) very good agreement with experimental results of selection between two chromosomal re-arrangements in an experimental *Drosophila* population, and less perfect, but still encouraging, agreement with another *Drosophila* experiment involving a mutant and wild-type alleles at an autosomal locus. Stuart's second paper (Barker 1958b) addressed the same issues but with a sex-linked locus.

We asked Stuart what actually was involved in working with SILLIAC, "Doing a SILLIAC run was the easy bit. One fed (or later had an operator feed) in a paper tape, watched dots flashing around on the valves, eventually took the paper tape output and put it through a tape reader to get a print out that hopefully was intelligible! Preparing the paper tape input was the really fun bit – the tape had a series of sprocket holes to drive it through the readers, and 5 possible holes across the tape – each hole actually representing one number (1, 2, 4, 8, 16). Errors were repaired by sticking little bits of black sticky tape over the offending holes. Getting a program to run first time (i.e. no black stickies) was a real but very rare EUREKA moment."

As a closing note on the SILLIAC story, it is worth noting that a FORTRAN version of the original 6-locus program, written at the University of California, Davis, by Donald Burnell under Fraser's supervision, and known as GS-D-1 (Fraser et al. 1965), was used as an invaluable teaching tool in quantitative genetics at the University of Sydney until the mid 1990s. Indeed, if someone could adapt the code to run with modern compilers, it would still be used – it provides a wonderful introduction to many of the essential concepts of selection for quantitative traits.

2.5 Dairy Genetics in the UK

As soon as his PhD was submitted, Stuart headed to the United Kingdom (UK) to work on dairy data with Alan Robertson (to whom he had been introduced, via mail, by Jim Rendel, who had also worked with Robertson on dairy cattle breeding programs in Edinburgh prior to moving to Australia). This was Stuart's first foray into quantitative genetics. The first eight months of 1961 were spent as Visiting Research Officer in the Breeding Research Division at the Thames Ditton headquarters of the Milk Marketing Board of England and Wales (MMB), on the south-western outskirts of London. Reflecting the interests of both countries in the results of this research, it was supported by grants from the MMB and the Australian Dairy Produce Board. Many, many hours were spent determining what useful records were available and having the relevant data punched onto cards. The overall strategy had been devised the previous year, via correspondence with Alan Robertson and Dr Joe Edwards, the Head of MMB's Production Division. A key element of the plan was to analyse records from more than one lactation on each animal. In Stuart's own words, "Before going, it was arranged that Alan would supervise the study, and early in the piece, I spent one day on his farm near Liverpool (helping him check animals and workers, collecting eggs and doing various chores – all the time discussing what would be done)". The analyses were conducted at the famous Rothamsted Experimental Station, near Harpenden (around 40 km north of London), on an Elliot 401 MK 1 computer made available by Frank Yates, then Head of Statistics, and a pioneer of statistics and statistical computing. As it so happens, this computer was the same make and model as one of the other very early computers to be used in Australia (by the Weapons Research Establishment in South Australia), which also commenced operations in 1956, just a month after SILLIAC. To return to Stuart's words, "Once the data were collected and punched onto cards, I then spent one day per week running the analyses on an Elliot 401 MK 1 computer there, using an hierarchical ANOVA program written by John Gower – in machine language. I travelled up by train, arriving a bit after 9.00 am. John Gower started a run each time before I got there, and I then sat with the machine for the next 24 hours! – starting new runs as necessary, watching to make sure that the computer did not blow up on me, and keeping myself awake using an electric calculator to join up bits of the hierarchy, and preparing mugs of coffee!! This I did for 12 or 13 weeks, and that only did part of the hierarchy for each analysis."

Sometime during these eight months, Stuart visited Edinburgh for three days to discuss the results with Alan Robertson. This was his first visit to one of the leading centres of animal quantitative genetics. Among other things, Stuart got to meet Douglas Falconer, and to learn of his pioneering textbook *Introduction to Quantitative Genetics* which had been published only the previous year (Falconer 1960). From then on, this was the textbook used by Stuart in his population and quantitative genetics course to the animal husbandry and genetics students back in Sydney.

Inevitably not all the analyses were completed during Stuart's stay in the UK, and the final analyses were completed on SILLIAC in 1962, after his return to Sydney.

The results of this work were published in two papers in 1966 (Barker and Robertson 1966; Robertson and Barker 1966). Among other things, the first paper showed the extent to which the then English method of defining contemporaries (within herds, within years) failed to allow for some important sources of variation. Importantly, by drawing on Robertson's (1966) innovative theoretical modelling of culling in dairy cattle (published in the previous issue of the same journal, *Animal Production*), the first Barker/Robertson paper was able to provide a far better explanation of the effect that selection on previous lactations has on the estimation of quantitative genetic parameters for later lactations. After taking account of this selection, in essence the analyses suggested that one could think in terms of first lactation having a genetic correlation of 0.8 with all subsequent lactations; later lactations having twice the phenotypic variance of first lactation, with genetic correlations of unity; and the phenotypic correlations amongst all pair-wise lactations being 0.4. Given this summary, the paper then made some very practical suggestions for taking account of all lactations in dairy-breeding programs. A feasible scheme would involve a selection index in which first lactation is given three times the weight of each subsequent lactation. The second paper investigated longevity, in the context of what had been (up until then), the common practice of selection based primarily on first lactation. The conclusion was very clear cut: "no evidence was found that selection on first lactation progeny test would reduce productive life in the dairy herd".

These two papers have been quite influential (total citations = 61). In highlighting the inadequacies of the contemporary-comparison approach in relation to allowing for identifiable sources of non-genetic variation and the effect of selection, their results contributed to the gradual realisation that mixed-model strategies had much to offer.

2.6 *With a* Drosophila *Pioneer in New York*

After his time at MMB, Stuart spent the last four months of 1961 in the Department of Zoology at Columbia University in New York, in Theodosius Dobzhansky's laboratory. This period was supported by a research grant from the U.S. Atomic Energy Commission and a Fulbright Travel Grant. By then in his early 60s, Dobzhansky had been a key figure in the "modern synthesis" of genetics with evolutionary theory, and, importantly for Stuart, Dobzhansky was also a pioneer of *Drosophila* field and laboratory studies, having by 1961 reached paper 30 (Dobzhansky et al. 1960) of his monumental series "Genetics of natural populations", commenced back in 1938 (Dobzhansky and Queal 1938). In addition, it had been a Dobzhansky paper (Dobzhansky and Pavlovsky 1953) that had provided the first experimental data tested in Stuart's first SILLIAC paper.

2.7 *Wearing Two Hats, Back in Sydney*

That sabbatical year of 1961 was seminal. It had provided Stuart with the opportunity for first-hand involvement in cutting-edge quantitative genetics to do with livestock, and it had exposed him to two pioneering *Drosophila* geneticists – the first (Alan Robertson) who was already using *Drosophila* as a model organism for quantitative-genetics studies, and Dobzhansky, the pioneer of evolutionary-population-genetics field studies in *Drosophila*. The interest in both population and quantitative genetics that Stuart had shown since commencing at the University of Sydney had been reinforced and extended by his year away. The die had been cast for a very productive research, teaching and advisory career, happily (and creatively) spanning both disciplines. In Stuart's own words, "My two hats of population and quantitative genetics have been worn ever since".

By now the Barker mould was more-or-less set: an overwhelming enthusiasm for experimental population genetics research using *Drosophila*; an extensive understanding of quantitative genetics applied to livestock (especially dairy cattle); considerable experience in, and enthusiasm for, the use of computers, especially for the analysis of what were then massive data sets; and enthusiasm for (and considerable ability in) teaching. In the laboratory, no experiment was too big, and these were so well organised they ran like clockwork!

3 A Foot in Both Camps

3.1 *Fitness in* Drosophila *Populations*

From the very beginning, and exceedingly relevant to this symposium, the first of these loves was concerned with the population genetics of fitness; witness the title of his PhD thesis ("The estimation of relative fitness of *Drosophila* populations and related studies in population genetics") (Barker 1960c), and the series titles of the papers from the thesis (Estimation of Relative Fitness of *Drosophila* Populations). A key result was the development of a model for estimating population fitness (by comparison with a standard population) and use of the model for estimating the fitness of a range of populations in a range of circumstances (Claringbold and Barker 1961; Barker 1962a,b,c; Barker 1963a,b). His deeper forays into fitness meant dissecting out the parameters for fitness components like fecundity, fertility and viability for the different strains and species investigated.

3.2 Quantitative Genetics Relating to Animal Improvement

At the same time as the dairy analyses were being completed, Stuart was also preparing a review entitled "Modern problems of population genetics in animal husbandry" (Barker 1967). This is actually a comprehensive summary of how quantitative genetics should be applied in practical livestock breeding programs, with particular emphasis on how best to create the foundation population, and how to maximise total selection response. The review draws heavily on relevant theoretical and experimental results, including (among the latter) the many results then emerging from Stuart's own first crop of postgraduate students. Stuart's summation of the need for research into the nature of quantitative genetic variation is just as relevant today as it was 40 years ago: "Future progress in the understanding of quantitative genetic variation... depends on analysis at the level of the gene, on knowledge of the effects of the genes through the proteins they control, and on understanding of the developmental and physiological pathways involved". After very little progress in this area during the last 40 years, we now, finally, in the wake of the Human Genome Project, have access to genomics research tools and strategies that are beginning to enable these issues to be addressed (e.g. Haley and de Koning 2007).

By the mid-60s, Stuart had gathered together a strongly-committed team of postgraduate students who were keen to build a bridge between Stuart's by now considerable experience with *Drosophila* research (primarily concerned with natural selection) and quantitative genetics of livestock (especially issues related to artificial selection). *Drosophila* was the ideal model organism to expeditiously and inexpensively test some of the important theoretical advances emerging from the same source as Falconer's by now influential textbook (i.e. the Institute of Animal Genetics in Edinburgh), especially the seminal paper by Robertson (1960) on limits to selection in finite populations.

In 1964 Stuart had obtained from Barrie Latter, then of the CSIRO Division of Plant Industry in Canberra, a sample from a *Drosophila melanogaster* population captured at Canberra in the summer of 1959. This so-called Canberra strain was the resource used in just about all of the quantitative genetics research subsequently conducted by Stuart's postgraduate students. As in practical animal breeding, the first requirement was to characterise one's base population. A comprehensive estimation of quantitative genetic parameters of this population was reported by Sheridan et al. (1968). The scale of this project was a sign of things to come for Stuart's students. For this paper alone, nearly 12,000 flies were scored for each of three bristle traits, and 8,000 of these flies were also scored for another two bristle traits. Furthermore, these bristle traits were not simple to score. The two most labour-intensive traits (abdominal bristle count) had averages of around 20 per fly; the other three had averages of around 10. This was not a project for the faint-hearted, nor the visually challenged!

The initial quantitative genetics work culminated in three key papers (Frankham et al. 1968a,b; Jones et al. 1968; which together have been cited 156 times) and two key quantitative geneticists (Les Jones and Dick Frankham) who have each made a major contribution (Les as a key geneticist in the Australian Dairy Herd Improvement Scheme for many years, and Dick as Professor of Biology at Macquarie University and a leading conservation geneticist). These papers provided confirmation of the essence of Robertson's (1960) theory, especially in relation to the interaction of effective population size and intensity of selection in determining response to selection. Above all, this work was instrumental in confirming that the simple additive infinitesimal model (that the genetic variation in quantitative traits is due to an infinite number of genes acting additively, each with a vanishing-small effect), despite all of its "unreality", actually does provide useful predictions of the results of real selection programs. Other postgraduate students also used *Drosophila* to test other aspects of the theory of artificial selection (Barker and Cummins 1969a,b; Barker and Karlsson 1974; Hollingdale and Barker 1971a,b; Rathie and Barker 1968; Sheridan and Barker 1974a,b; Hammond 1974; Yoo 1974). Some of these projects involved long-term selection, e.g. 66 generations for Hollingdale and Barker (1971b). The work culminated in a series of five papers published in 1980 by Brian Yoo (Yoo 1980a,b,c,d; Yoo et al. 1980), who selected for 69 generations. Considered together, the publications of Stuart's quantitative genetics postgraduate students at the University of Sydney represent a comprehensive experimental testing of the major theoretical issues of artificial selection for quantitative traits.

3.3 Natural Selection in the Laboratory and in Nature

Meanwhile, Stuart maintained his experimental interests in natural selection, which soon had been extended to include inter-specific competition (Barker and Podger 1970a,b).

Like his contemporaries, Stuart was greatly excited in the mid-60s when it became evident that electrophoresis of proteins was exposing hitherto unimagined levels of genetic variation at loci throughout the genome of all species from humans (Harris 1966) to *Drosophila* (Hubby and Lewontin 1966). Spending nine months of sabbatical leave in Chicago with Dick Lewontin in 1968/69 had provided Stuart with ample encouragement, and soon after, a chance observation by CSIRO's Alan Dyce of young *Drosophila* emerging from rotting cladodes of prickly pear (*Opuntia* spp.), was communicated to Stuart, who quickly appreciated that a novel and very powerful natural experimental system had been presented to him – an opportunity to apply the new technologies for quantifying genetic variation in a *Drosophila* species that occupies a well-defined ecological niche of known distribution that is tailor-made for evolutionary-genetic investigations. *Drosophila buzzatii* live only on the rotting cladodes of the prickly pear, whose distribution in Australia is exceedingly patchy, thanks to a well-documented biological control program using the *Cactoblastis cactorum* moth (Mann 1970; Cameron 2007). This well-defined natural experimental system comprised a powerful resource for investigating major evolutionary questions concerned with adaptation and migration, and it provided Stuart with a powerful resource for ecological genetics research – a resource which would have been very pleasing to his mentor, Theodosius Dobzhansky.

An ARC grant was secured, an extraordinary research assistant and part-time Masters student (John Mulley) was employed (in February 1970), and a great adventure began. A portent of the effort that was to be devoted to this model system is there for all to see in their first paper (Barker and Mulley 1976), which reported sampling from 35 localities throughout the length and breadth of the cactus distribution, from near Sydney to northern Queensland, and from the coast to 400 km inland – an area of approximately 160,000 square km, corresponding to more than half of the whole of eastern Australia! From just eight of these localities, 3,416 flies were collected, each of which was genotyped for 29 loci. With a mean of 427 flies genotyped per locality, the authors were certainly justified in concluding that for the 21 loci found to be monomorphic, "the probability of undetected segregation was negligible"! More than 5000 flies were collected from the 35 localities, and each of these was genotyped for the eight polymorphic loci. In total, the number of genotypes reported in this single study was around 112,000! And all of this was done manually. These data were elegantly manipulated by John James through a raft of multivariate investigations in a search of correlations between, on the one hand, allele frequency and heterozygosity, and on the other, climatic parameters (mean meteorological values as well as their variability) from each collection site (Mulley et al. 1979).

An unexpected bonus of the *buzzatii* field trips was the discovery, for the first time in Australia, of *Drosophila aldrichi*, another member of the *mulleri* subgroup to which *buzzatii* belongs. Similar in appearance, it was first detected in the laboratory after some tracks on the electrophoretograms displayed bands that had not been seen before (in *buzzatii*). *D. aldrichi*, like *buzzatii*, had evidently been an inadvertent immigrant and escapee from the cladodes originally brought into Australia with other organisms as potential biological control agents. Alternatively, both drosophilids could have been stowaways with the original *Opuntia* introductions, brought in by travellers as botanical curiosities.

Complementary to this new, exciting field work, another hard-working postgraduate student, Jeffrey Moth, was beavering away during the early 1970s with extensive laboratory investigations of inter-specific competition (Moth and Barker 1976, 1977, 1981). He gathered data on fitness parameters in *D. melanogaster* and *D. simulans* in order to ultimately computer simulate their competitive interaction.

The crowning achievement of the early *Drosophila*-cactus work was a paper in *Nature* (Barker and East 1980), showing for the first time evidence of natural selection at allozyme loci (following artificial perturbation of a natural population in the field, in the prickly-pear system). It is difficult for people who weren't researchers in the 1970s to realise the importance of this clear indication that natural selection does operate at allozyme loci (the allele frequencies changed in a repeatable manner following perturbation by artificial migration). For many, this was strong support for the selectionist (if not the pan-selectionist) view of the world. Others tended to the view that such examples still left room for substantial neutral evolution.

With the realisation that many different natural strains of yeast form a major component of the diet of cactophilic *Drosophila*, and that these yeast strains could be cultured in the laboratory, the *Drosophila*-cactus system was soon extended to the *Drosophila*-cactus-yeast system, which has proved to be an extraordinarily rich resource of both field and laboratory experiments aimed primarily at identifying the selective forces maintaining genetic polymorphisms in natural populations. So rich was this system that it soon embraced an international team of researchers, including Bruce Weir (University of North Carolina), John Oakeshott, Geoff Chambers and John Gibson (ANU), John James (University of NSW), Fabio Sene (Brazil), Bob Sokal and Neil Oden (Stony Brook University), Freddy Christiansen, Volker Loeschcke and Joergen Bundgaard (University of Aarhus), Tim Prout and Mel Green (University of California, Davis), Phil Hedrick (then at University of Kansas), Richard Halliburton (Western Connecticut State University), Tom Starmer (Syracuse University), Jim Fogleman (University of Denver), Forbes Robertson (University of Edinburgh), Bryan Clarke (University of Nottingham), Esteban Hasson (University of Buenos Aires), Alfredo Ruiz (Autonoma University, Barcelona), and Ian Davies (UNE).

Among the 75 or so papers that have arisen from this research, highlights include the one by Barker et al. (1986a) which showed that genetic variation at allozyme loci was associated with differences in yeast-strain preference by egg-laying females. Another study published in the same year (Barker et al. 1986b) indicated the extent of data gathering that exemplifies Stuart's approach to research. A site on a property called "Yarrawonga" in the Hunter Valley of NSW was visited every month for four years, with a total of 15,313 flies being captured and subsequently genotyped – each at six loci! Whilst inevitably not all the results were clear-cut, a very clear conclusion was drawn: "variation in allele frequencies over time cannot be attributed solely to drift". Furthermore, "microspatial heterogeneity . . . coupled with habitat selection . . . could be a major factor promoting diversifying selection and the maintenance of polymorphism".

3.4 Epistasis

As evidence of his continuing involvement in both experimental population/evolutionary genetics and applied animal breeding, Stuart published a review of experimental evidence for epistasis (Barker 1979), showing himself to be equally at home in discussing evidence for epistasis as a source of quantitative genetic variation in (**a**) fitness in natural populations and (**b**) production traits in domesticated animals and plants. In this, Stuart was following in the worthy footsteps of none other than Sewall Wright, who made so many major contributions in both evolutionary genetics and the genetics of animal and plant breeding. In his 1979 review, Stuart raised the issue of Wright's (1922, 1977) shifting balance theory of evolution, based on the concept of "multiple-peak" epistasis. In the context of animal or plant improvement, Wright had argued that if epistasis is an important source of genetic variation, it follows that splitting a selection line into a series of sublines and selecting within each line, with occasional selection and crossing between sublines, would ultimately be more successful than selection within a single larger line. The failure of experimental tests of this theory (e.g. Rathie and Nicholas 1980) is not so much evidence against epistasis, but rather is an indication of the over-riding importance of additive genetic variation (and of the infinitesimal model describing it) for most quantitative traits. As the genomics revolution bears fruit in future years, no doubt we will become much more knowledgeable about epistasis; we may even begin to understand why such a ubiquitous phenomenon remains a second-order issue in terms of both natural and artificial selection.

3.5 Time for a Change

By 1979, Stuart had been at the University of Sydney for 23 years, and had been promoted through the ranks to Associate Professor. During all of this time he had maintained much the same heavy undergraduate teaching load that he had been given when he first started, At a very rough count, he had taught around 1200 veterinarians and 250 students majoring in animal science or animal genetics; and he had averaged around 180 contact hours per year, the majority of these on the Camden campus, involving 26 round trips each year, with each trip comprising roughly 120 kms, much of it very demanding driving along suburban streets.

In August 1979, Stuart moved to the University of New England (UNE), Armidale, to become Professor of Animal Science. In that same month, as Foundation President of the newly-formed Australian Association for Animal Breeding and Genetics (AAABG),[1] Stuart hosted the association's first conference, held at the UNE.

Stuart also continued his very active work as a consultant to two very different practical animal breeding programs. For many years before and after his move to UNE, he provided substantial advice for the Steggles poultry breeding operation in the Hunter Valley, and he was heavily involved in providing advice for the Wright family's pioneering development of Beefmakers (a Hereford/Simmental synthetic) at Wallamumbi, near Armidale.

Not surprisingly, Stuart's *Drosophila* research (both field and laboratory) continued unabated at the UNE – the busy fly-room in Sydney was replicated in Armidale. But moving to Armidale provided Stuart with the opportunity to become involved in different aspects of the genetics of domesticated animals. Among

[1] In 1997, the name, but not the initials, of the association was changed to Association for the Advancement of Animal Breeding and Genetics Incorporated

other things he stimulated further investigations into the quantitative genetics of resistance to internal parasites in sheep, and, in particular, investigations of what became known as the Golden Ram – a ram with offspring showing exceptionally-high resistance to *Haemonchus contortus* (Albers et al. 1987). Building on the ideas of Clunies-Ross (1932), Stuart and his colleagues distinguished between resistance to infection and resistance to the effects of infection (resilience), the latter measured as the difference between performance for a production trait with and without infection. Although the heritability of resilience proved to be frustratingly low, it was concluded that selection for performance when infected could be very effective, especially if combined with direct selection for resistance. The importance of this paper is indicated by its being Stuart's most-cited paper, with 109 citations.

3.6 Management of Domestic Animal Genetic Diversity

Prior to his move to Armidale, Stuart had became involved in conservation genetics, first via his appointment in 1978 as chair of the Expert Committee on Animal Genetic Resources of the Society for the Advancement of Breeding Researches in Asia and Oceania (SABRAO), followed in 1983 by appointment as a member of the FAO/UNEP Expert Panel on Animal Genetic Resources Conservation and Management, a position he held until 1992. In that year, he played a key role in an FAO Expert Consultation on the Management of Global Animal Genetic Resources. The following year, he chaired an FAO Working Group on "An Integrated Global Program to Establish the Genetic Relationships among the Breeds of each Domestic Animal Species". For four years from 1996 he chaired FAO's "Panel of Experts on the Development of the Global Strategy for the Management of Farm Animal Genetic Resources", and he remained on that panel until 2001. These FAO activities involved Stuart working closely with another former postgraduate student (one of the present authors – KH) who for ten years led FAO's work on management of domestic animal diversity. Among many other things, Stuart assisted with the technical development of the first detailed survey on the State of the World's Farm Animal Genetic Resources – an activity directed at capacity building for sustainable management of livestock diversity. The final report (FAO 2007) was negotiated in Interlaken, in early September 2007, when delegates from 109 countries adopted a Global Plan of Action for Animal Genetic Resources, the first internationally agreed framework directed at improved characterisation, sustainable use and development utilisation, and conservation of farm animal genetic resources.[2]

In addition to this expert advisory work on managing animal genetic resources, Stuart also became heavily involved in the use of genetic markers for quantifying domestic animal diversity and for inferring phylogenies. In particular, he developed extensive collaborations throughout Asia, using firstly protein electrophoresis and later DNA polymorphisms to document genetic variation in a wide range of animal

[2] http://www.fao.org

populations, starting with Tan et al. (1993) and including the much-cited ($n = 61$) Barker et al. (1997), which compared protein-coding loci (primarily allozymes) and microsatellites as indicators of genetic variation in Asian water buffalo, concluding very firmly in favour of the latter. Inevitably, there was also a *Drosophila* paper on conservation genetics: an analysis of a set of long-term "captive" *Drosophila* populations showed a gradual loss of genetic variation over time, and an effective population size just a few per cent of actual population size (Briscoe et al. 1992). Long after retirement, Stuart is still characterising genetic variation in a conservation context (Flamand et al. 2003).

3.7 Natural Selection Continued

Meanwhile, the *Drosophila*-cactus-yeast work also continued. In yet another happy marriage between evolutionary genetics and the tools-in-trade of animal breeders, the first of two papers with Tim Prout (Prout and Barker 1989) reported that heritability for thorax length in the natural environment is far lower than in the laboratory. With the ecology of the natural environment in this particular case being so well documented and understood in detail, this was a very illuminating study, showing how the potential effectiveness of natural selection (as determined by heritability) is so dependent on the environment/ecology in which the selection occurs. A few years later, in another pioneering study, Prout and Barker (1993) estimated F_{ST} from both a quantitative trait (thorax length) and allozyme loci on the same individuals, with the former being significantly larger than the latter, which was taken to indicate natural selection on the quantitative trait. Several years later, Barker and Starmer (1999) showed that oviposition site preference (OSP) in *Drosophila buzzatii* is a heritable quantitative trait, and concluded that "OSP for yeast species would seem to be a powerful force for the maintenance of genetic variation, and not only at loci affecting the choice of oviposition sites".

Barely pausing to mark his retirement on 1st July 1998 (42 years after his first appointment as temporary lecturer at the University of Sydney), Emeritus Professor Stuart Barker has continued researching and publishing in *Drosophila* and conservation genetics. Indeed, 24 papers have been published since his official retirement! In one of these, Wilson et al. (2006) presented the first report of male recombination and neo-sex-chromosome formation in *Drosophila* – in this particular case, *Scaptodrosophila hibisci*, a native Australian *Drosophila* specific to a number of Hibiscus species.

How does one encapsulate Stuart's apparently unbounded energy? We can think of no better way than to quote a limerick penned for Stuart by Bryan Clarke FRS, now Emeritus Professor of Genetics at the University of Nottingham:

Old hyperactivity Stuart
during work, is aflame to get through it.
He embodies the notion
of Brownian motion;
I don't know how Stuart can do it.

4 Conclusion

Stuart Barker has spent nearly half a century exploring, understanding, advancing and teaching the science of genetics of animal populations and their evolution, and of livestock populations and their characterisation, utilisation and conservation. In so doing, he has mentored a large number of diverse postgraduates (in total, 23 PhDs and 12 Masters) who have gone on to make substantial contributions in evolutionary/conservation biology and animal breeding. To mention just one example, the development of viable industry-based national genetic evaluation and improvement programs in dairy cattle, beef cattle and sheep in Australia is due in no small part to former postgraduate students of Stuart.

The breadth of Stuart's research has been greater than that of most geneticists, to the extent that the present authors have been able only to scratch the surface of some of his contributions, and those of his many collaborators (including post-docs and students).

In concluding, we hope we have been able to provide some indication of Stuart's many talents and his most significant scientific contributions to knowledge of fitness, evolution and the quantitative genetics of animal breeding.

Acknowledgments We are very grateful to Stuart for providing us with answers to our many questions (including the quotations included in this paper). Inevitably, in a review of this nature, there is much work that has not been mentioned; we apologise in advance to the many colleagues of Stuart (including post-docs and students) whose work has not been specifically discussed. We thank Alison Ward for her detective work in the archives of the University of Sydney. We also thank the many colleagues and former students who provided us with background information. In particular, we thank Bryan Clarke for permission to reproduce his limerick. We especially thank John Mulley and Jerry Taylor for their detailed feedback on a draft. In addition, we are very grateful to John James, Lisa Mascord, Rob Banks, Jeff Moth, Ken Phillips, Richard Thomas, Dick Frankham, and Brian Kinghorn for their constructive suggestions. Wherever possible, we have incorporated all suggestions into the final version.

References

Albers, G.A.A., Gray, G.D., Piper, L.R., Barker, J.S F., Le Jambre, L.F. and Barger I.A. 1987. The genetics of resistance and resilience to haemonchus-contortus infection in young Merino sheep. *International Journal for Parasitology* **17**: 1355–1363.

Barker, J.S.F. 1955. The problem of environment and breed improvement. *Queensland Agricultural Journal* **80**: 241–245.

Barker, J.S.F. 1956. Are egg-laying competitions outmoded? *Queensland Agricultural Journal* **82**: 43–46.

Barker, J.S.F. 1957a. The breed structure and genetic analysis of the pedigree cattle breeds in Australia. I. The Jersey. *Australian Journal of Agricultural Research* **8**: 561–586.

Barker, J.S.F. 1957b. Expansion and distribution of the pedigree Jersey breed in Australia. *Australian Journal of Agricultural Research* **8**: 547–560.

Barker, J.S.F. 1958a. Simulation of genetic systems by automatic digital computers. III. Selection between alleles at an autosomal locus. *Australian Journal of Biological Sciences* **11**: 603–612.

Barker, J.S.F. 1958b. Simulation of genetic systems by automatic digital computers. IV. Selection between alleles at a sex-linked locus. *Australian Journal of Biological Sciences* **11**: 613–625.

Barker, J.S.F. 1959. The breed structure and genetic analysis of the pedigree cattle breeds in Australia. I. The Jersey: a correction. *Australian Journal of Agricultural Research* **10**: 769–770.

Barker, J.S.F. 1960a. Yet another population cage. *Drosophila Information* Service **34**: 113.

Barker, J.S.F. 1960b. An adaptation of the population bottle of Reed and Reed (1948). *Drosophila Information Service* **34**: 113–114.

Barker, J.S.F. 1960c. PhD Thesis *The estimation of relative fitness of Drosophila populations and related studies in population genetics*. University of Sydney, Australia.

Barker, J.S.F. 1962a. Estimation of generation interval in experimental populations of *Drosophila*. *Genetical Research* **3**: 388–404.

Barker, J.S.F. 1962b. Sexual isolation between *Drosophila melanogaster* and *Drosophila simulans*. *American Naturalist* **96**: 105–115.

Barker, J.S.F. 1962c. Studies of selective mating using yellow mutant of *Drosophila melanogaster*. *Genetics* **47**: 623–640.

Barker, J.S.F. 1963a. Estimation of relative fitness of *Drosophila* populations. 2. Experimental evaluation of factors affecting fitness. *Evolution* **17**: 56–71.

Barker, J.S.F. 1963b. The estimation of relative fitness of *Drosophila* populations. 3. The fitness of certain strains of *Drosophila melanogaster*. *Evolution* **17**: 138–146.

Barker, J.S.F. 1967. Modern problems of population genetics in animal husbandry. *Der Zuchter* **37**: 309–323.

Barker, J.S.F. 1979. Inter-locus interactions: A review of experimental evidence. *Theoretical Population Biology* **16**: 323–346.

Barker, J.S.F. and Allingham, D.H. 1959. Migration of pedigree Jersey cattle in Australia. *Australian Journal of Agricultural Research* **10**: 749–766.

Barker, J.S.F. and Burton, H.W. 1956. Report on the Queensland random sample poultry production trial. *Queensland Agricultural Journal* **82**: 456–466.

Barker, J.S.F. and Cummins, L.J. 1969a. Disruptive selection for sternopleural bristle number in *Drosophila melanogaster*. *Genetics* **61**: 697–712.

Barker, J.S.F. and Cummins, L.J. 1969b. Effect of selection for sternopleural bristle number on mating behaviour in *Drosophila melanogaster*. *Genetics* **61**: 713–719.

Barker, J.S.F. and Davern, C.I. 1956. The mutagenic action of formaldehyde on *Drosophila melanogaster*. *Australian Journal of Biological Sciences* **8**: 382–399.

Barker, J.S.F. and Davey, G.P. 1960. The breed structure and genetic analysis of the pedigree cattle breeds in Australia. II. The Poll Hereford. *Australian Journal of Agricultural Research* **11**: 1072–1100.

Barker, J.S.F. and East P.D. 1980. Evidence for selection following perturbation of allozyme frequencies in a natural population of *Drosophila*. *Nature* **284**: 166–168.

Barker, J.S.F., East, P.D. and Weir, B.S. 1986b. Temporal and microgeographic variation in allozyme frequencies in a natural population of *Drosophila buzzatii*. *Genetics* **112**: 577–611.

Barker, J.S.F. and Karlsson, L.J. 1974. Effects of population-size and selection intensity on responses to disruptive selection in *Drosophila melanogaster*. *Genetics* **78**: 715–735.

Barker, J.S.F., Moore, S.S., Hetzel, D.J.S., Evans, D., Tan, S.G. and Byrne, K. 1997. Genetic diversity of Asian water buffalo (*Bubalus bubalis*): Microsatellite variation and a comparison with protein-coding loci. *Animal Genetics* **28**: 103–115.

Barker, J.S.F. and Mulley, J.C. 1976. Isozyme variation in natural populations of *Drosophila buzzatii*. *Evolution* **30**: 213–233.

Barker, J.S.F. and Podger, R.N. 1970a. Interspecific competition between *Drosophila melanogaster* and *Drosophila simulans* – Effects of larval density and shortterm adult starvation on fecundity, egg hatchability and adult viability. *Ecology* **51**: 855–864.

Barker, J.S.F. and Podger, R.N. 1970b. Interspecific competition between *Drosophila melanogaster* and *Drosophila simulans* – Effects of larval density on viability, developmental period and adult body weight. *Ecology* **51**: 170–189.

Barker, J.S.F. and Robertson, A. 1966. Genetic and phenotypic parameters for first 3 lactations in Friesian cows. *Animal Production* **8**: 221–240.

Barker, J.S.F. and Starmer, W.T. 1999. Environmental effects and the genetics of oviposition preference for natural yeast substrates in *Drosophila buzzatii*. *Hereditas* **130**: 145–175.

Barker, J.S.F., Vacek, D.C., East, P.D. and Starmer, W.T. 1986a. Allozyme genotypes of *Drosophila buzzatii*: feeding and oviposition preferences for microbial species, and habitat selection. *Australian Journal of Biological Sciences* **39**: 47–58.

Briscoe, D.A., Malpica, J.M., Robertson, A., Smith, G.J., Frankham, R., Banks, R.G. and Barker, J.S.F. 1992. Rapid loss of genetic variation in large captive populations of *Drosophila* flies – Implications for the genetic management of captive populations. *Conservation Biology* **6**: 416–425.

Cameron, M.M. 2007. *Cactoblastis cactorum the Miracle Bug*. M.M. Cameron, Brisbane.

Claringbold, P.J. and Barker, J.S.F. 1961. Estimation of relative fitness of *Drosophila* populations. *Journal of Theoretical Biology* **1**: 190–203.

Clunies-Ross, I. 1932. Observations on the resistance of sheep to infestations by the stomach worm *Hameonchus contortus*. *Journal of the Council for Scientific and Industrial Research* **5**: 73–80.

Dobzhansky, Th., Krimbas, C. and Krimbas, M.G. 1960. Genetics of natural populations. XXX. Is the genetic load in *Drosophila pseudoobscura* a mutational or a balanced load? *Genetics* **45**: 741–753.

Dobzhansky, Th. and Pavlovsky, O. 1953. Indeterminate outcome of certain experiments on *Drosophila* populations. *Evolution* **7**: 198–210.

Dobzhansky, Th. and Queal, M.L. 1938. Genetics of natural populations. I. Chromsome variation in populations of *Drosophila pseudoobscura*. *Genetics* **23**: 239–251.

Falconer, D.S. 1960. *Introduction to Quantitative Genetics*. Oliver and Boyd, Edinburgh.

FAO. 2007. *The State of the World's Animal Genetic Resources for Food and Agriculture* (B. Rischkowsky and D. Pilling, eds.). Food and Agriculture Organisation of the United Nations, Rome.

Flamand, J.R.B., Vankan, D., Gairhe, K.P., Duong, H. and Barker, J.S.F. 2003. Genetic identification of wild Asian water buffalo in Nepal. *Animal Conservation* **6**: 265–270.

Frankham R., Jones L.P. and Barker, J.S.F. 1968a. Effects of population size and selection intensity in selection for a quantitative character in *Drosophila*. 1. Short-term response to selection. *Genetical Research* **12**: 237–248.

Frankham R., Jones L.P. and Barker, J.S.F. 1968b. Effects of population size and selection intensity in selection for a quantitative character in *Drosophila*. 3. Analyses of lines. *Genetical Research* **12**: 267–283.

Franklin, I., Grigg, G. and Mayo, O. 2004. James Meadows Rendel 1915–2001. *Historical Records of Australian Science* **15**: 269–284.

Fraser, A.S. 1957a. Simulation of genetic systems by automatic digital computers. I. Introduction. *Australian Journal of Biological Sciences* **10**: 484–491.

Fraser, A.S. 1957b. Simulation of genetic systems by automatic digital computers. II. Effects of linkage on rates of advance under selection. *Australian Journal of Biological Sciences* **10**: 492–499.

Fraser, A.S. 1958. Monte Carlo analyses of genetic models. *Nature* **181**: 208–209.

Fraser, A., Miller, D. and Burnell, D. 1965. Polygenic balance. *Nature* **206**: 114.

Haley, C.S. and de Koning, D.J. 2007. Towards *in vitro* genetics. *Trends in Genetics* **23**: 382–386.

Hammond, K. 1974. *Drosophila* as an experimental model in animal breeding research. *Proceedings of the First World Congress on Genetics Applied to Livestock Production* **1**: 425–437.

Harris H. 1966. Enzyme polymorphism in man. *Proceedings of the Royal Society Series B* **164**: 298–310.

Hollingdale, B. and Barker, J.S.F. 1971a. Selection for increased abdominal bristle number in *Drosophila melanogaster* with concurrent irradiation. 1. Populations derived from an inbred line. *Theoretical and Applied Genetics* **41**: 208–215.

Hollingdale, B. and Barker, J.S.F. 1971b. Selection for increased abdominal bristle number in *Drosophila melanogaster* with concurrent irradiation. 2. Populations derived from an outbred cage population. *Theoretical and Applied Genetics* **41**: 263–274.

Hubby, J.L. and Lewontin R.C. 1966. A molecular approach to the study of genic heterozygosity in natural populations. I. The number of alleles at different loci in *Drosophila pseudoobscura*. *Genetics* **54**: 577–594.

Jones, L.P., Frankham, R. and Barker, J.S.F. 1968. Effects of population size and selection intensity in selection for a quantitative character in *Drosophila*. 2. Long-term response to selection. *Genetical Research* **12**: 249–266.

Mann, J. 1970. *Cacti naturalised in Australia and their Control*, S.G. Reid, Government Printer Brisbane.

McCann, D.A. and Batterham, P. 1993. Australian genetics: a brief history. *Genetica* **90**: 81–114.

Moth, J.J. and Barker, J.S.F. 1976. Interspecific competition between *Drosophila melanogaster* and *Drosophila simulans* – Reduction in fecundity and destruction of eggs when medium is inhabited by larvae. *Oecologia* **23**: 151–164.

Moth, J.J. and Barker, J.S.F. 1977. Interspecific competition between *Drosophila melanogaster* and *Drosophila simulans* – Effects of adult density on adult viability. *Genetica* **47**: 203–218.

Moth, J.J. and Barker, J.S.F. 1981. Interspecific competition between *Drosophila melanogaster* and *Drosophila simulans* – Effects of adult density, species frequency, light, and dietary P-32 on fecundity. *Physiological Zoology* **54**: 28–43.

Mulley, J.C., James, J.W. and Barker, J.S.F. 1979. Allozyme genotype-environment relationships in natural populations of *Drosophila buzzatii*. *Biochemical Genetics* **17**: 105–126.

Prout, T. and Barker, J.S.F. 1989. Ecological aspects of the heritability of body size in *Drosophila buzzatii*. *Genetics* **123**: 803–813.

Prout, T. and Barker, J.S.F. 1993. F-statistics in *Drosophila buzzatii*: selection, population size and inbreeding. *Genetics* **134**: 369–375.

Provine, W.B. 1986. *Sewall Wright and Evolutionary Biology*, University of Chicago Press, Chicago.

Rathie, K.A. and Barker, J.S.F. 1968. Effectiveness of regular cycles of intermittent artificial selection for a quantitative character in *Drosophila melanogaster*. *Australian Journal of Biological Sciences* **21**: 1187–1213.

Rathie, K.A. and Nicholas, F.W. 1980. Artificial selection with differing population structures. *Genetical Research* **36**: 117–131.

Robertson, A. 1953. A numerical description of breed structure. *Journal of Agricultural Science* **43**: 334–336.

Robertson A. 1960. A theory of limits to artificial selection. *Proceedings of the Royal Society Series B* **153**: 235–249.

Robertson, A. 1966. A mathematical model of culling in dairy cattle. *Animal Production* **8**: 95–108.

Robertson, A. and Asker, A.A. 1951. The genetic history and breed structure of British Friesian cattle. *Empire Journal of Experimental Agriculture* **19**: 1–13.

Robertson, A. and Barker, J.S.F. 1966. Correlation between first lactation milk production and longevity in dairy cattle. *Animal Production* **8**: 241–252.

Sheridan, A.K., Frankham, R., Jones, L.P., Rathie, K.A. and Barker, J.S.F. 1968. Partitioning of variance and estimation of genetic parameters for various bristle number characters of *Drosophila melanogaster*. *Theoretical and Applied Genetics* **38**: 179–187.

Sheridan, A.K. and Barker, J.S.F. 1974a. 2-trait selection and genetic correlation. 1. Prediction of responses in single-trait and in 2-trait selection. *Australian Journal of Biological Sciences* **27**: 75–88.

Sheridan, A.K. and Barker, J.S.F. 1974b. 2-trait selection and genetic correlation. 2. Changes in genetic correlation during 2-trait selection. *Australian Journal of Biological Sciences* **27**: 89–101.

Tan, S G., Barker, J.S.F., Selvaraj, O.S., Mukherjee, T.K. and Wong, Y.F. 1993. Genetic-studies of water-buffalo blood markers. 1. Red-cell acid-phosphatase, albumin, catalase, red-cell alpha-esterase-3, group-specific component, and protease inhibitor. *Biochemical Genetics* **31**: 223–230.

Wilson, A.C.C., Sunnucks, P., Bedo, D.G. and Barker, J.S.F. 2006. Microsatellites reveal male recombination and neo-sex chromosome formation in *Scaptodrosophila hibisci* (Drosophilidae). *Genetical Research*. **87**: 33–43.

Wright, S. 1922. The effects of inbreeding and crossbreeding in guinea pigs. III. Crosses between highly inbred families. *USDA Bulletin* **1121**: 1–60.

Wright, S. 1977. *Evolution and the Genetics of Populations*. Vol. 3, *Experimental Results and Evolutionary Deductions*. University of Chicago Press, Chicago.

Yoo, B.H. 1974. Correlated responses of different *scute* genotypes to long-term selection for increased abdominal bristle number in *Drosophila melanogaster. Australian Journal of Biological Sciences* **27**: 205–218.

Yoo, B.H. 1980a. Long-term selection for a quantitative character in large replicate populations of *Drosophila melanogaster*. 1. Response to selection. *Genetical Research* **35**: 1–17.

Yoo, B.H. 1980b. Long-term selection for a quantitative character in large replicate populations of *Drosophila melanogaster*. 2. Lethals and visible mutants with large effects. *Genetical Research* **35**: 19–31.

Yoo, B.H. 1980c. Long-term selection for a quantitative character in large replicate populations of *Drosophila melanogaster*. 3. The nature of residual genetic variability. *Theoretical and Applied Genetics* **57**: 25–32.

Yoo, B H. 1980d. Long-term selection for a quantitative character in large replicate populations of *Drosophila melanagaster*. 5. The inbreeding effect of selection. *Australian Journal of Biological Sciences* **33**: 713–723.

Yoo, B H., Nicholas, F.W. and Rathie, K.A. 1980. Long-term selection for a quantitative character in large replicate populations of *Drosophila melanogaster*. 4. Relaxed and reverse selection. *Theoretical and Applied Genetics* **57**: 113–117.

Index